Rare-earth-free Ferrimagnetic Mn_4N Spintronics

Online at: https://doi.org/10.1088/978-0-7503-5477-6

Rare-earth-free Ferrimagnetic Mn_4N Spintronics

Takashi Suemasu
Department of Applied Physics, Institute of Pure and Applied Science, University of Tsukuba, Tsukuba, Ibaraki 305-8573, Japan

IOP Publishing, Bristol, UK

ISBN 978-0-7503-5477-6 (ebook)
ISBN 978-0-7503-5475-2 (print)
ISBN 978-0-7503-5478-3 (myPrint)
ISBN 978-0-7503-5476-9 (mobi)

DOI 10.1088/978-0-7503-5477-6

Version: 20240201

IOP ebooks

British Library Cataloguing-in-Publication Data: A catalogue record for this book is available from the British Library.

Published by IOP Publishing, wholly owned by The Institute of Physics, London

IOP Publishing, No.2 The Distillery, Glassfields, Avon Street, Bristol, BS2 0GR, UK

US Office: IOP Publishing, Inc., 190 North Independence Mall West, Suite 601, Philadelphia, PA 19106, USA

Contents

Preface

Since the perpendicular magnetic anisotropy of Mn_4N thin films was reported in 2012, Mn_4N has attracted much attention as a new rare-earth-free spintronic material. Mn_4N epitaxial films exhibit magnetic compensation and angular momentum compensation at room temperature upon impurity doping. Recently, record speeds of current-induced magnetic domain wall motion at room temperature driven purely by spin-transfer torque (STT) have been achieved in Mn_4N and Ni-doped compensated Mn_4N, attracting considerable attention. Mn_4N also exhibits negative thermal expansion, which is a very unusual property. In addition, a great variety of mixed crystals have been formed. Spintronics, a rapidly growing field in solid-state physics, has already been the subject of many technical books and the results of cutting-edge research have been published in many scientific journals. This book is intended to provide graduate students and researchers interested in ferrimagnetic Mn_4N-based mixed crystals with a broad overview of the history of this research.

Chapter 1 discusses the origin of the magnetic moment of atoms, the principle of magnetic domain walls, STT, spin-orbit torque (SOT), and magnetic domain wall motion, and concludes with a summary of the characteristics of Mn_4N. In chapter 2, we briefly discuss the history of Mn_4N-based mixed crystals. Note that there were many comprehensive studies of these materials in bulk form in the 1950s–1970s. However, due to space limitations, this book will focus on relatively recently published results, including epitaxial thin films. Chapter 3 focuses mainly on molecular beam epitaxy of Mn_4N thin films and introduces their crystallinity and magnetic properties. In addition, x-ray magnetic circular dichroism is introduced as a means of determining the presence or absence of a magnetic compensation composition. In chapter 4, we describe ultrafast current-induced domain wall motion demonstrated in Mn_4N epitaxial thin films. In chapter 5, we introduce Ni or Co-doped Mn_4N films that compensate for magnetization at room temperature. In chapter 6, we introduce Mn_4N films doped with nonmagnetic elements to pursue compensation of magnetization and discuss the conditions under which magnetic compensation occurs. In chapter 7, we offer Ni-doped Mn_4N microstrips that provide room temperature magnetic compensation and ultrafast magnetic domain wall motion of up to 3000 m s^{-1}. Chapter 8 provides an overview of Mn_4N compounds doped with impurities such as Fe, Co, Ni, Cu, Zn, Al, In, Ga, Ge, Sn, Pd, Ag, Sb, Pt, Au, Gd, and Dy. Chapter 9 presents the latest reports such as SOT switching and the Dzyaloshinskii–Moriya interaction measurements.

Acknowledgements

I am grateful to all those who helped me in the preparation of this book. In particular, I would like to thank E. Tapp and her colleagues at IOP Publishing for the opportunity to write this book, Taku Hirose for collecting the figures in chapters 2 and 8, past and present graduate students in my laboratory who have consistently worked on thin film growth and characterization, and colleagues at the University of Tsukuba for their help with various measurements. Most of the content of chapters 3–7 is the result of collaboration between graduate students in my laboratory and researchers in Grenoble, France. I am especially grateful to Dr L. Vila, Dr J. P. Attané, Dr S. Pizzini, and their colleagues in Grenoble, France. The ultrafast current-induced magnetic wall motion in chapters 4 and 7 was achieved through the exchange of doctoral students (Dr Keita Ito, Dr Toshiki Gushi, Dr Taro Komori, Dr Sambit Ghosh, and PhD student Mr Tomohiro Yasuda as of 1 September 2023) under the double degree program (DDP) between the University of Tsukuba and Université Grenoble Alpes. Special emphasis should be given to Professor Emeritus Katsuhiro Akimoto (University of Tsukuba) and Professor Etienne Gheeraert (Université Grenoble Alpes) for their kind support regarding the DDP program. Finally, I thank my wife Hiroka and my children for encouraging me to pursue my work.

Author biography

Takashi Suemasu

Takashi Suemasu received his BS, MS, and PhD degrees in electrical and electronics engineering from Tokyo Institute of Technology, Japan, and became a full professor at the University of Tsukuba, Japan, in 2010. In 2014, he became a visiting researcher at the Institute for Photovoltaics, Universität Stuttgart, Germany. In April 2021, he was appointed Chair of the Department of Applied Physics, Institute of Pure and Applied Sciences, University of Tsukuba. He has worked on thin film growth and device applications of semiconducting silicide and nitride based spintronic materials. In 1996, he began research on semiconducting β-$FeSi_2$ with Professor Fumio Hasegawa of the University of Tsukuba (now Professor Emeritus), and succeeded in the room temperature operation of p-Si/β-$FeSi_2$/n-Si light emitting diodes with an emission wavelength near 1.6 μm in 2000. He also has been interested in Si compound semiconductors with controllable forbidden bandwidth. He began working on $BaSi_2$ in 2001 and demonstrated solar cell operation with p-$BaSi_2$/n-Si heterojunction diodes in 2016 and $BaSi_2$-pn homojunction diodes in 2019. Spintronics is another of his research interests. Since the Fe-rich phase Fe_3Si is ferromagnetic among Fe–Si compounds, he started research on magnetic materials in 2004 and has gradually expanded his research targets to 3*d* nitride thin films such as Fe_4N, Co_4N, Ni_4N, and Mn_4N and their mixed crystals. In 2019–2021, ultrafast domain wall motion purely driven by spin-transfer torques was demonstrated in Mn_4N and compensated $Mn_{4-x}Ni_xN$ with colleagues working at the Université Grenoble Alpes, Spintec, and Institut Néel in France. He enjoys swimming a couple of days a week and playing golf on the large field near Tsukuba more than anything else.

List of symbols

Experimental techniques used for preparation and characterization of samples

CVD	chemical vapor deposition
EDX	energy dispersive x-ray spectroscopy
FMR	ferromagnetic resonance
MBE	molecular beam epitaxy
NPD	neutron powder diffraction
PAMBE	plasma-assisted MBE
PLD	pulsed laser deposition
RHEED	reflection high-energy electron diffraction
SD	sputtering deposition
SSR	solid-state reaction
TEM	transmission electron microscopy
XRD	x-ray diffraction
XPS	x-ray photoelectron spectroscopy

Substrates used for thin-film growth

STO	$SrTiO_3$
BTO	$BaTiO_3$
LSAT	$(LaAlO_3)_{0.3}(SrTaAl_2O_6)_{0.7}$
MAO	$MgAl_2O_4$
YSZ	yttria-stabilized zirconia

Magnetic order and transition temperature

σ_{AH}	anomalous Hall conductivity
ρ_{xx}	longitudinal resistivity
ρ_{xy}	transverse resistivity
AFM	anti-ferromagnetism
C_p	heat capacity
FiM	ferrimagnetism
FM	ferromagnetism
IMA	in-plane magnetic anisotropy
PM	paramagnetism
PMA	perpendicular magnetic anisotropy
T_C	Curie temperature
T_N	Néel temperature

Measurement methods for magnetic and magneto-transport properties

AHE	anomalous Hall effect
ANE	anomalous Nernst effect
DSC	different scanning calorimetry
FE	ferroelectric
L-TCR	low temperature coefficient of resistivity
LTE	low thermal expansion
ME	magneto-electric effect

M–H curves	magnetization versus magnetic field curves
MOKE	magneto-optical Kerr effect
MPMS	magnetic property measurement system
MVE	magneto volume effect
NTE	negative thermal expansion
PME	piezomagnetic effect
PPMS	physical properties measurement system
SOT	spin-orbit torque
STT	spin-transfer torque
TCR	temperature coefficient of resistivity
VSM	vibrating sample magnetometer

IOP Publishing

Rare-earth-free Ferrimagnetic Mn_4N Spintronics

Takashi Suemasu

Chapter 1

Introduction

1.1 Origin of magnetic moment

Magnetization appears as a collection of magnetic moments in the atoms and ions that make up materials. The origin of the magnetic moment is the spin angular momentum $\boldsymbol{S}$ and orbital angular momentum $\boldsymbol{L}$ of the electrons. In this book, we do not go into the magnetic moment of the nucleus, which also has a spin magnetic moment, because it is much smaller than the magnetic moment of electrons.

1.1.1 Orbital magnetic moment

According to quantum mechanics, the orbital angular momentum $\boldsymbol{L}$ of an individual electron is expressed using the orbital quantum number l and the principal quantum number n,

$$\boldsymbol{L}^2 = l(l+1)\hbar^2, \; l = 0, 1, 2, \ldots n-1. \tag{1.1}$$

$$|\,\boldsymbol{L}\,| = \sqrt{l(l+1)}\,\hbar. \tag{1.2}$$

'$l + 1$' in equation (1.1) rather than just l originates from the uncertainty principle. $l = 0, 1, 2, 3$ corresponds to the orbital named s (sharp), p (principal), d (diffuse), and f (fundamental). Therefore, the s electrons do not contribute to the orbital angular momentum. The orientation of the orbital angular momentum along the magnetic field in the z axis direction is quantized and is labeled by the magnetic quantum number m_l, The component of the orbital angular momentum along the field in the z axis direction is given by equation (1.3).

$$L_z = m_l\hbar, \; m_l = 0, \pm 1, \ldots \pm l. \tag{1.3}$$

In the case of d electrons ($l = 2$), m_l is allowed to be 0, ±1 and ±2, meaning that L_z has five orientations ($-2\hbar, -\hbar, 0, \hbar, 2\hbar$) and is always smaller than $|\,\boldsymbol{L}\,|$, as illustrated in figure 1.1. This suggests that the orbital angular momentum never points directly

doi:10.1088/978-0-7503-5477-6ch1

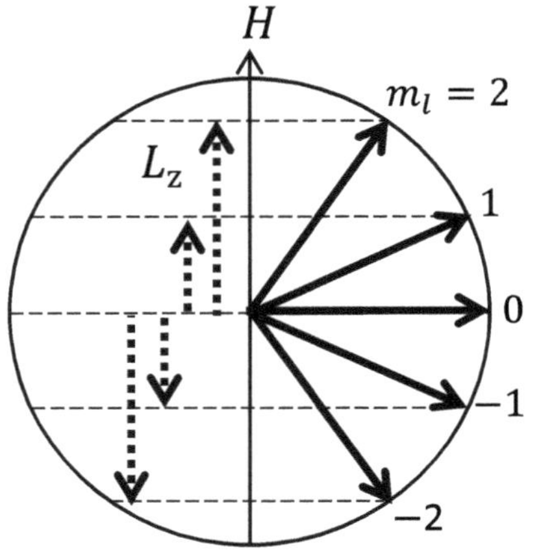

Figure 1.1. Z component of angular momentum of an electron (L_z) along the magnetic field H for d orbital ($l = 2$). $|\boldsymbol{L}| = \sqrt{6}\hbar$.

along the magnetic field but precesses instead in a cone around it. Assuming an electron orbiting with velocity v at a distance r from the nucleus in an atom, the following orbital magnetic moment m_{orbit} arises,

$$|\boldsymbol{m}_{\text{orbit}}| = IA. \tag{1.4}$$

Here, I is the circular current and A is the area of the orbit. Since the current is the charge passing per unit time, it can be expressed as,

$$I = -ev/2\pi r. \tag{1.5}$$

The negative sign occurs because the charge on the electron is negative.

The energy of a magnetic moment $\boldsymbol{m}_{\text{orbit}}$ in a magnetic field H is expressed as,

$$E = -\mu_0 \boldsymbol{m}_{\text{orbit}} \cdot \boldsymbol{H}. \tag{1.6}$$

Here, μ_0 is the magnetic permeability of vacuum. The angular momentum of an electron with mass m_e orbiting in a circle with radius r is $m_e vr$. This value is equal to the angular momentum projected onto the magnetic field $m_l\hbar$. Therefore, the value of v is given by,

$$v = m_l\hbar/m_e r. \tag{1.7}$$

Thus, equation (1.6) gives the energy,

$$E = \mu_0 \frac{e\hbar}{2m_e} m_l H = \mu_0 \mu_B m_l H. \tag{1.8}$$

Here, μ_B is the Bohr magneton expressed as equation (1.9), and it is the natural unit of the magnetic moment of an electron.

$$\mu_B = e\hbar/2m_e. \tag{1.9}$$

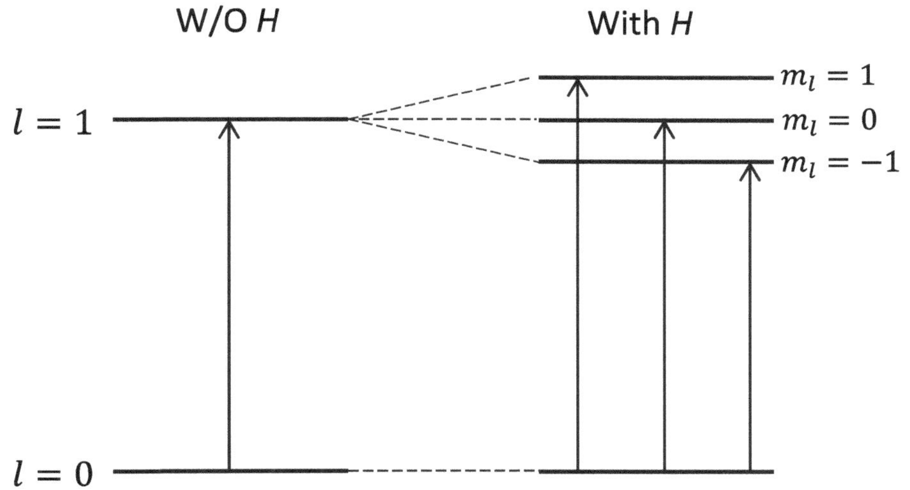

Figure 1.2. Example of a *s-p* transition in the normal Zeeman effect without and with an external magnetic field.

Therefore, the magnetic moment along the field axis is expressed as,

$$m_{\text{orbit}} = -\mu_{\text{B}} m_l. \tag{1.10}$$

It is noted that the orbital magnetic moment is antiparallel to the orbital angular momentum. The energy of s electron does not split in the presence of a magnetic field. In contrast, the energy of p electrons splits into three, corresponding to $m_l = -1$, 0, and 1. Therefore, three lines appear in the normal Zeeman spectrum as a result, as shown in figure 1.2.

1.1.2 Spin magnetic moment

The spin of an electron is derived from the Dirac equation. The magnitude of the spin angular momentum of an individual electron, $|\boldsymbol{S}|$, is expressed by using the spin quantum number s as,

$$|\boldsymbol{S}| = \sqrt{s(s+1)}\hbar = \frac{\sqrt{3}}{2}\hbar. \tag{1.11}$$

The orientation of the spin angular momentum along the magnetic field in the z axis direction is quantized and is labeled by m_{S}, which is the spin analog to the magnetic quantum number m_l. It is given by $m_{\text{S}}\hbar = \pm\hbar/2$, meaning that $m_{\text{S}}\hbar$ is always smaller than S, as illustrated in figure 1.3. We see that the spin angular momentum vector can never align along the magnetic field direction and precesses instead in a cone around it. Therefore, the spin magnetic moment along the field axis is expected to be $-\mu_{\text{B}} m_{\text{S}}$. Actually, the theory of quantum mechanics gives us,

$$m_{\text{spin}} = -g_e \mu_{\text{B}} m_{\text{S}}. \tag{1.12}$$

Here, g_e is a g-factor of the electron.

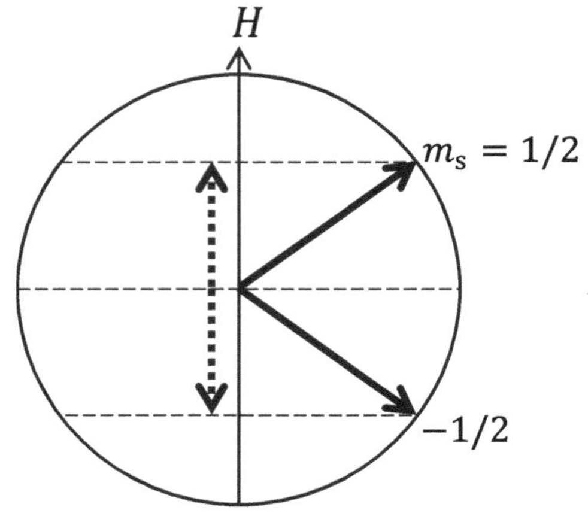

Figure 1.3. Z component of angular momentum of an electron along the magnetic field H is either $\hbar/2$ or $-\hbar/2$. $|\boldsymbol{S}| = \sqrt{3}\hbar/2$.

1.1.3 Spin−orbit interaction

Electrons with an orbital magnetic moment behave like a circulating electric current. This indicates the presence of a spin magnetic moment in an associated magnetic field, and thus the magnetic moment from the electron's spin is supposed to interact with the magnetic field caused by its orbital motion. This interaction is the spin–orbit interaction and increases as the atomic number Z increases. This can be understood by thinking of an electron fixed in space and a nucleus orbiting around it. The current produced by the circulating nucleus is greater when the nuclear charge is larger. In fact, the spin–orbit interaction increases in proportion to Z^4 [1].

Once the s and l quantum numbers of the individual electrons are given, the way to calculate the angular momentum of all of the electrons in an atom depends on the relative magnitudes of the orbit–orbit, spin–orbit, and spin–spin couplings. Two different schemes for calculating the total angular momentum of an atom are provided here. This is very important because the magnetic moment of an atom is determined by the total angular momentum of the electrons.

(a) Russell–Saunders coupling

In light atoms, the spin–orbit interaction is weak. Thus, the coupling between the individual spins and the individual orbital angular momenta is greater than the spin–orbit interaction. Therefore, the best way to calculate the total angular momentum is first to sum up the orbital angular momenta of all of the individual electrons to get the total orbital momentum. Their spin angular momenta are then added to obtain the total spin momentum. The total spin and angular components are next added to get the total angular momentum. In the case of l quantum numbers of the electrons, the rules for determining the allowed values of the total orbital quantum number are rather complicated. In this case, the allowed L is given by,

$$L = l_1 + l_2, l_1 + l_1 - 1, \ldots\ldots, |l_1 - l_2|. \tag{1.13}$$

We will take an example here for an atom with just two electrons with orbital quantum numbers l_1 and l_2. Suppose that there are two electrons, one with $l = 1$, and the other with $l = 2$, the L values are allowed to be 1, 2, and 3. By analogy with the m_l values for individual electrons, we define a total M_L for atoms that ranges from $-L$, $-L+1$, $-L+2$,…., L, and give the total orbital angular momentum projected onto a specific direction. The number of M_L is $(2L+1)$. In the same way, the spins are combined into a total spin quantum number,

$$S = s_1 + s_2, s_1 - s_2. \tag{1.14}$$

The total spin S is thus, $M_S = -S, -S+1, \ldots, S$. The number of S is $(2S+1)$. In the case of a two-electron atom, the allowed S values are 0 or 1. Thus, M_S values are, respectively, −1, 0, and +1. The total angular momentum quantum number J is thus determined by vector addition of the total atomic spin and total orbital angular momenta as,

$$J = L + S, L + S - 1, \ldots\ldots, |L - S|. \tag{1.15}$$

The corresponding total J values, M_J, are −4, −3, −2, −1, 0, 1, 2, 3, 4. The magnitude of the total atomic angular momentum is then given by,

$$|\boldsymbol{J}| = \sqrt{J(J+1)}\hbar. \tag{1.16}$$

The component along the magnetic field is $M_J\hbar$. This scheme is known as Russell–Saunders coupling [2]. There are two important points, as shown below.

1. The difference in energy between states with different J values but the same S and L values are small compared with those between levels having different L or S values.
2. For a filled shell of electrons, L, S, and J are zero. Thus, there is no net angular momentum, and thus the permanent magnetic dipole moment is also zero. We can state that only the incomplete outer shells have to be considered in calculating J, L, and S values. Atoms with no incomplete shells are called diamagnetic.

(b) Hund's rules

The German scientist Friedrich Hund developed the following three rules to identify the lowest-energy configuration for the electrons in an incomplete shell. These rules are based on the assumption that angular momentum states are described by Russell–Saunders coupling.

First rule: Electrons occupy the shells so that they maximize the total spin S. As a result, the electrons occupy orbitals with one electron per orbital, and the spins are all parallel until all the orbitals are occupied by one electron. For more electrons, they are paired up in orbitals with an opposite spin. This can be interpreted qualitatively by the Pauli exclusion principle. Therefore, the repulsive Coulomb energy becomes smaller between electrons of the same spin, and the energy is lower.

Second rule: The configuration with the greatest total atomic orbital angular momentum L lies lowers in energy for a given spin arrangement. The basis for this rule is that electrons orbiting in the same direction meet less frequently than those orbiting in the opposite directions. Therefore, their repulsion is on average less when L is large. Table 1.1 summarizes the electron configuration of 3d and 4f orbitals according to the above two rules. They can accommodate 10 and 14 electrons, respectively.

Third rule: For atoms with less than half-filled shells, the lowest-energy electronic configuration is the one with the lowest value of $J = | L - S |$. Meanwhile, when the shell is more than half-filled, the opposite rule applies, i.e., the arrangement with the highest value of $J = | L + S |$ has the lowest energy. The origin of the rule is the spin–orbit coupling and is related to the fact that antiparallel dipole moments have lower energy than parallel dipole moments.

Table 1.2 shows the calculated quantum numbers for various 3d ions and those obtained experimentally. Let us look at Mn^{2+} as an example. The configuration of electrons in a Mn^{2+} ion is $(1s)^2(2s)^2(2p)^6(3s)^2(3p)^6(3d)^5(4s)^0$, while that in a Mn atom is $(1s)^2(2s)^2(2p)^6(3s)^2(3p)^6(3d)^5(4s)^2$. Thus, the incomplete shell is only $(3d)^5$. Since there are five 3d orbitals, the electrons occupy five orbitals individually with parallel spins. Thus, S = 5/2, corresponding to m_l = 2, 1, 0, −1, and −2. As a result, L must be 0.

Table 1.1. Configuration of 3d and 4f electrons.

	m_l						m_l						
	2	1	0	−1	−2		3	2	1	0	−1	−2	−3
$(3d)^1$	↑					$(4f)^1$	↑						
$(3d)^2$	↑	↑				$(4f)^2$	↑	↑					
$(3d)^3$	↑	↑	↑			$(4f)^3$	↑	↑	↑				
$(3d)^4$	↑	↑	↑	↑		$(4f)^4$	↑	↑	↑	↑			
$(3d)^5$	↑	↑	↑	↑	↑	$(4f)^5$	↑	↑	↑	↑	↑		
$(3d)^6$	↑↓	↑	↑	↑	↑	$(4f)^6$	↑	↑	↑	↑	↑	↑	
$(3d)^7$	↑↓	↑↓	↑	↑	↑	$(4f)^7$	↑	↑	↑	↑	↑	↑	↑
$(3d)^8$	↑↓	↑↓	↑↓	↑	↑	$(4f)^8$	↑↓	↑	↑	↑	↑	↑	↑
$(3d)^9$	↑↓	↑↓	↑↓	↑↓	↑	$(4f)^9$	↑↓	↑↓	↑	↑	↑	↑	↑
$(3d)^{10}$	↑↓	↑↓	↑↓	↑↓	↑↓	$(4f)^{10}$	↑↓	↑↓	↑↓	↑	↑	↑	↑
						$(4f)^{11}$	↑↓	↑↓	↑↓	↑↓	↑	↑	↑
						$(4f)^{12}$	↑↓	↑↓	↑↓	↑↓	↑↓	↑	↑
						$(4f)^{13}$	↑↓	↑↓	↑↓	↑↓	↑↓	↑↓	↑
						$(4f)^{14}$	↑↓	↑↓	↑↓	↑↓	↑↓	↑↓	↑↓

Table 1.2. Calculated values of S, L, J, g_J, the magnitude of total angular momentum, the magnitude of total spin angular momentum, and experimentally reported magnetic moment for various $3d$ ions.

		S	L	J	g_J	$g_J\sqrt{J(J+1)}\,[\mu_B]$	$2\sqrt{S(S+1)}\,[\mu_B]$	Expt. $[\mu_B]$
Ti^{3+}, V^{4+}	$(3d)^1$	1/2	2	3/2	4/5	1.55	1.73	1.7
Ti^{2+}, V^{3+}	$(3d)^2$	1	3	2	2/3	1.63	2.83	2.8
V^{2+}, Cr^{3+}	$(3d)^3$	3/2	3	3/2	2/5	0.78	3.87	3.8
Cr^{2+}, Mn^{3+}	$(3d)^4$	2	2	0	0	0	4.90	4.9
Mn^{2+}, Fe^{3+}	$(3d)^5$	5/2	0	5/2	2	5.92	5.92	5.9
Fe^{2+}, Co^{3+}	$(3d)^6$	2	2	3	3/2	6.71	4.90	5.4
Co^{2+}, Ni^{3+}	$(3d)^7$	3/2	3	9/2	4/3	6.63	3.87	4.8
Ni^{2+}	$(3d)^8$	1	3	4	5/4	5.59	2.83	3.2
Cu^{2+}	$(3d)^9$	1/2	2	5/2	6/5	3.55	1.73	1.9

Table 1.3. Calculated values of S, L, J, Landé g-factor g_J, the magnitude of total angular momentum, the magnitude of total spin angular momentum, and experimentally reported magnetic moment for various 4f ions.

		S	L	J	g_J	$g_J\sqrt{J(J+1)}\,[\mu_B]$	$2\sqrt{S(S+1)}\,[\mu_B]$	Expt. $[\mu_B]$
Ce^{3+}	$(4f)^1$	1/2	3	5/2	6/7	2.54	6.93	2.5
Pr^{3+}	$(4f)^2$	1	5	4	4/5	3.85	10.95	3.5
Nd^{3+}	$(4f)^3$	3/2	6	9/2	8/11	3.52	12.96	3.4
Pm^{3+}	$(4f)^4$	2	6	4	3/5	2.68	12.96	–
Sm^{3+}	$(4f)^5$	5/2	5	5/2	2/7	0.85	10.95	1.7
Eu^{3+}	$(4f)^6$	3	3	0	0	0	6.93	3.4
Gd^{3+}	$(4f)^7$	7/2	0	7/2	2	7.94	0	8.9
Tb^{3+}	$(4f)^8$	3	3	6	3/2	9.72	6.93	9.8
Dy^{3+}	$(4f)^9$	5/2	5	15/2	4/3	10.65	10.95	10.6
Ho^{+}	$(4f)^{10}$	2	6	8	5/4	10.61	12.96	10.4
Er^{3+}	$(4f)^{11}$	3/2	6	15/2	6/5	9.58	12.96	9.5
Tm^{3+}	$(4f)^{12}$	1	5	6	7/6	7.56	10.95	7.6
Yb^{3+}	$(4f)^{13}$	1/2	3	7/2	8/7	4.53	6.93	4.5

Since the shell is more than half filled, this makes J equal to S, and thus $J = 5/2$. It is noted in table 1.2 that the experimentally obtained magnetic moments of $3d$ ions are almost determined not by J but by S. This means that the orbital angular momentum L vanishes due to the itinerant nature of the $3d$ electrons.

Table 1.3 shows the calculated quantum numbers for various $4f$ ions and those obtained experimentally. Let us take a look at Ce^{3+} as an example. The configuration of electrons in a Ce^{3+} ion is $(4f)^1(5s)^2(5p)^6(5d)^0(6s)^0$, while that in a Ce atom is $(4f)^1(5s)^2(5p)^6(5d)^2(6s)^1$. Thus, the incomplete shell is only $(4f)^1$. Since there are seven $4f$ orbitals, the $4f$ electron is able to occupy only one orbital with up-spin. Thus, $S = 1/2$, corresponding to $m_l = 3$. As a result, L must be 3. Since the shell is less than half filled, this makes J equal to $| L - S |= 5/2$. We can see from table 1.3 that the experimentally obtained magnetic moments of $4f$ ions are almost determined not by S but by J, in contrast to $3d$ ions. This means that $4f$ electrons are orbiting in an inner shell, and are thus rather isolated from $3d$ electrons. Note that $\hbar = 1$ is used in tables 1.2 and 1.3.

It is also noted that, strictly speaking, the existence of spontaneous magnetic moments is not necessarily determined by shells alone but must be considered in terms of electronic theory, such as the band structure of metals.

(c) *jj* coupling

The Russell–Saunders coupling scheme does not explain the electron configuration in heavy atoms such as actinides. This is because their spin and orbital angular momenta of individual electrons couple strongly. The orbital and spin angular momenta of each electron i are combined, resulting in a total angular momenta per electron as,

$$j_i = l_i + s_i. \tag{1.17}$$

The resultant j_i interact weakly via electrostatic coupling of their electron distributions and forms a resultant total angular momentum that is expressed as,

$$J = \sum j_i. \tag{1.18}$$

(d) Anomalous Zeeman effect

The normal Zeeman effect shown in figure 1.2 occurs for atoms with a total spin angular momentum being equal to zero. Otherwise, more complex transition lines in the spectrum (called an anomalous Zeeman effect) occur. This is a result of spin–orbit coupling. The additional complexity is brought about because the splitting of the upper and lower levels of the transition are not equal. The reason for the unequal splitting is the anomalous g-factor of the electrons.

Here we deduce the g-factor in the following. The total magnetic moment of electrons $\boldsymbol{m}$ is the sum of the orbital magnetic moment and the spin magnetic moment, and is given by,

$$\boldsymbol{m} = -(\boldsymbol{L} + 2\boldsymbol{S})\mu_B. \tag{1.19}$$

Meanwhile, the total angular momentum $\boldsymbol{J}$ is expressed as,

$$\boldsymbol{J} = \boldsymbol{S} + \boldsymbol{L}. \tag{1.20}$$

When there is an interaction between the orbital motion of the electron and its spin (spin–orbit coupling), represented by $\lambda \boldsymbol{L} \cdot \boldsymbol{S}$, the total angular momentum $\boldsymbol{J}$ is conserved, while an exchange of angular momentum components occurs between orbital and spin angular momentum. Thus, the spin angular momentum $\boldsymbol{m}$ will be in a precessional motion around the total angular momentum $\boldsymbol{J}$. On a time scale longer than the period of the precession, only the component of the total magnetic moment parallel to the total angular momentum, $\boldsymbol{m}_z$, remains as,

$$\boldsymbol{m}_z = -g\mu_B \boldsymbol{J}. \tag{1.21}$$

Therefore, the relation

$$\boldsymbol{m}_z \cdot \boldsymbol{J} = \boldsymbol{m} \cdot \boldsymbol{J} \tag{1.22}$$

should be satisfied.

$$\boldsymbol{m}_z \cdot \boldsymbol{J} = -g\mu_B \boldsymbol{J}^2 = -g\mu_B J(J+1). \tag{1.23}$$

$$\begin{aligned}
\boldsymbol{m} \cdot \boldsymbol{J} &= -\mu_B(\boldsymbol{L} + 2\boldsymbol{S}) \cdot (\boldsymbol{L} + \boldsymbol{S}) \\
&= -\mu_B(\boldsymbol{L}^2 + 3\boldsymbol{L} \cdot \boldsymbol{S} + 2\boldsymbol{S}^2) \\
&= -\mu_B\left(\boldsymbol{L}^2 + 2\boldsymbol{S}^2 + \frac{3}{2}[(\boldsymbol{L} + \boldsymbol{S})^2 - \boldsymbol{L}^2 - \boldsymbol{S}^2]\right) \\
&= -\mu_B\left[\frac{3}{2}\boldsymbol{J}^2 + \frac{1}{2}\boldsymbol{S}^2 - \frac{1}{2}\boldsymbol{L}^2\right] \\
&= -\mu_B\left[\frac{3}{2}J(J+1) + \frac{1}{2}S(S+1) - \frac{1}{2}L(L+1)\right].
\end{aligned} \tag{1.24}$$

From equations (1.23) and (1.24), g is expressed as,

$$g = \frac{3}{2} + \frac{S(S+1) - L(L+1)}{2J(J+1)}. \tag{1.25}$$

In the case of $S = 0$, $g = 1$ from equation (1.25). The magnetic moment is thus independent of L and we can see the normal Zeeman effect, such as in figure 1.2. Meanwhile, at $S \neq 0$, the g value depends on both S and L. Thus, the upper and lower levels in the spectroscopic transition are split by different amounts, as illustrated in figure 1.4. It is noted that the allowed transitions are further limited by the angular momentum conservation law $\Delta M_J = 0$ or ± 1. This corresponds to the emission of linearly polarized light or circularly polarized light, respectively. If the applied field is very strong, then the coupling between S and L can be broken in favor of their direct coupling to the magnetic field. S and L then precess independently about the field direction. The electromagnetic field inducing the electronic transition

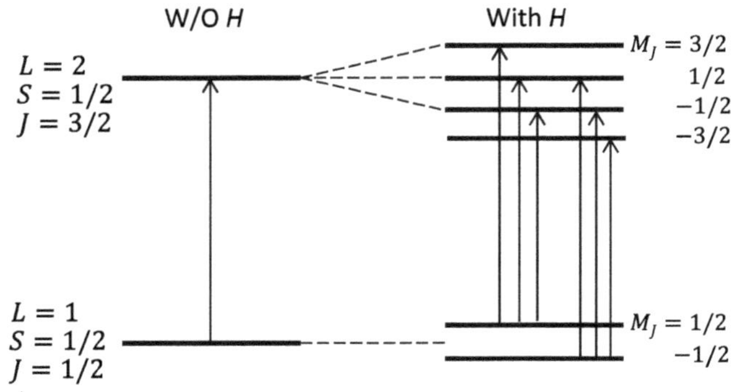

Figure 1.4. Example of a transition in the anomalous Zeeman effect for a transition between *p* and *d* orbitals without and with an external magnetic field.

couples only to the orbital distribution of the electrons. Thus, the presence of the spin does not show up in the transition.

1.2 Magnetic properties of materials

In the previous section, we examined the origin of the magnetic moment of single atoms and ions. In this section, we first classify materials by the way in which the magnetic moments of the electrons associated with the atoms and ions that make up the materials are arranged. The magnetic domain is then brought out to see how the magnetization changes with respect to the external magnetic field.

1.2.1 Classification of magnetic materials [3]

(a) Diamagnetism

A material that has the property of slightly magnetizing in the opposite direction of the magnetic field when a magnetic field is applied is called a diamagnetic material. This is due to the fact that the Lorentz force applied to the electrons by the applied magnetic field changes their orbital motion, resulting in the induction of a slight magnetic moment in the opposite direction of the magnetic field. In accordance with Lenz's law, it is considered that electron motion is induced in the direction that prevents an increase in the magnetic field. Diamagnetism is a property of all matter. However, it is so small that it is hidden from view in materials with large atomic magnetic moments. Diamagnetism can only be observed in materials where the electrons form a closed shell structure and have no atomic magnetic moment (figure 1.5).

(b) Paramagnetism

If there is not a very strong interaction between the magnetic moments, then they are oriented in disparate directions and wobble around, constantly changing direction in time due to thermal fluctuations. Such a state is called paramagnetism (figure 1.6).

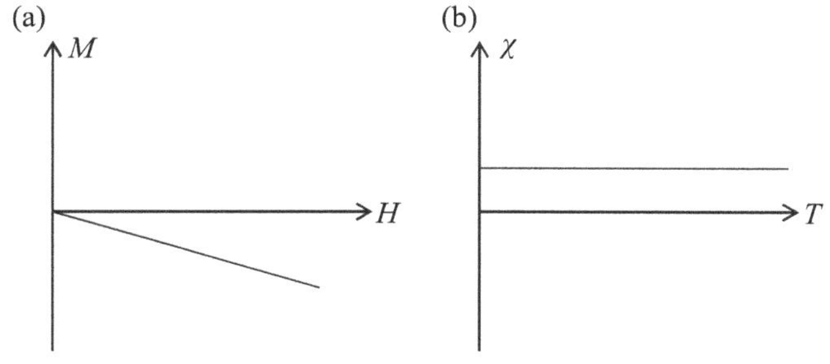

Figure 1.5. (a) *M–H* curve and (b) temperature dependence of magnetic susceptibility for diamagnetic materials.

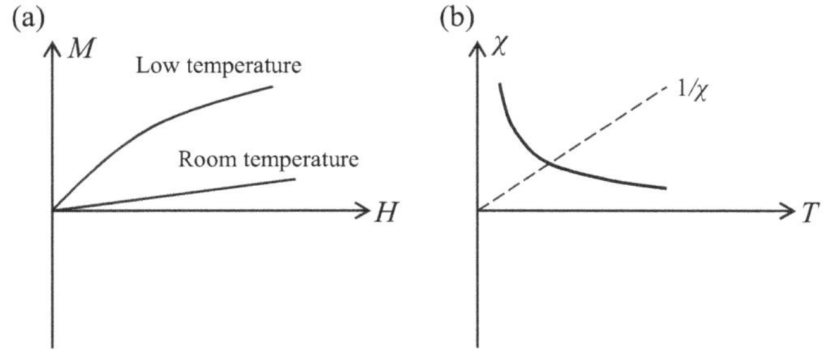

Figure 1.6. (a) *M–H* curve and (b) temperature dependence of magnetic susceptibility for paramagnetic materials.

Magnetization increases in proportion to the magnetic field. At low temperatures and high magnetic fields, magnetization tends to saturate. The magnetic susceptibility can be expressed by the following equation (1.26) according to Curie's law.

$$\chi = \frac{C}{T}. \tag{1.26}$$

(c) Ferromagnetism

The interaction between the magnetic moments acts to align them parallel to each other. As a result, the magnetic moments align and a material with spontaneous magnetization is called ferromagnet. The magnitude of the spontaneous magnetization of ferromagnetic materials depends on temperature. Even in ferromagnetic materials, at high temperatures thermal fluctuations overcome the interaction to align magnetic moments and the magnetic moment becomes disjointed, and spontaneous magnetization does not occur. As the temperature is lowered, spontaneous magnetization occurs at a certain temperature. This temperature is called the Curie temperature and is denoted by T_C. The magnetic susceptibility can be expressed by the following equation (1.27) according to Curie–Weiss law.

$$\chi = \frac{C}{T - T_C}. \tag{1.27}$$

In addition to Fe, Co, and Ni, ferromagnetic materials include the rare-earth metal Gd, which has a Curie temperature of 293 K. Gd does not exhibit spontaneous magnetization until its temperature falls below room temperature. Tb and Dy are also ferromagnetic at low temperatures (figure 1.7).

(d) Antiferromagnetism

In ferromagnetic materials all of the magnetic moments are aligned parallel to each other in the same direction. In contrast, in antiferromagnetic materials, neighboring magnetic moments are aligned antiparallel to each other, indicating that the material as a whole has zero spontaneous magnetization with magnetic moments that cancel each other out. Antiferromagnetism is manifested in various compounds, such as FeO, CoO, NiO, MnO, and MnF_2. The magnitude of the magnetic susceptibility is of the same order for paramagnetic materials. It differs from paramagnetic materials in that the magnetic moments are aligned due to strong interactions between them. At high temperatures, as is the same with ferromagnets, the magnetic moment of antiferromagnetic materials is scattered due to thermal fluctuations. At low temperatures, however, the magnetic moment begins to align at a certain temperature. This temperature is called the Néel temperature and is denoted by T_N. The magnetic susceptibility can be expressed by the following equation (1.28) according to the Curie–Weiss law (figure 1.8).

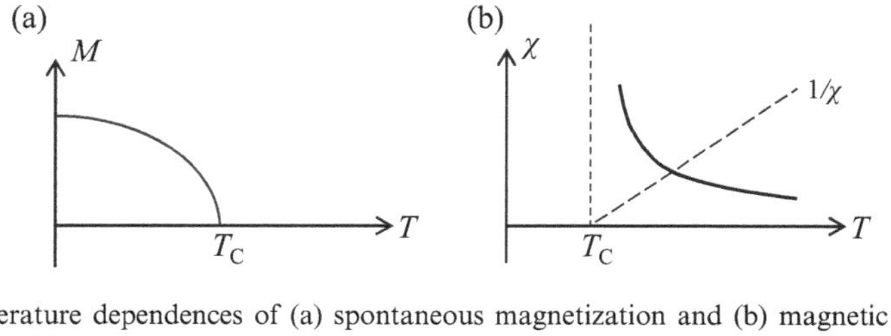

Figure 1.7. Temperature dependences of (a) spontaneous magnetization and (b) magnetic susceptibility for ferromagnetic materials.

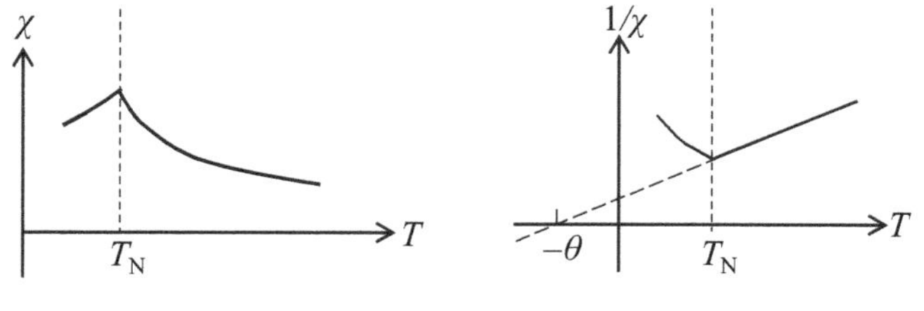

Figure 1.8. Temperature dependences of (a) magnetic susceptibility and (b) inverse of magnetic susceptibility for antiferromagnets.

$$\chi = \frac{C}{T + \theta}. \tag{1.28}$$

(e) Ferrimagnetism

The interaction of neighboring magnetic moments so that their orientations are antiparallel is the same as that of antiferromagnets. However, the magnetic moments do not cancel each other as a whole, resulting in spontaneous magnetization. Ferrimagnetism is mechanically the same as antiferromagnetism but is the same as ferromagnetism in that it has spontaneous magnetization. Even in the case of ferrimagnetism, magnetic moments are oriented in disparate directions at high temperatures due to thermal fluctuations. Below a certain temperature, the magnetic moments align and spontaneous magnetization occurs. This temperature is called the Curie temperature because of its similarity to ferromagnetism (figure 1.9). The magnetic susceptibility can be expressed by the following equation (1.29).

$$\frac{1}{\chi} = \frac{T}{C} + \frac{1}{\chi_0} - \frac{\sigma}{T - \theta}. \tag{1.29}$$

However, the temperature dependence of magnetization depends somewhat differently from that of a ferromagnet. Consider a ferrimagnet that is composed of two sublattices, A and B, in which the total magnetic moment of sublattice A (M_A) is antiparallel to that of sublattice B (M_B) with $| \boldsymbol{M}_A | > | \boldsymbol{M}_B |$. Since the temperature dependence of $| \boldsymbol{M}_A |$ is different from that of $| \boldsymbol{M}_B |$, the temperature dependence of the net magnetization $| \boldsymbol{M}_A - \boldsymbol{M}_B |$ is specific to materials, such as in figures 1.10(a), 1.10(b), and 1.10(c). In figure 1.10(b), $| \boldsymbol{M}_B |$ decreases more rapidly with increasing temperature than $| \boldsymbol{M}_A |$. Therefore, the net magnetization $| \boldsymbol{M}_A - \boldsymbol{M}_B |$ increases as the temperature increases, and reaches a maximum before falling into zero at T_C. This behavior can be seen in the cubic spinel $NiOCr_2O_3$ [4]. Another example is shown in figure 1.10(c). The spontaneous magnetization

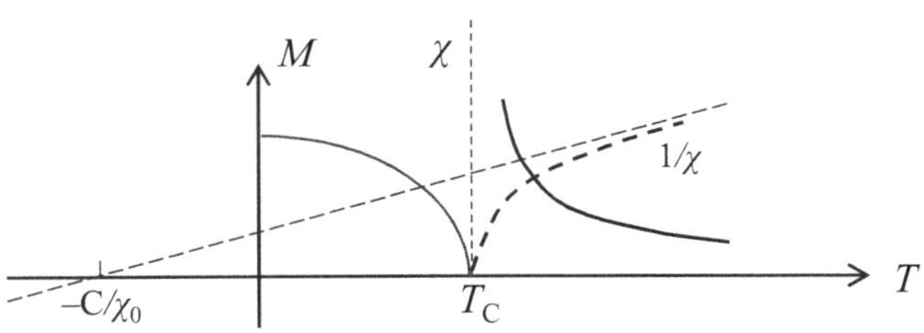

Figure 1.9. Temperature dependences of (a) spontaneous magnetization and (b) magnetic susceptibility for ferrimagnets.

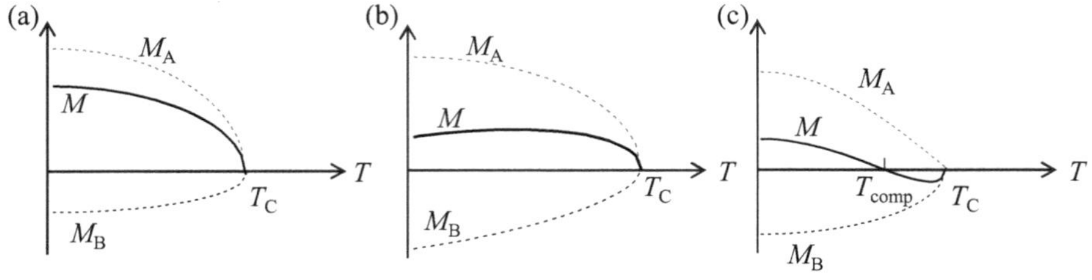

Figure 1.10. Schematics of the spontaneous magnetization of sublattices A and B (broken curves) and resultant magnetization (solid curves) in (a) and (b) typical ferrimagnetic materials, and (c) unusual ferrimagnetic material.

decreases to zero at a certain temperature ($< T_C$), and a spontaneous magnetization points in the opposite direction. This temperature is called the compensation point (T_{comp}). In addition, the $|\, \boldsymbol{M_A} \,|$ and $|\, \boldsymbol{M_B} \,|$ are exactly the same, and thus the net magnetization becomes zero. One example that shows such a behavior is $Li_{0.5}Fe_{1.25}Cr_{1.25}O_4$[5]. If a ferrimagnet contains more than two sublattices, then a more complicated temperature dependence of the spontaneous magnetization may occur (figure 1.10).

The most technologically important ferrimagnets thus far are ferrites [6, 7]. They are ferrimagnetic oxides, and therefore are electrically insulating. As a result, they are used in applications where the electrical conductivity exhibited by most ferromagnetic materials is unfavorable. For example, they have been used in high-frequency applications because an alternating field does not induce undesirable eddy currents in an insulating material. Ferrites are usually manufactured using ceramic fabrication techniques. For example, powdered NiO and Fe_2O_3 are mixed together to produce $NiOFe_2O_3$, and are pressed into shape and then heated. This method has the advantage that the shape of the magnet can easily be controlled by the choice of mold.

1.2.2 Magnetization process of materials

When an external magnetic field ($\boldsymbol{H}$) is applied to a ferromagnetic material, the magnetization of the material develops as shown in figure 1.11. This is called the M–H curve. The magnetization is the vector addition of the total magnetic moment per unit volume. Since a magnetic material is an aggregate of many magnetic moments, the H dependence of M depends on the fact that a torque expressed as $\boldsymbol{N} = \boldsymbol{m} \times \boldsymbol{H}$ acts when an external magnetic field is applied to the magnetic moment. From point O to point P, the work (W) done by the external magnetic field in changing the magnetization corresponds to the gray area and can be expressed as,

$$W = \mu_0 \int H \mathrm{d}M. \qquad (1.30)$$

An actual ferromagnet is a collection of magnetizations oriented in several different directions, as shown in figure 1.12. Regions with magnetization in the

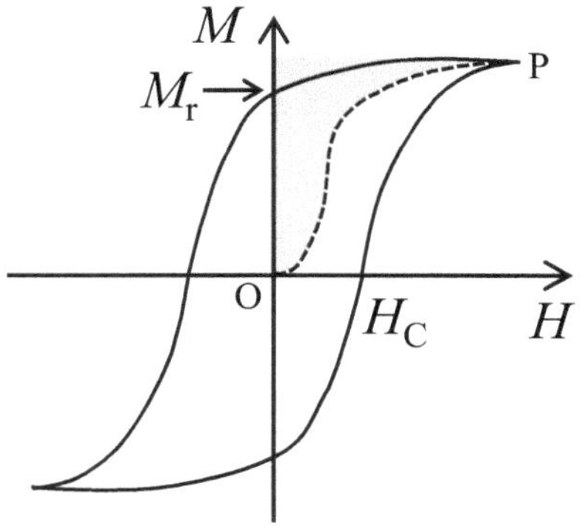

Figure 1.11. Schematics of the *M–H* curve.

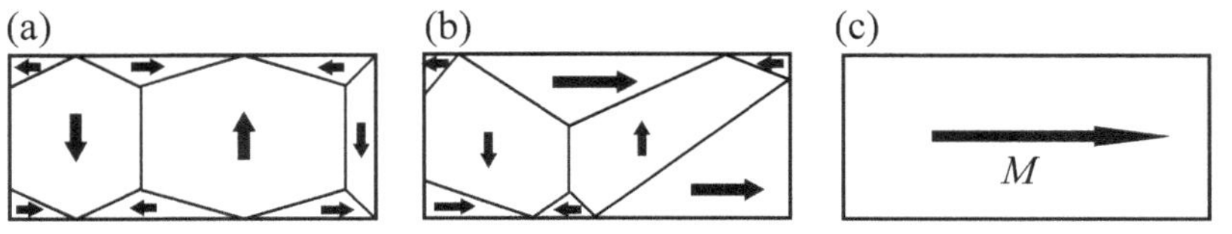

Figure 1.12. Schematics of (a) magnetic domains at $H = 0$, (b) at small H applied in the right direction, and (c) sufficient H for M to saturate.

same direction are called magnetic domains. At the boundaries between magnetic domains, the magnetic moment rotates. This boundary region is called the magnetic domain wall (DW). Magnetic moments in a crystal have a tendency to orient in a certain crystal axis direction, as shown in figure 1.12, which is called magnetic anisotropy. The spin–orbit interaction determines which crystal axis direction the magnetization will face. Most of the magnetic moment is carried by the electron spin. The angular momentum of the electron orbitals is also associated with the symmetry of the crystal. Due to the spin–orbit interaction, the magnetic moment is related to the crystal orientation and magnetic anisotropy appears. When an external magnetic field is applied, the size of the magnetic domain changes, as shown in figure 1.12(b). Magnetic domains with magnetization in a direction different from the magnetic field gradually narrow and eventually become only magnetic domains with magnetization oriented in the direction of the magnetic field, as shown in figure 1.12(c), meaning the saturation of magnetization. Then, as the magnetic field is weakened from this state and the direction of the field is reversed passing through the zero-field state, the magnetization does not follow the same magnitude and reverse direction but rather follows a hysteresis curve. The reversal of magnetization within one magnetic domain occurs by the introduction of magnetic walls. To introduce a magnetic wall into a magnetic domain, the direction of magnetization must be changed against magnetic anisotropy. Extra energy is required to do this, and reversing the magnetic field does not immediately reverse the magnetization. The magnetization at $H = 0$ is the remanent magnetization, and the magnetic field necessary for $M = 0$ is the coercivity. When we see the *M–H* curve in figure 1.11, it should be noted whether demagnetizing field correction is performed or not. In the case of M–H measurements with H applied perpendicular to the plane, the demagnetizing field is $-M$.

1.2.3 Energies in ferromagnets

The balance of several magnetic energies determines the spatial orientation of the magnetization in a magnetic system.

The Zeeman energy E_Z, expressed below, is the interaction between the magnetization $\boldsymbol{M}$ and the magnetic field $\boldsymbol{H}$,

$$E_Z = -\mu_0 \boldsymbol{M} \cdot \boldsymbol{H}. \tag{1.31}$$

Magnetic anisotropy has many origins, including shape anisotropy, crystalline anisotropy, or interfacial anisotropy. Among them, the magnetocrystalline anisotropy energy E_C depends on the angle between a crystalline axis and magnetization. We will next have a look at a system with a uniaxial anisotropy. The E_C is expressed as,

$$E_C = K_u \sin^2 \theta. \tag{1.32}$$

Here, K_u is the uniaxial anisotropy constant, and θ is the relative angle between the magnetization and a crystalline axis.

Using the interfacial anisotropy constant K_I and the magnetic film thickness t, the interfacial anisotropy energy E_I is given by,

$$E_I = \frac{K_I}{t} \sin^2 \theta. \tag{1.33}$$

Equation (1.33) shows that the E_I becomes high with decreasing t.

The origin of ferromagnetism, antiferromagnetism, and ferrimagnetism is the exchange interaction between magnetic moments, and its energy is expressed using two neighboring spins S_1 and S_2 as,

$$\varepsilon_{ex} = -J S_1 \cdot S_2. \tag{1.34}$$

Here, J is the exchange integral. A negative (positive) value of J leads to ferrimagnetism or antiferromagnetism (ferromagnetism).

The magnetic dipolar energy E_d originates from all of the interactions of magnetic dipoles in a system. When the shape of the system is ellipsoidal and the magnetization is homogeneous, E_d can be expressed using a diplolar field H_d acting on the magnetization as,

$$E_d = -\frac{\mu_0}{2} \boldsymbol{M} \cdot \boldsymbol{H}_d, \quad \boldsymbol{H}_d = -\overline{\overline{N}} \boldsymbol{M}. \tag{1.35}$$

Here, $\overline{\overline{N}}$ is the demagnetizing tensor. The demagnetizing tensor of a ferromagnetic thin film is diagonal with only one non-zero demagnetizing factor, equal to 1 along the z axis. Thereby, E_d can be written as,

$$E_d = -\frac{\mu_0}{2} M_S^2 \sin^2 \theta = K_d \sin^2 \theta. \tag{1.36}$$

Here, K_d is the shape anisotropy, θ is the angle of the magnetization relative to the z axis, and M_S is the spontaneous magnetization. The out-of-plane magnetized state has a higher energy and is not stable. Therefore, the magnetization tends to be oriented in the in-plane direction in thin films. This interaction is thereby called shape anisotropy. It is noted that the expression of E_d is similar to that of E_C. The uniaxial anisotropy K_u must overcome K_d to obtain perpendicularly magnetized films. The difference between these two components is called the effective uniaxial anisotropy K_u^{eff}. If the difference ($K_u^{eff} = K_u - K_d$) is positive, then the magnetization is in the out-of-plane direction. Since the energy of the magnetic wall is the exchange interaction energy and the magnetic anisotropy energy, the energy of the magnetic material is the sum of the magneto static energy, the exchange interaction energy, the magnetic anisotropy energy, and the elastic energy due to magnetostriction, which determines the magnetic domain structure to minimize it.

1.2.4 Magnetic domain and magnetic domain wall

As shown in figure 1.12, several magnetic domains exist. This happens because if there are no magnetic domains and all magnetic moments are perfectly oriented in the same direction, then magnetic dipoles would be generated on the surface and the diamagnetic field associated with those dipoles would destabilize the magnetization. In other words, it can be said that the energy of the ferromagnetic material becomes higher and more unstable by the energy of the shape magnetic anisotropy. Such energy produced with the generation of magnetic poles is called magnetostatic energy. When the material has some magnetic domains, such as in figures 1.13(b) and (c), the magnetostatic energy decreases. However, it is not the case that energy will go down if as many magnetic domains as possible can be created. This happens because there are magnetic DWs between magnetic domains, wherein the magnetic moment is twisted between the two adjacent magnetic domains as shown in figure 1.14. This means that such twisted magnetic moments are unstable from the viewpoint of exchange interaction, whose energy is optimized if neighboring magnetic moments are parallel or as close to parallel as possible to each other. In other words, magnetic DW has energy, and thus too much increase in magnetic

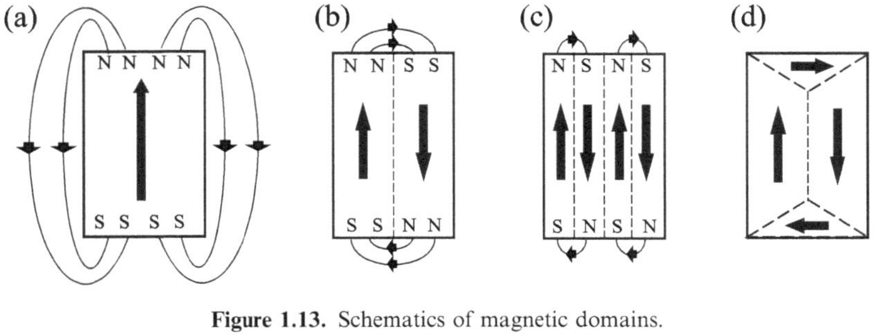

Figure 1.13. Schematics of magnetic domains.

walls is also undesirable. The magnetic domains shown in figure 1.13(d) are called closure domains, wherein magnetic poles are not generated. However, the magnetization of each magnetic domain is oriented in different directions, and thus such a domain structure does not respond well to expansion and contraction of the magnetic material due to magnetorestriction. In addition, when the material is uniaxial, there are magnetic domains with their magnetization orienting in the hard-axis direction, increasing the energy. Thus, the question of how magnetic domains are formed is not simple to answer.

The width of the DW is again determined by a balance between competing energy contributions. The exchange energy is optimized once neighboring magnetic moments are aligned in parallel. For this reason, wide walls are preferred so that the angular variation of magnetic moments between adjacent atomic planes is as small as possible. However, the magnetocrystalline anisotropy is optimized when the moments are as closely aligned to the easy axes as possible. This favors narrow walls with a sharp transition between the domains so that few moments with unfavorable crystalline alignment exist in the transition region. In practice, a compromise is reached where the total energy across the boundaries is minimized. The most energetically favorable type of magnetic DW is one that does not generate magnetic poles in the material, and thus does not introduce a demagnetizing field. One such magnetic DW is a Bloch wall, illustrated for a 180° boundary. In this case, the magnetization perpendicular to the boundary does not change across the wall. Thus, no magnetic poles or demagnetizing fields occurs. Another type of magnetic DW is called a Néel wall and occurs in thin films of magnetic materials. In Néel walls, the spins rotate in the plane normal to the DW, rather than around an axis normal to the DW. The spin rotations in a Bloch wall and in a Néel wall are shown in figures 1.14(a) and (b), respectively. Néel walls have an energetic advantage in thin films because the free poles are formed on the wall's surface rather than on the film's surface, which reduces the magnetostatic energy.

The width of the DW (Δ) is given by equation (1.37) using the exchange stiffness constant (A) and the uniaxial magnetic anisotropy constant (K_u) as,

$$\Delta = \sqrt{A/K_u}. \tag{1.37}$$

Here, the magnitude of A is expressed as,

$$A = \frac{S^2 J_{ex}}{a}, \tag{1.38}$$

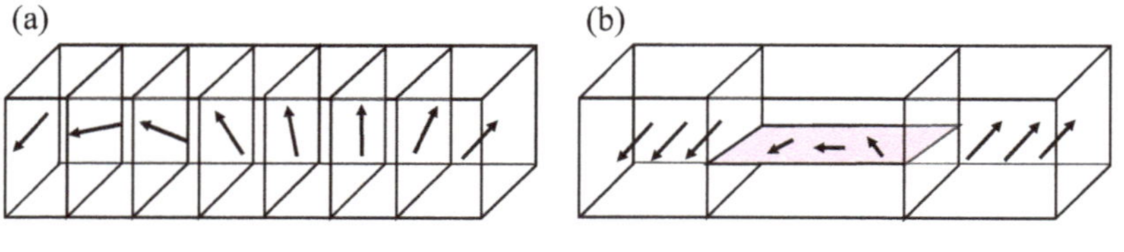

Figure 1.14. Schematics of rotation of magnetic moments in (a) Bloch wall and (b) Néel wall.

where S is the spin angular moment, a is the lattice constant, and J_{ex} is the exchange integral. The DW energy (E_{ex}) caused by exchange interaction is therefore expressed as,

$$E_{ex} = \frac{2A}{\Delta}. \tag{1.39}$$

Meanwhile, the DW energy due to the anisotropy (E_a) is,

$$E_a = 2K_u\Delta. \tag{1.40}$$

Equations (1.39) and (1.40) show that the summation of E_{ex} and E_a reaches the minimum at a certain value of Δ. For example, in the case of $A{\sim}10^{-11}$ J m^{-1}, the DW thickness is around 100 nm for soft ferromagnets ($K_u \sim 10^3$ J m^{-3}), and around 10 nm for a hard ferromagnets ($K_u \sim 10^5$ J m^{-3}).In the case of thin films, when the film thickness decreases and becomes comparable to the DW width, the standard Bloch walls cannot be maintained. Here, the wall has a smaller energy when the spins in the wall are rotating in the plane normal to the wall, using an in-plane rotation rather than the standard Bloch wall.

1.3 Magnetic domain wall dynamics

1.3.1 LLG equation

The dynamics of magnetization in the absence of spin transfer torque (STT) are described by the Landau–Lifshitz–Gilbert (LLG) equation as [8, 9],

$$\frac{d\boldsymbol{m}}{dt} = \gamma_0 \boldsymbol{H}_{\text{eff}} \times \boldsymbol{m} + \alpha \boldsymbol{m} \times \frac{d\boldsymbol{m}}{dt}. \tag{1.41}$$

Here, $\boldsymbol{m}$ is the unit vector along the magnetization, γ_0 is the gyromagnetic ratio, and α is the damping constant. $\boldsymbol{H}_{\text{eff}}$ is the effective field, which is the sum of the external fields and any other effects acting on the magnetization such as the magnetic anisotropy, and is given by the first derivative of the total energy E with respect to $\boldsymbol{m}$,

$$\boldsymbol{H}_{\text{eff}} = -\frac{1}{\mu_0 M_S}\frac{dE}{d\boldsymbol{m}}. \tag{1.42}$$

The first term of equation (1.41) shows the Lamor precession of the magnetic moment around the $\boldsymbol{H}_{\text{eff}}$. This precession's angular velocity is proportional to $\boldsymbol{H}_{\text{eff}}$ and γ_0. The second term is the damping term, corresponding to the relaxation of the magnetic moment. A schematic of the torques acting on the magnetization is shown in figure 1.15.

1.3.2 Spin-transfer torque (STT)

The STT due to spin-polarized current has been proposed by theory in 1996 [10, 11]. When the conduction electrons go through a region such as a DW, wherein the magnetic moment gradually changes, the spin polarization of the conduction electrons becomes misaligned with that of the local magnetic moment there. Such

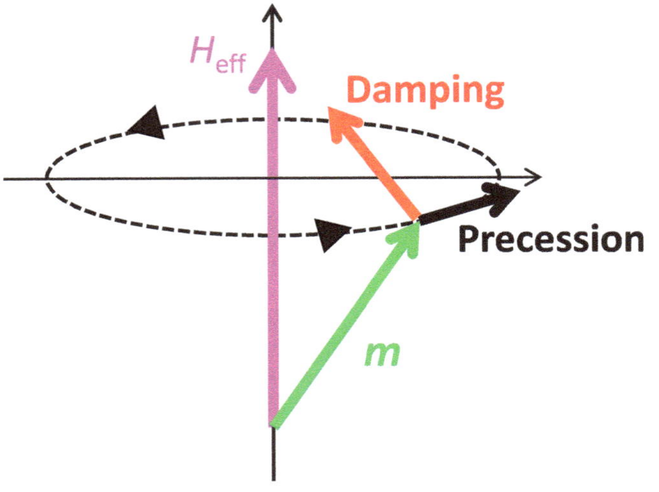

Figure 1.15. Torques acting on magnetization.

a misalignment generates a torque on the local magnetic moment. The effective field causes two different torques: (i) an adiabatic torque resulting in spin relaxation towards the effective field and (ii) a non-adiabatic torque resulting in spin precession around the effective field. By adding these two torques, the LLG equation is given as [12, 13],

$$\frac{d\boldsymbol{m}}{dt} = \gamma_0 \boldsymbol{H}_{eff} \times \boldsymbol{m} + \alpha \boldsymbol{m} \times \frac{d\boldsymbol{m}}{dt} - (\boldsymbol{u} \cdot \nabla)\boldsymbol{m} + \beta \boldsymbol{m} \times [(\boldsymbol{u} \cdot \nabla)\boldsymbol{m}]. \tag{1.43}$$

$$\boldsymbol{u} \approx \frac{g\mu_B P}{2eM_S}\boldsymbol{J}. \tag{1.44}$$

Here, $\boldsymbol{u}$ is the spin-drift velocity, β is the non-adiabatic factor, P is the spin polarization of the electric current, g is the g-factor, e is the elementary charge, and $\boldsymbol{J}$ is the current density flowing in the direction of electrons. The third term in equation (1.43) is the adiabatic or damping-like torque. The fourth term is the non-adiabatic or field-like torque, whose direction is the same as the first term, and therefore assists the magnetization's precessional motion. It tends to align $\boldsymbol{m}$ with the spins of the incoming electrons. The four torques acting on the magnetization are shown in figure 1.16.

As the conduction electron passes through the DW and enters the next magnetic domain, its spin moment rotates by 180° and aligns with the local magnetic moments. This is related to the fact that the angular momentum of the conduction electrons is transferred to the magnetization of the localized electrons and acts as a torque. Spin torque from spin-polarized currents faces an important practical upper limit in efficiency because the strength of the torque is limited by the amount of angular momentum carried by an individual charge carrier, and the efficiency factors are typically smaller than 1. Spin torque due to a spin-polarized current is

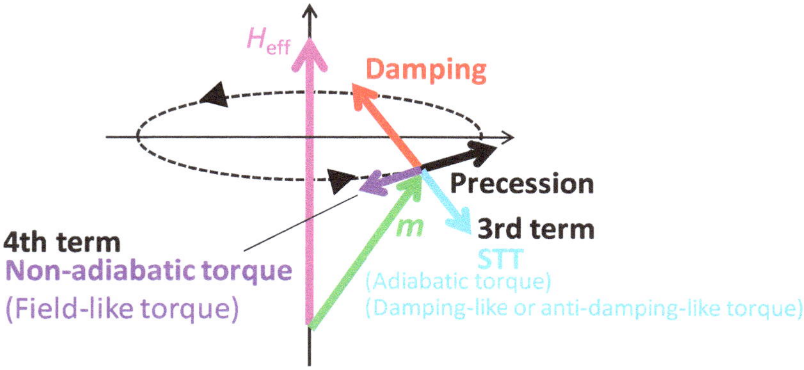

Figure 1.16. Four torques acting on the magnetization.

also not energy efficient due to the associated charge current. In a typical magnetic tunnel junction used for memory devices, each electron dissipates several tenths of eV on average. In contrast, the energy required to excite a magnon and contribute to magnetic dynamics is only about 10 μeV. Therefore, creating ultra-dense embedded magnetic memory devices will likely require a different and more efficient mechanism for controlling magnetic devices. This can be done by spin–orbit torque (SOT).

1.3.3 Spin–orbit torque (SOT)

Current-induced torques with efficiencies greater than 1 can be generated by using SOTs to transform a flowing charge current into a pure spin current in the direction perpendicular to the charge current. In the case of STTs, the spin polarization of conduction electrons is inevitable, while spin current or spin accumulation at the interface between heavy metal and magnetic layers is responsible for SOTs. In such heterostructures, the spin current is generated in the heavy metal by 3D bulk mechanisms, such as the spin Hall effect, and 2D interfacial mechanisms that produce an inverse spin galvanic effect, known as the Rashba-Edelstein effect. These spin currents and/or the spin accumulation act as an effective field and exert torques on the magnetic moments in magnetic layers, leading to the magnetization reversal and the DW motion. Let us now discuss SOTs by spin Hall effect. In heavy metals such as Pt and W, which have a strong spin–orbit coupling (SOC), the charge current in such heavy metal layers can be converted into a pure spin current flowing vertically toward the magnetic layer due to scatterings. When this spin current is absorbed by the magnetic layer in contact with the heavy metal layer, it creates a SOT in the magnetic layer. The SOT created by the spin Hall effect is called a damping-like or an anti-damping-like torque, which is in the same direction as the adiabatic STT in equation (1.43). The resulting torque to be added into the LLG equation is expressed as,

$$T_{\text{spin Hall effect}} = \gamma_0 \frac{|g| \mu_B \theta_{\text{SH}}}{2eM_s t}[\boldsymbol{m} \times (\boldsymbol{m} \times \boldsymbol{\sigma})]. \tag{1.45}$$

Here, t is the magnetic layer thickness, θ_{SH} is the spin Hall angle, which is the ratio of the vertically flowing spin current density to the applied in-plane charge current density in dimensionless units, and $\boldsymbol{\sigma}$ is the direction of the polarization of the spin current. The strength of the spin Hall effect is usually parametrized by θ_{SH}. The sign of θ_{SH} is positive for Pt, but negative for Ta and W [14]. Figure 1.17 shows a schematic of the current-induced effective field by SOTs in a system with a spin Hall effect [15].

We next move on to the SOTs by Rashba effect. This is called a field-like torque and is the same direction as that of the non-adiabatic STTs in equation (1.43). The term to be added into the LLG equation is given as,

$$T_{\text{Rashba}} = \gamma_0 \frac{\alpha_P P}{\mu_0 \mu_B e M_s} (\boldsymbol{m} \times \boldsymbol{\sigma}). \tag{1.46}$$

Here, α_P is the Rashba coefficient representing the strength of the Rashba SOC. In fact, both damping and field like contribution can arise from spin Hall effect or Edelstein effect. Thus, the situation is actually more complicated.

Here we see how the magnetic moment in ferromagnetic layers (CoFeB) flips by SOTs when the anti-damping-like torque is dominant [15]. Provided that charge

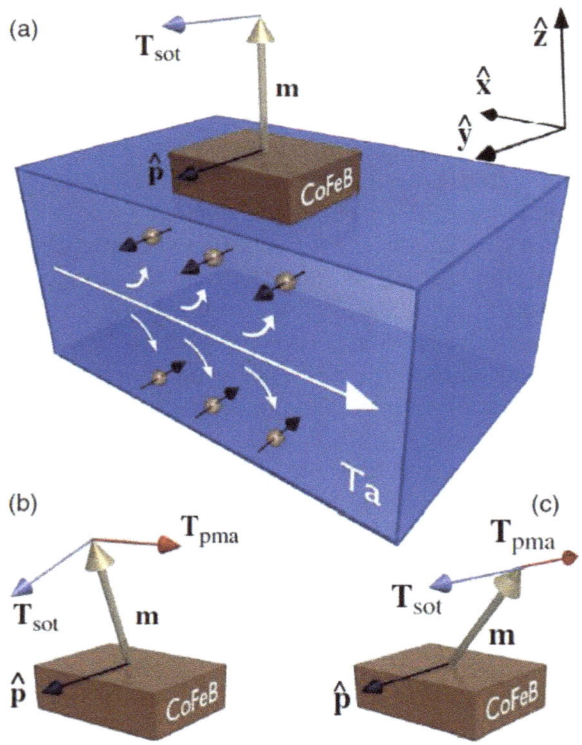

Figure 1.17. Schematic of the current-induced effective fields by SOTs acting on the magnetization. Reprinted figure with permission from [15], Copyright (2017) by the American Physical Society.

current $\boldsymbol{j}$ flows in the x direction in Ta layers, which is a strongly spin–orbit-coupled nonmagnetic metal, it creates a bulk spin Hall effect and thus gives rise to y-polarized spin moments, shown with small dark blue arrows, deflected along the z direction. This mechanism creates a spin current with spin moments pointing along the y direction that flows along the z axis into the CoFeB with PMA and magnetization $\boldsymbol{m}$ and travels some distance before losing spin polarization, causing spin–orbit torque T_{sot} (blue vector) that acts on $\boldsymbol{m}$ (gold vector). In other words, when the pure spin current is absorbed by the ferromagnet, it creates a spin transfer torque. In response to the onset of current and spin flow, $\boldsymbol{m}$ tilts in the direction of T_{sot} i.e., toward the y axis, shown in (b). As $\boldsymbol{m}$ tilts away from the z axis, T_{sot} also tilts as shown in (b). As a result, a new effective field-like torque develops due to PMA, T_{pma} (red vector). Below the critical current, which depends on the strength of anisotropy, $\boldsymbol{m}$ precesses with decreasing amplitude depending on the damping constant α and reaches a steady state, wherein $\boldsymbol{m}$ is tilted in the $x–z$ plane where the two torques cancel, as shown in (c).

From the viewpoint of spin torque efficiency, SOT is superior to STT. We see this by taking a spin Hall effect in a bilayer structure consisting of heavy metal (HM) and magnetic layers (ML) in figure 1.18. The efficiency of the spin torque generated by the spin Hall effect in total is the sum of the rate at which angular momentum is absorbed by the magnetic layer per unit charge current. Since the charge current density J and the spin current density go through regions of very different areas, which are wt for the charge current density and wL for the spin current density, the torque efficiency for a spin Hall structure is given by [15],

$$\frac{T_{\mathrm{sot}} wL}{Jwt} \propto \theta_{SH} \frac{L}{t} \frac{g\mu_B}{e}. \tag{1.47}$$

The ratio L/t can be 30 or more even for a very small structure. Therefore, the torque caused by the spin Hall effect is more efficient than one quantized spin per unit charge by more than one order of magnitude, which far exceeds conventional STTs. From a microscopic perspective, the in-plane charge current of each electron is used many times to transfer torque to the ferromagnetic material, as shown in figure 1.18(b) [16]. Electrons become spin-polarized in the HM by spin–orbit coupling when they are deflected toward the ML. At the interface, they transfer angular momentum to the local magnetic moments through the same mechanism as for the STTs. However, because there is no net electron flow into the ML, the

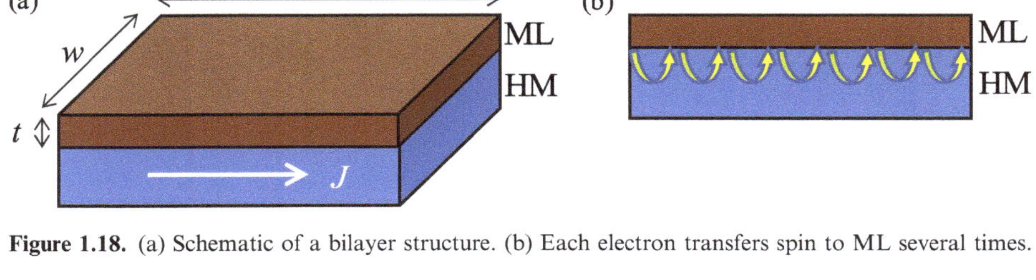

Figure 1.18. (a) Schematic of a bilayer structure. (b) Each electron transfers spin to ML several times.

electrons eventually diffuse back into the heavy metal, and this process is repeated many times. The relevant length scale is the spin diffusion length in the HM (~2 nm). Thus, many cycles of spin transfer occur despite in a compact magnetic structure. This means that large θ_{SH} occurs in materials with large electrical resistivity, typically Pt alloys.

1.4 Spintronics applications of ferrimagnet

Spintronics is an expanding field of science and technology that utilizes not only electric charge but also electron spin. The pioneering work of Fert and Grünberg on the giant magnetoresistance (GMR) is considered the birth of spintronics. In 2007, they were awarded the Nobel prize in Physics [17, 18]. GMR is observed in a layered structure, wherein a nonmagnetic layer is sandwiched by two magnetic layers. When the current flows perpendicular or parallel to the plane, the total resistance of the stack changes depending on the relative angle of magnetizations between the two magnetic layers, with the parallel magnetization direction having a smaller resistance than the antiparallel one. The difference in resistance between the antiparallel and parallel states is important (figure 1.19). The GMR value reaches up to 40% in Co/Cu multilayers [19]. Later even higher magnetoresistances were obtained, up to 1800% at 4 K [20] and 604% at room temperature [21] in a stacked structure by replacing the nonmagnetic metal layer with an insulating barrier layer. Such a large difference in magnetoresistance is called tunneling magnetoresistance (TMR) [22, 23], and was discovered originally by Jullière in 1975 [24]. The mechanism of TMR involves electrons tunneling through the insulating barrier between the two magnetic layers, resulting in a dependence of resistance on the magnetization orientations of the two electrodes. Beyond the fundamental interest associated with the discovery of spin transport phenomena in materials, this discovery has enabled the development of a new generation of TMR and GMR read-heads for magnetic hard disks drives [25–27].

Spintronics has created several potential applications in memory and logic devices, including magnetic random-access memories (MRAMs) [28], DW MRAMs [29], and DW logic devices [30–34]. Among these devices, MRAMs have become a commercial memory product [35], while TMR and GMR are used

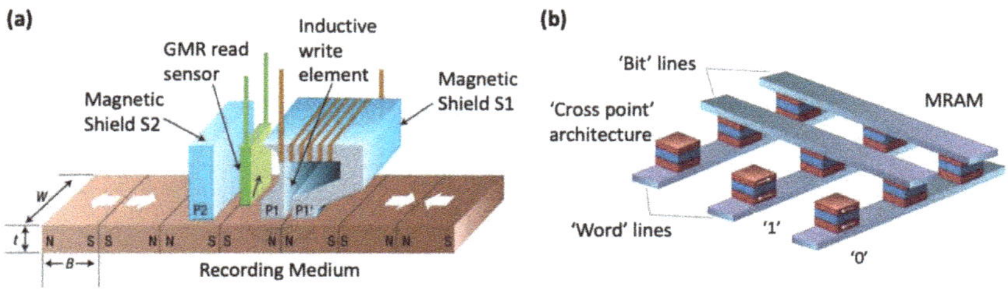

Figure 1.19. (a) GMR head for hard disk, consisting of a writing element followed by the GMR read sensor. (b) MRAM in the basic cross point architecture. Reprinted from [27], Copyright (2007), with permission from Springer Nature.

for memory and sensor applications [28, 36]. The advantage of spintronics over the current CMOS technology is the non-volatility, i.e., no power consumption during data retention. In fact, data can be stored almost indefinitely in the magnetized state of nanomagnets. Thus, spintronics could lead to the next generation of ultrafast, high-density, and low power-consumption nanoelectronic devices.

1.4.1 Devises using domain wall motion

Both ferromagnetic and ferrimagnetic materials consist of magnetic domains, wherein the magnetization is homogeneous in each magnetic domain. The boundary between these magnetic domains is a magnetic DW. Controlling the movement of these DWs will enable us to manipulate the magnetization of spintronics nanoelements. For example, in perpendicularly magnetized thin films, the magnetization in the DWs rotates by 180° from one side of the DW to another side [37–39]. Such magnetic textures are very interesting because they can be moved with the help of magnetic fields [40–42] and/or currents [43–45]. Current-induced DW motion by STTs was theoretically predicted by Berger in 1978 [46]. Since then, especially in the past two decades, it has been a major focus in spintronics for the development of applications such as racetrack memories [47, 48], DW MRAM [29], and DW based logic devices [30–32, 49]. Racetrack memory uses currents to move a row of magnetic domains along the magnetic wire, which passes through a read/write element, as shown in figure 1.20. The red and blue represent magnetic domains, and writing is done using the magnetic field generated by an electric current. Improved magnetic detection capabilities based on the development of spintronic magnetoresistive sensors have made it possible to achieve much higher bit densities using much smaller magnetic domains. MRAM can replace dynamic RAM (DRAM) because of its low-power, high speed, and non-volatility [28]. The development of racetrack memories could lead to potential HDD replacement candidates and

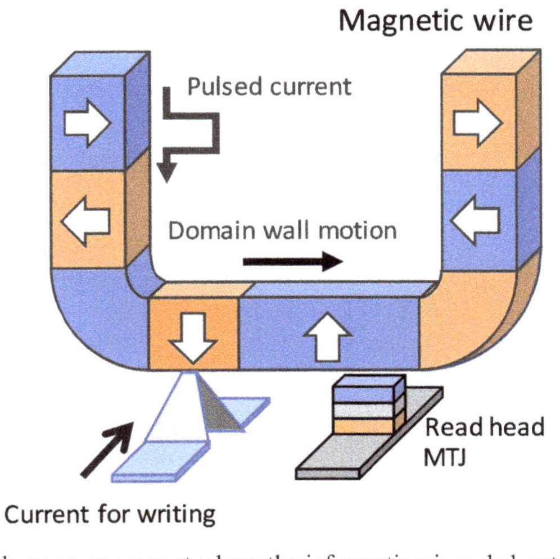

Figure 1.20. Racetrack memory concept where the information is coded onto magnetic domains.

DW-based logic could replace conventional transistors in logic functions. STTs [13, 50] and SOTs [51] are the two major mechanisms for current-induced DW motion. In the case of the STTs acting on a DW, the charge current is spin-polarized by the local magnetization through the *s-d* interaction [50, 52]. This incoming spin-polarized current then provides the local magnetic moments of the DW with its angular momentum, resulting in the DW motion [13, 53]. In the case of SOTs, a heavy metal layer in contact with the ferromagnetic layer is required to generate a spin-polarized current by spin Hall effect [54–56]. The Rashba-Edelstein effect can also cause SOTs [57, 58]. The spin current generated in the heavy metal accumulates and diffuses into the ferromagnetic layer, generating SOT that drives the DW motion [59–63].

1.4.2 DW motion in ferrimagnets

DW motion has been studied extensively in ferromagnetic materials, but recently interest has been growing in DW motion in ferrimagnetic materials. Ferrimagnetic materials consist of two sublattices with magnetic moments oriented in the opposite directions, wherein the magnetization of one sublattice is higher than that of the other one. As a result, the magnetization of ferrimagnet is much smaller than that of ferromagnet. Because the ferrimagnetic sublattice is composed of different elements and atoms, the net angular momentum is also lower. Low magnetization and low angular momentum are important to increase a DW velocity. In fact, STTs and SOTs work more efficiently with smaller magnetization [13, 53].

In some ferrimagnetic materials, the magnetization and the angular momentum can be controlled by changing the temperature of the sample or adjusting the composition of the material, reducing the magnetic moment and angular momentum of the two sublattices to the point where they completely cancel out each other [64], which are called the magnetic compensation point (MCP) and the angular momentum compensation point (ACP), respectively [65, 66]. Such compensated ferrimagnets are intermediate cases between antiferromagnets and ferromagnets and are very interesting to study. They can be also useful to investigate spin transport and dynamics in anti-ferromagnetically coupled systems, which are the focus of attention in antiferromagnetic spintronics [67]. In fact, such ferrimagnets have a significant advantage over antiferromagnets because ferrimagnets have finite Zeeman couplings, which can be more easily manipulated by magnetic fields, as can spin textures such as DWs and skyrmions. In principle, the spin polarization makes it possible to polarize and detect the spin current. This is essential for the study of spin dynamics (figure 1.21).

Figure 1.22 summarizes current-induced DW motion (CIDWM) velocities achieved in various materials. According to recent studies, the precessional torque is negligible at the ACP, resulting in a very large DW velocity [68–70], and very fast and efficient magnetization switching [71, 72]. Early experiments were conducted with ferromagnetic materials using STTs [73–75], but gradually the focus has shifted to ferrimagnetic materials. Most experiments up to now have been done using SOTs without [76] or with field assistance. Because the efficiency of STTs turned out to be

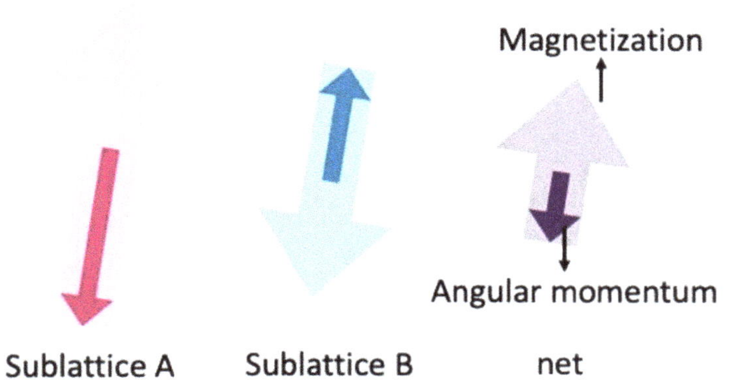

Figure 1.21. Diagrams of magnetization and angular momentum in ferrimagnet. The light red (light blue) arrow shows the magnetization, while the dark red (dark blue) arrow shows the angular momentum of sublattice A(B). The light and dark purple arrows represent the net magnetization and angular momentum, respectively.

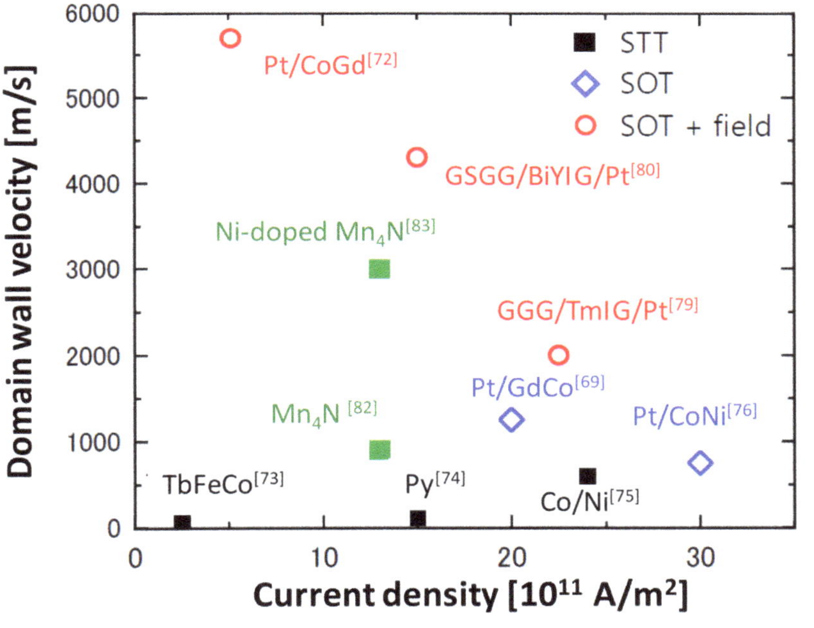

Figure 1.22. CIDW velocity achieved in materials. Record velocity was achieved in $(Mn,Ni)_4N$ purely driven by STTs.

quite disappointing, SOTs have dominated scientific activity on DWs for the last decade [60, 62]. Recent experiments rely on ferrimagnets containing rare earths such as GdCo [68, 69], TbCo [77, 78], TmYIG [79], and BiYIG [80]. However, the current densities for DW propagation are still high in such material systems, and thus the assistance of an external in-plane field is necessary for DW propagation. Some of these materials are insulating ferrimagnets that complicate the electrical readout.

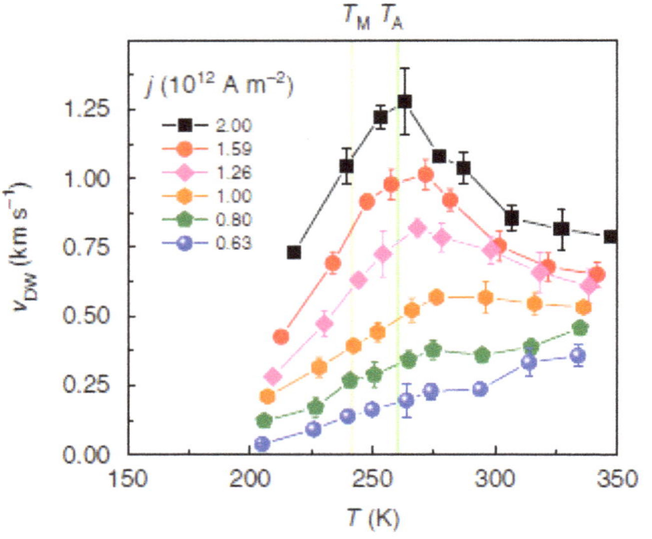

Figure 1.23. Temperature dependence of CIDW velocity at an angular momentum compensation temperatures (T_A). Reproduced with permission from reference. Reprinted by permission from Springer Nature [69], Copyright (2018).

In addition, the use of rare earths is costly from an ecological point of view and important from a geopolitical point of view [81]. Quite recently, ultrafast CIDWM driven purely STTs have been demonstrated in rare-earth free Mn_4N [82]and Ni-doped compensated Mn_4N strips [83].

1.5 Theory of current-induced DW motion

1.5.1 Current induced DW motion

Here we focus on the theory of current-induced DW motion (CIDWM) in magnetic strips. CIDWM is a form of STT-driven magnetization reversal [84]. We consider a flow of electrons through the DW. Figures 1.24(a) and (b) are schematics of a simplified CIDWM mechanism, wherein we only consider the damping-like STT [85]. In reality, magnetic moments are also subjected to the precession torque, the field-like STT, and the damping torque, leading to a more complex dynamics that includes DW motion and magnetization precession within the DW [86].

Let us review the characteristics of DW dynamics under spin-polarized current and magnetic field, wherein the 1D model proposed by Malozemoff and Slonczewski is used [87]. Figure 1.25 illustrates a schematic of the perpendicularly magnetized 1D model. The distribution of magnetization is assumed to be uniform in the y and z axis directions, *i.e.* $\frac{\partial \boldsymbol{m}}{\partial y} = \frac{\partial \boldsymbol{m}}{\partial z} = 0$. To describe the DW motion in the 1D model, we choose two collective coordinates, which are the DW position, q, and the azimuth of the magnetization in the DW, φ. The DW profile is expressed in spherical coordinates as,

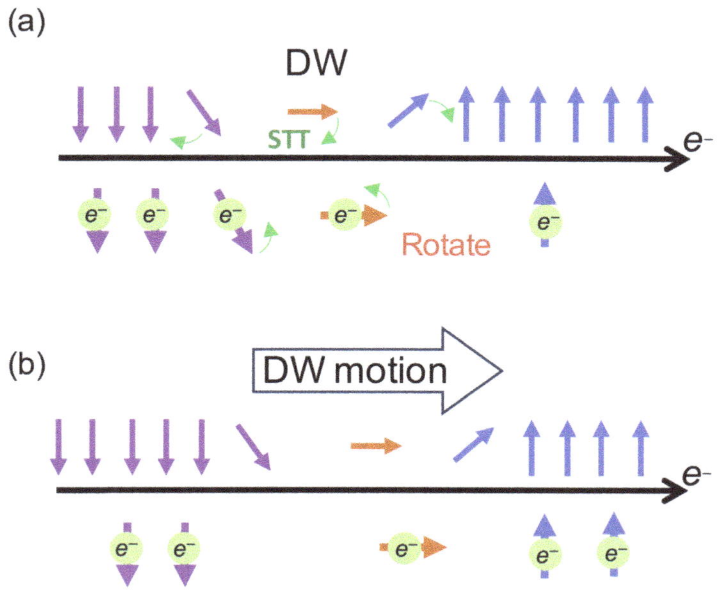

Figure 1.24. Diagrams of CIDWM. Arrows indicate magnetic moments. Electrons flow from left to right. (a) As electrons pass through the DW, the electron spins rotate to align along the localized magnetic moment. The STT acts as a counteraction on the localized moment. (b) When the STT acts on the localized moment, the DW moves with the electron flow in this model. Reproduced from [85]. CC BY 4.0.

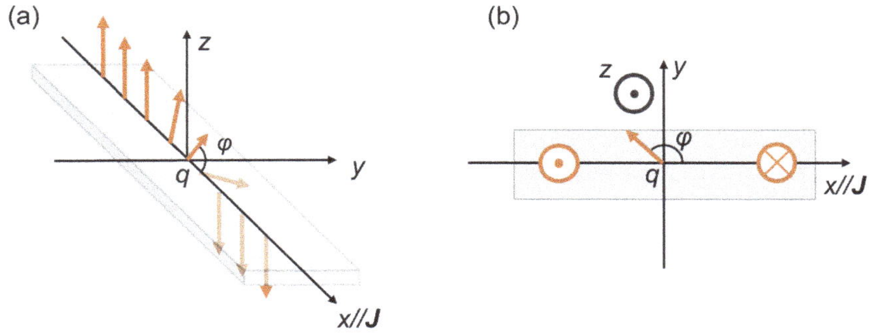

Figure 1.25. 1D magnetic strip model. (a) Distribution of the magnetization around the DW. The orange arrows denote local magnetic moments. (b) Top view of (a). q and φ are the position of DW and the azimuth of magnetic moment in DW, respectively. Reproduced from [85]. CC BY 4.0.

$$\theta(x, t) = 2 \arctan \exp\left(\frac{x - q(t)}{\Delta}\right), \quad \phi(x, t) = \varphi(t). \tag{1.48}$$

Here, $m = (\sin\theta\cos\phi, \sin\theta\sin\phi, \cos\theta)$. By a Lagrangian approach [88], we can find the equations of motion under a current flowing in the x direction and a magnetic field in the easy-axis z direction as,

$$\frac{1}{\Delta}\dot{q} - \alpha\dot{\varphi} = \frac{u}{\Delta} + \frac{\mu_0\gamma H_K}{2}\sin 2\phi,\ \dot{\varphi} + \frac{\alpha}{\Delta}\dot{q} = \beta\frac{u}{\Delta} + \mu_0\gamma H_{\text{app}}. \tag{1.49}$$

Here, Δ is the DW width, given by $\Delta(\varphi) = \sqrt{\frac{A}{K_{\text{u}} + K_{\text{D}}(\varphi)}}$ for 1 D model, where A is the exchange stiffness, $K_D = \frac{\mu_0 M_S^2}{2}(N_x \sin^2\varphi + N_y \cos^2\varphi - N_z)$, where N_x and N_y are the demagnetizing coefficients. $H_k = \frac{2K_D}{\mu_0 M_S}$ is the DW demagnetizing field [13].

When a magnetic field drives a DW, the DW moves in the steady regime at a constant value of φ up to a field called the Walker field (H_W). At above H_W, the DW moves by transforming continuously, i.e., $\dot{\varphi} \neq 0$, from transverse to vortex DW for in-plane systems and from Bloch to Néel DW in out-of-plane systems. This continuous transformation causes the velocity to decrease until it reaches a second linear v-H region with a reduced mobility.

When the DW is driven by a spin-polarized current, in the adiabatic limit ($\beta = 0$) the DW moves continuously only when it can initiate a precession motion and align with the incoming spin polarization. This occurs above a critical current density J_c given by,

$$J_C = \frac{\mu_0 e}{\hbar P}\Delta(\varphi) M_S H_K. \tag{1.50}$$

According to equation (1.50), J_c depends on the direction of magnetization. In an in-plane magnetized system, the transverse anisotropy is expressed as,

$$K_D = \left| K_z - K_y \right| \approx |\, K_z \,| = \frac{1}{2} M_S^2 \mu_0 - K_u. \tag{1.51}$$

Introducing typical values yields a transverse anisotropy constant of the order of 1 MJ m^{-3}. Meanwhile, the transverse anisotropy constant is small in thin films with perpendicular magnetization, which is associated with the energy difference between Bloch and Néel DWs. In this case $K_D = \left| K_x - K_y \right|$. For a strip of small thickness t and width W, the DW can be modeled as an ellipse, and thus the demagnetizing factors can be expressed as [53],

$$K_x \approx \frac{1}{2} M_S^2 \mu_0 \left(\frac{t}{t + \Delta}\right),\ K_y \approx \frac{1}{2} M_S^2 \mu_0 \left(\frac{t}{t + W}\right). \tag{1.52}$$

Therefore, w obtain a transverse anisotropy constant in the order of some 10^4 J m^{-3}, considering parameters of typical perpendicular ferrimagnetic strips. This means that the critical current density is much lower in these systems than those with an in-plane magnetized system [89].

In the adiabatic limit, when the current density exceeds J_C, the DW begins to precess and moves at a velocity expressed as,

$$v_{\text{precession}} = \frac{1}{1 + \alpha^2}\sqrt{u^2 - u_C^2}. \tag{1.53}$$

Here, u_C is the spin-drift velocity at $J = J_C$. When J_C is very small as is the case for systems with PMA, the expression below can be approximated.

$$v_{\text{precession}} \approx \frac{1}{1+\alpha^2} \mid u \mid = \frac{1}{1+\alpha^2}\frac{g\mu_B}{2eM_S}PJ. \tag{1.54}$$

Therefore, the DW mobility, v_{DW}/J, is proportional to the ratio between the spin polarization P and the spontaneous magnetization M_S in the precessional regime. It can therefore be stated that perpendicularly magnetized materials with small M_S together with large P are attractive for CIDWM devices.

1.5.2 Spin-transfer torque in ferrimagnets

The LLG equation for ferrimagnetic materials must be modified to account for magnetic moments in the sublattices. Several modifications of the LLG equation have already been reported in Refs. [90–92]. In a ferrimagnet consisting of two sublattices, we can express an individual LLG equation for these two sublattices, 1 and 2, as follows.

$$\dot{\boldsymbol{m}}_1 = \gamma_1(\boldsymbol{H}_1 + H_0\boldsymbol{m}_2) \times \boldsymbol{m}_1 + \alpha_1\boldsymbol{m}_1 \times \dot{\boldsymbol{m}}_1. \tag{1.55}$$

$$\dot{\boldsymbol{m}}_2 = \gamma_2(\boldsymbol{H}_2 + H_0\boldsymbol{m}_1) \times \boldsymbol{m}_2 + \alpha_2\boldsymbol{m}_2 \times \dot{\boldsymbol{m}}_2. \tag{1.56}$$

Here, $\boldsymbol{m}_i$ is the unit vector along the magnetic moment, γ_i is the gyromagnetic ratio, $\boldsymbol{H}_i$ is the effective magnetic field, and α_i is the damping coefficient for sublattice i. H_0 stands for the effective field due to the exchange interaction between the two sublattices. Assuming that $\boldsymbol{m}_1$ and $\boldsymbol{m}_2$ are aligned antiparallel (*i.e.* H_0 is sufficiently large), the following equation can be derived by combining equations (1.55) and (1.56) because $\boldsymbol{m}_1 \times \boldsymbol{m}_2$ is zero.

$$\dot{\boldsymbol{m}} = \gamma_{\text{eff}}\boldsymbol{H}_{\text{eff}} \times \boldsymbol{m} + \alpha_{\text{eff}}\boldsymbol{m} \times \dot{\boldsymbol{m}}. \tag{1.57}$$

Here, $\boldsymbol{m} = \boldsymbol{m}_1 = -\boldsymbol{m}_2$ is the unit vector along the total magnetization, and γ_{eff}, $\boldsymbol{H}_{\text{eff}}$, and α_{eff} are the effective parameters defined by,

$$\gamma_{\text{eff}} = \frac{M_1 - M_2}{\frac{M_1}{\gamma_1} - \frac{M_2}{\gamma_2}} = \frac{M_S}{A_{\text{tot}}}. \tag{1.58}$$

$$\boldsymbol{H}_{\text{eff}} = \frac{M_1H_1 - M_2H_2}{M_1 - M_2} = \frac{M_1H_1 - M_2H_2}{M_S}. \tag{1.59}$$

$$\alpha_{\text{eff}} = \frac{\alpha_1\frac{M_1}{\gamma_1} + \alpha_2\frac{M_2}{\gamma_2}}{\frac{M_1}{\gamma_1} - \frac{M_2}{\gamma_2}} = \frac{\alpha_1A_1 - \alpha_2A_2}{A_{\text{tot}}}. \tag{1.60}$$

Here, M_i is are the spontaneous magnetization and $A_i = M_i/\gamma_i$ is the angular momentum at sublattice i. M_S is the total magnetization and A_{tot} is the total angular momentum. These equations indicate the existence of two singularities at compensation points: $A_{tot} = 0$ and $M_S = 0$. At the angular momentum compensation point ($A_1 = A_2$), both γ_{eff} and α_{eff} diverge, and this leads to the instantaneous alignment of magnetization to the local effective field [91]. The divergence of α_{eff} has been experimentally confirmed by ferromagnetic resonance measurements in the all-optical pump and probe technique [92]. At the magnetization compensation point ($M_1 = M_2$), γ_{eff} becomes zero and H_{eff} diverges. However, according to equation (1.57), is multiplied by γ_{eff}, canceling the M_S term and thus keeping the product finite.

A new modification including an adiabatic STT term in CPP-GMR stack is suggested in reference [93]. Figure 1.26 shows the assumed geometry. A positive current means that the electrons flow from layer 1 to pinned layer 2. They formulize the STT acting on the magnetic moment at sublattice i in the layer 1 as,

$$\boldsymbol{\Gamma}_i = \pm \frac{\gamma_i \hbar}{2eV} \frac{I_{e,\,i} g_i^{\pm}}{M_i} \boldsymbol{M}_i \times (\boldsymbol{m}_i \times \boldsymbol{m}_3). \tag{1.61}$$

Here, V is the volume of layer 1, $I_{e,\,i}$ is the electric current flowing in sublattice i, $g_i^{\pm}$ is the coefficient for sublattice i, which depends on the spin polarization, and $\boldsymbol{m}_3$ is the unit vector along the magnetic moment of layer 2. The combination of this torque for each sublattice introduces the effective LLG equation as follows:

$$\dot{\boldsymbol{m}} = \gamma_{eff} \boldsymbol{H}_{eff} \times \boldsymbol{m} + \alpha_{eff} \boldsymbol{m} \times \dot{\boldsymbol{m}} \pm \alpha_{I_{eff}}^{\pm} \boldsymbol{m} \times (\boldsymbol{m} \times \boldsymbol{m}_3). \tag{1.62}$$

Here, $\alpha_{I_{eff}}^{\pm} = \frac{\hbar}{eV} \frac{I_{e,\,1} g_1^{\pm} + I_{e,\,2} g_2^{\pm}}{A_{tot}}$. It can be stated that the magnitude of STT term diverges at the ACP. The expression of CPP-GMR stack is optimized in reference [93], and thus the vector of STT is not aligned to $(\nabla \cdot \boldsymbol{u})\boldsymbol{m}$ but to $\boldsymbol{m} \times (\boldsymbol{m} \times \boldsymbol{m}_3)$. Thus, the expression of STT in the ferrimagnetic wire is in the context of semiclassical adiabatic description [44] due to the balance between the supplied and accepted angular momentum transitions.

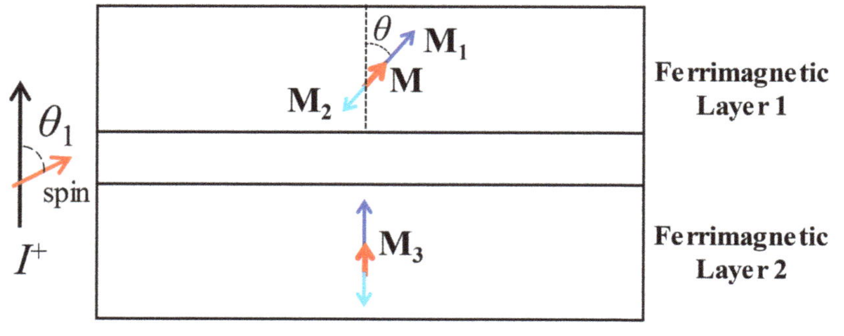

Figure 1.26. Schematic representation of a ferrimagnetic multilayer containing two strongly coupled antiparallel magnetic subnetworks. At positive bias, electrons flow from layer 1 to layer 2. Reprinted from [93], with the permission of AIP Publishing.

Figure 1.27 shows a schematic of the 1D model that we discuss here. Consider the unit cell of length dx, the cross section S, and the unit vector of magnetization $\boldsymbol{m}(x)$. Angular momentum supplied per unit time is given as the product of four terms: the spin angular momentum $\hbar/2$, the number of electrons flowing per unit time I/e, the spin polarization of current P, and the relative vector of magnetizations. It is expressed as,

$$\text{(Supplied)} = \frac{\hbar I}{2e}P \times [\boldsymbol{m}(x - dx) - \boldsymbol{m}(x)] = \frac{\hbar I}{2e}P \times \left(-\frac{\partial \boldsymbol{m}}{\partial x}dx\right). \tag{1.63}$$

Meanwhile, angular momentum accepted per unit time dt is given as,

$$\text{(Accepted)} = \left(\frac{M_1}{\gamma_1} - \frac{M_2}{\gamma_2}\right)Sdx\frac{\boldsymbol{m}(t + dt) - \boldsymbol{m}(t)}{dt} = A_{\text{tot}}Sdx\frac{\partial \boldsymbol{m}}{\partial t}. \tag{1.64}$$

STT is derived as the time derivative of magnetization vector in the following,

$$\frac{\partial \boldsymbol{m}}{\partial t} = \frac{\hbar PJ}{2eA_{\text{tot}}}\left(-\frac{\partial \boldsymbol{m}}{\partial x}\right) = -u_{\text{eff}}\frac{\partial \boldsymbol{m}}{\partial x}. \tag{1.65}$$

Therefore, the spin-drift velocity u_{eff} in a ferrimagnet is given by,

$$u_{\text{eff}} = \frac{\hbar PJ}{2eA_{\text{tot}}} = \frac{\hbar \gamma_{\text{eff}} PJ}{2eM_S} = \frac{g_{\text{eff}}\mu_B PJ}{2eM_S}. \tag{1.66}$$

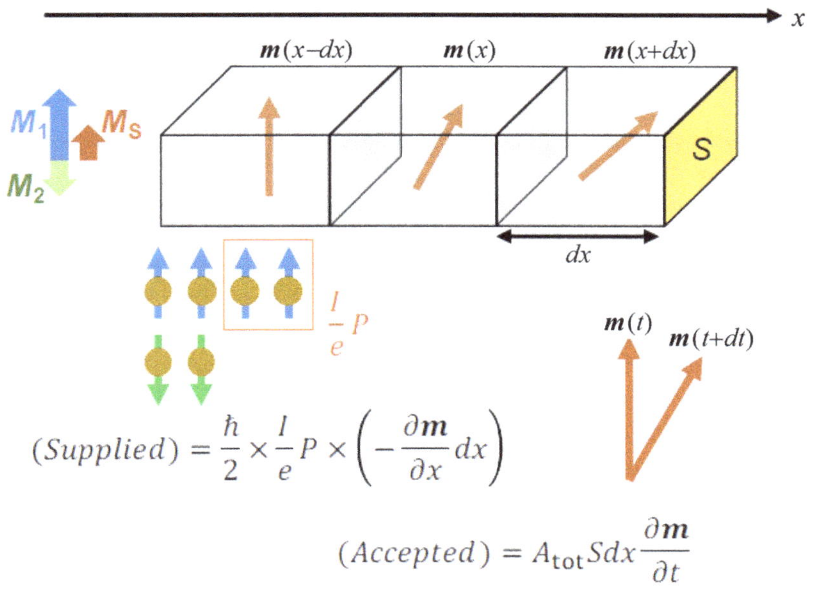

Figure 1.27. Schematic illustration of a 1D magnetic wire. Reproduced from [85]. CC BY 4.0.

Here, $g_{\text{eff}} = \frac{\hbar\gamma_{\text{eff}}}{\mu_B}$. The spin-drift velocity u_{eff} diverges not at the MCP but at the ACP in ferrimagnetic materials, at which extremely efficient STT can be expected.

The SOT terms were added into ferrimagnet-LLG equations in a very recent report, resulting in ultrafast CIDWM at angular momentum compensation temperature T_A at 260 K [69], as shown in figure 1.23. The DW velocities in ferromagnet (v_{FM}) and ferrimagnet (v_{FiM}) are derived by replacing γ and α by γ_{eff} and α_{eff}, respectively,

$$v_{\text{FM}} = \frac{\gamma}{M_S}\frac{\pi}{2}\frac{Dj_{\text{HM}}}{\sqrt{j_{\text{HM}}^2 - \alpha j_0^2}}, \quad \left(j_0 = \frac{2etD}{\hbar\theta_{\text{SH}}^{\text{eff}}\Delta}\right). \tag{1.67}$$

$$v_{\text{FiM}} = \frac{\pi}{2}\frac{Dj_{\text{HM}}}{\sqrt{\left(A(T)_{\text{tot}}j_{\text{HM}}\right)^2 - \left(A_0 j_0\right)^2}}, \quad (A_0 = \alpha_1 A_1 + \alpha_2 A_2) \tag{1.68}$$

Here, D is the DMI parameter, j_{HM} is the current density flowing in the heavy metal layer, t is the magnetic layer thickness, $\theta_{\text{SH}}^{\text{eff}}$ is the effective spin Hall angle, and Δ is the DW width. Thus, v_{FiM} saturates to be $\gamma\pi D/2M_S$. It can be stated that ferrimagnetic materials possess singularities of γ, α, and spin-drift velocity u at the ACP, and also result in a high DW velocity driven by SOT.

1.6 Features of Mn_4N

As described in later chapters, the interest of Mn_4N films for spintronics arises from the following several features.

(i) The combination of relatively large PMA and small M_S results in the appearance of perpendicular magnetization, or out-of-plane easy magnetization axis. This perpendicular magnetization facilitates the downscaling of spintronic devices, as well as the storage of data permanently using a bistable magnetization state. PMA has been reported in Mn_4N films formed on various substrates such as glass, Si(001), MgO(001), $SrTiO_3$[STO](001), $(LaAlO_3)_{0.3}(Sr_2TaAlO_6)_{0.7}$[LSAT](001), and $LaAlO_3$(001) [94–101], and the tensile strain in the in-plane direction has been proposed to be responsible for PMA [97, 101]. In this sense, Mn_4N thin films are quite attractive because they have a relatively small $M_S \sim 100$ kA m^{-1} [99, 102, 103], with an anisotropy field value $\mu_0 H_K > 2.5$ T [97, 104]. The interesting fact is that Mn_4N films are under tensile stress even when grown on LSAT(001) with a smaller lattice constant than Mn_4N [100], and thus compressive stress is likely to be induced in the Mn_4N layers. The uniaxial magnetic anisotropy constant (K_u) of Mn_4N thin films has been reported to be of the order of 0.1 MJ m^{-3} [99, 102, 105, 106]. Recently, a theoretical study of the in-plane tensile strain of Mn_4N and the resulting contribution of N to the PMA was performed [106]. Mn_4N thin films are perpendicularly magnetized because the PMA is sufficiently large to exceed the demagnetizing field energy.

(ii) The effective spin polarization in Mn_4N based on the DW velocity measurements is relatively high ($P = 0.8$) [82]. This can enhance magneto-transport effects used in spintronics such as the anomalous Hall effect, the giant magnetoresistance, thereby suggesting the possibility of using Mn_4N as a magnetic soft layer in a magnetic tunnel junction stack.

(iii) The magnetization reversal is caused by the nucleation and the motion of Bloch DWs. The combination of a relatively high P and a small M_S results in high DW velocities due to the STT [82]. The current-induced DW velocity due to the adiabatic torque is given by,

$$v \approx \frac{1}{1+\alpha^2}\frac{g\mu_B}{2eM_S}PJ \propto \frac{P}{M_S}. \tag{1.69}$$

Note that the advantage of small magnetization is not limited to the case of pure STT on DWs: the torque efficiency is supposed to vary as the inverse of the free layer magnetization for both STTs and SOTs.

(iv) The combination of high PMA and moderate exchange constant results in relatively sharp Bloch walls. Sharp DWs could lead to an enhanced spin transfer effect through the non-adiabatic contribution [107] and thereby contribute to faster DW motion at smaller currents [53, 89]. The DW width (Δ) in the nanowire is given by equation (1.37) [108]. With the increase of K_u, Δ decreases, resulting in the reduction of the threshold current density required to move the DWs [53, 89]. In this respect, materials possessing strong PMA are beneficial for CIDWM devices. Furthermore, the DW width needs to be sufficiently small to scale DW-based logic devices [109] and three terminal or racetrack memories [110] down to the ultimate technology nodes.

(v) Mn_4N does not require chemical elements such as heavy metals and rare earths, which are considered important from a geopolitical and environmental perspective.

Bibliography

[1] Atkins P W 1999 *Molecular Quantum Mechanics* (Oxford: Oxford University Press)
[2] Russel H N and Saunders F A 1925 *Atrophys. J.* **61** 38
[3] Kittel C 1981 *Introduction to Solid State Physics* 8th edn (New York: Wiley)
[4] Tomiyasu K and Kagomiya I 2004 *J. Phys. Soc. Jpn.* **73** 2539
[5] Rais A, Yousif A A, Gismeiseed A M and Elzain M 2004 *Hyperine Interact.* **156** 229
[6] Narang S B and Pubby K 2021 *J. Mag. Mag. Mater.* **167** 167163
[7] Kefeni K K, Msagati T A M and Mamba B B 2017 *Mater. Sci. Eng.* B **215** 37
[8] Landau L and Lifshitz E 1935 *Physik. Zeits. Sowjetunion* **8** 153
[9] Gilbert T L 2004 *IEEE Trans. Magn.* **40** 3443
[10] Slonczewski J C 1996 *J. Magn. Magn. Mater.* **159** L1
[11] Berger L 1996 *Phys. Rev.* B **54** 9353
[12] Zhang S and Li Z 2004 *Phys. Rev. Lett.* **93** 127204

[13] Thiaville A, Nakatani Y, Miltat J and Suzuki Y 2005 *Europhys. Lett.* **69** 990

[14] Schreier M *et al* 2015 *J. Phys.* D **48** 025001

[15] Hellman F *et al* 2017 *Rev. Mod. Phys.* **89** 025006

[16] Pai C-F, Liu L, Li Y, Tseng H W, Ralph D C and Buhrman R A 2012 *Appl. Phys. Lett.* **101** 122404

[17] Baibich M N *et al* 1988 *Phys. Rev. Lett.* **61** 2472

[18] Binasch G, Grünberg P, Saurenbach F and Zinn W 1989 *Phys. Rev.* B **39** 4828

[19] Youssef J B *et al* 1997 *J. Mag. Mag. Mater.* **165** 288

[20] Bowen M *et al* 2003 *Appl. Phys. Lett.* **82** 233

[21] Ikeda S *et al* 2008 *Appl. Phys. Lett.* **93** 082508

[22] Miyazaki T and Tezuka N 1995 *J. Mag. Mag. Mater.* **139** L231

[23] Moodera J S, Kinder L R, Wong T M and Meservey R 1995 *Phys. Rev. Lett.* **74** 3273

[24] Jullière M 1975 *Phys. Lett.* A **54** 225

[25] Seagate reaches One Terabit per Square Inch Milestone in Hard Drive Storage with New Technology Demonstration ~ News Archive ~ Seagate Seagate.com 2022 (https://seagate.com/in/en/news/news-archive/terabit-milestone-storage-seagate-master-pr/) .

[26] 2014: HDD areal density reaches 1 terabit/sq. in. ~ The Storage Engine ~ Computer History Museum computerhistory.org 2022 (https://computerhistory.org/storageengine/hdd-areal-density-reaches-1-terabit-sq-in/).

[27] Chappert C, Fert A and Van Dau F N 2007 *Nat. Mater.* **6** 813

[28] Dieny B *et al* 2020 *Nat. Electron.* **3** 446

[29] Brataas A, Kent A D and Ohno H 2012 *Nat. Mater.* **11** 372

[30] Allwood D A 2002 *Science* **296** 2003

[31] Currivan J A, Youngman J, Mascaro M D, Baldo M A and Ross C A 2012 *IEEE Mag. Lett.* **3** 3000104

[32] Manipatruni *et al* 2019 *Nature* **565** 35

[33] Noël P *et al* 2020 *Nature* **580** 483

[34] Li X *et al* 2023 *Nano Lett.* **23** 6845

[35] Samsung Electronics Starts Commercial Shipment of eMRAM Product Based on 28nm FD-SOI Process 2022 (https://news.samsung.com/global/samsung-electronics-starts-commercial-shipment-of-emram-product-based-on-28nm-fd-soi-process).

[36] Zhu J-G and Park C 2006 *Mater. Today* **9** 36

[37] LaBonte A E 1969 *J. Appl. Phys.* **40** 2450

[38] Hubert A and Schäfer R 1998 *Magnetic Domains: The Analysis of Magnetic Microstructures* (New York: Springer) p 696

[39] Coey J M D 2003 *J. Appl. Phys.* **93** 8224

[40] Ono T *et al* 1999 *Science* **284** 468

[41] Beach G S D, Nistor C, Knutson C, Tsoi M and Erskine J L 2005 *Nat. Mater.* **4** 741

[42] Fukumoto K *et al* 2006 *Phys. Rev. Lett.* **96** 097204

[43] Tsoi M, Fontana R E and Parkin S S P 2003 *Appl. Phys. Lett.* **83** 2617

[44] Yamaguchi A *et al* 2004 *Phys. Rev. Lett.* **92** 077205

[45] Kläui M *et al* 2005 *Phys. Rev. Lett.* **94** 106601

[46] Berger L 1978 *J. Appl. Phys.* **49** 2156

[47] Parkin S S P, Hayashi M and Thomas L 2008 *Science* **320** 190

[48] Parkin S S P and Yang S-H 2015 *Nat. Nanotechnol.* **10** 195

[49] Luo Z *et al* 2020 *Nature* **579** 214

[50] Slonczewski J 1996 *J. Mag. Mag. Mater.* **159** L1
[51] Manchon A *et al* 2019 *Rev. Mod. Phys.* **91** 035004
[52] Freitas P P and Berger L 1985 *J. Appl. Phys.* **57** 1266
[53] Mougin A, Cormier M, Adam J P, Metaxas P J and Ferré J 2007 *Europhys. Lett.* **78** 57007
[54] Kato Y K, Myers R C, Gossard A C and Awschalom D D 2004 *Science* **306** 1910
[55] Wunderlich J, Kaestner B, Sinova J and Jungwirth T 2005 *Phys. Rev. Lett.* **94** 047204
[56] Sinova J, Valenzuela S O, Wunderlich J, Back C H and Jungwirth T 2015 *Rev. Mod. Phys.* **87** 1213
[57] Meier L *et al* 2007 *Nat. Phys.* **3** 650
[58] Manchon A, Koo H C, Nitta J, Frolov S M and Duine R A 2015 *Nat. Mater.* **14** 871
[59] Khvalkovskiy A V *et al* 2009 *Phys. Rev. Lett.* **102** 067206
[60] Mihai Miron I *et al* 2010 *Nat. Mater.* **9** 230
[61] Haazen P P J *et al* 2013 *Nat. Mater.* **12** 299
[62] Emori S, Bauer U, Ahn S-M, Martinez E and Beach G S D 2013 *Nat. Mater.* **12** 611
[63] Ryu K-S, Thomas L, Yang S-H and Parkin S S P 2013 *Nat. Nanotechnol.* **8** 527
[64] Imai M *et al* 2018 *Appl. Phys. Lett.* **113** 052402
[65] Stanciu C D *et al* 2006 *Phys. Rev.* B **73** 220402
[66] Binder M *et al* 2006 *Phys. Rev.* B **74** 134404
[67] Baltz V *et al* 2018 *Rev. Mod. Phys.* **90** 015005
[68] Kim K-J *et al* 2017 *Nat. Mater.* **16** 1187
[69] Caretta L *et al* 2018 *Nat. Nanotechnol.* **13** 1154
[70] Haltz E, Krishnia S, Berges L, Mougin A and Sampaio J 2021 *Phys. Rev.* B **103** 014444
[71] Han J *et al* 2017 *Phys. Rev. Lett.* **119** 077702
[72] Cai K *et al* 2020 *Nat. Electrnon.* **3** 37
[73] Awano H *et al* 2015 *J. Mag. Mag. Mater.* **383** 50
[74] Hayashi M *et al* 2007 *Phys. Rev. Lett.* **98** 037204
[75] Shobitz M *et al* 2019 *Phys. Rev. Lett.* **123** 217201
[76] Yang S-H *et al* 2015 *Nat. Nanotechnol.* **10** 221
[77] Siddiqui S A, Han J, Finley J T, Ross C A and Liu L 2018 *Phys. Rev. Lett.* **121** 057701
[78] Je S-G *et al* 2018 *Appl. Phys. Lett.* **112** 062401
[79] Avci C O *et al* 2019 *Nat. Nanotechnol.* **14** 561
[80] Caretta L *et al* 2020 *Science* **370** 1438
[81] Palomino A *et al* 2021 *Sustain. Mater. Technol.* **28** e00270
[82] Gushi T *et al* 2019 *Nano Lett.* **19** 8716
[83] Gosh S *et al* 2021 *Nano Lett.* **21** 2580
[84] Berger L 1984 *J. Appl. Phys.* **55** 1954
[85] Gushi T 2019 Mn_4N thin films for spintronics applications based on current-induced domain wall motion *PhD Thesis* University of Tsukuba
[86] Koyama T *et al* 2011 *Nat. Mater.* **10** 194
[87] Malozemoff A P and Slonczewski J C 1972 *Phys. Rev. Lett.* **29** 952
[88] Nasseri S A, Sarma B, Durin G and Serpico C 2015 *Phys. Proc.* **75** 974
[89] Emori S and Beach G S D 2011 *Appl. Phys. Lett.* **98** 132508
[90] Wangsness R K 1953 *Phys. Rev.* **91** 1085
[91] Giles R and Mansuripur M 1991 *J. Magn. Soc. Jpn* . **15** 299
[92] Stanciu C D, Kimel A V, Hansteen F, Tsukamoto A, Itoh A, Kirilyuk A and Rasing T 2006 *Phys. Rev.* B **73** 220402(R)

[93] Canizo-Cabrera A, Garcia-Vazquez V and Te-ho Wu 2006 *J. Appl. Phys.* **99** 08G512
[94] Ching K M, Chang W D, Chin T S, Duh J G and Ku H C 1994 *J. Appl. Phys.* **76** 6582
[95] Ching K M, Chang W D and Chin T S 1995 *J. Alloys Compd.* **222** 184
[96] Tsunoda M and Kabara K 2012 Presented at Int. Conf. of the Asian Union of Magnetics Societies (ICAUMS) 2pPS-47, Nara, Japan, 2012
[97] Yasutomi Y, Ito K, Sanai T, Toko K and Suemasu T 2014 *J. Appl. Phys.* **115** 17A935
[98] Shen X, Chikamatsu A, Shigematsu K, Hirose Y, Fukumura T and Hasegawa T 2014 *Appl. Phys. Lett.* **105** 072410
[99] Kabara K and Tsunoda M 2015 *J. Appl. Phys.* **117** 17B512
[100] Hirose T, Komori T, Gushi T, Toko K and Suemasu T 2020 *J. Cryst. Growth* **535** 125566
[101] Hirose T, Komori T, Gushi T, Anzai A, Toko K and Suemasu T 2020 *AIP Adv.* **10** 025117
[102] Anzai A, Takata F, Gushi T, Toko K and Suemasu T 2018 *J. Cryst. Growth* **489** 20
[103] Takei W J, Heikes R R and Shirane G 1962 *Phys. Rev.* **125** 1893
[104] Ito K, Yasutomi Y, Kabara K, Gushi T, Higashikozono S, Toko K, Tsunoda M and Suemasu T 2016 *AIP Adv.* **6** 056201
[105] Komori T, Anzai A, Gushi T, Toko K and Suemasu T 2019 *J. Cryst. Growth* **507** 163
[106] Isogami S, Masuda K and Miura Y 2020 *Phys. Rev. Mater.* **4** 014406
[107] Burrowes C *et al* 2010 *Nat. Phys.* **6** 17
[108] Thiaville A, García J M and Miltat J 2002 *J. Magn. Magn. Mater.* **242–245** 1061
[109] Parveen F, He Z, Angizi S and Fan D 2017 *IEEE Computer Society Annual Symp. on VLSI (ISVLSI)* p 152
[110] Fukami S and Ohno H 2017 *Jpn. J. Appl. Phys.* **56** 0802A1

Takashi Suemasu

Chapter 2

Research history of Mn_4N-based mixed crystals: several research topics

It is important to have an overview of the history of Mn_4N-based mixed crystals before going into more detail in chapter 3 and beyond. A wide variety of detailed studies are summarized in reference [1]. Two different studies have been conducted independently, bisecting the left-hand and right-hand sides of the periodic table, as shown in figure 2.1. On the right-hand side of the periodic table, noncollinear antiferromagnetic structures and various magnetic orders in Mn_3AN have been studied, wherein A is a metallic element. For example, many physically interesting phenomena have been reported in Mn_3GaN, such as spin-glass structures [2], magnetostriction effects in Mn_3CuN [3], and phase separation in Mn_3ZnN in the low-temperature region [4]. Furthermore, negative thermal expansion (NTE) phenomena and anomalous behavior of electrical resistivity have been observed due to such complex magnetic ordering. In particular, the development of NTE materials based on Mn_3CuN, Mn_3ZnN, and Mn_3GaN is still being actively pursued. In addition, a phenomenon in which the electrical resistivity becomes constant regardless of temperature in a certain temperature range has been observed in the materials of Mn_3NiN, Mn_3CuN, and Mn_3AgN. Since the temperature coefficient of resistivity (TCR) becomes remarkably small in this case, such materials are called low TCR (L-TCR) materials. Meanwhile, on the left-hand side of the periodic table, Mn_4N mixed-crystal thin films have been developed using 3*d* transition metals and rare metals for application to spintronics devices. In the early stage, the introduction of ferromagnetic elements such as Fe, Co, and Ni has been aimed at developing tunnel-magneto-resistance (TMR) devices that combine the PMA of Mn_4N thin films with the high spin polarization. Following the observation of ultrafast magnetic domain wall motion in ferrimagnetic Mn_4N films, studies on magnetic domain wall motion devices have been actively pursued in addition to the development of TMR devices. In this chapter, we give an overview of Mn_4N crystals mixed with elements on the right-hand side of the periodic table.

doi:10.1088/978-0-7503-5477-6ch2

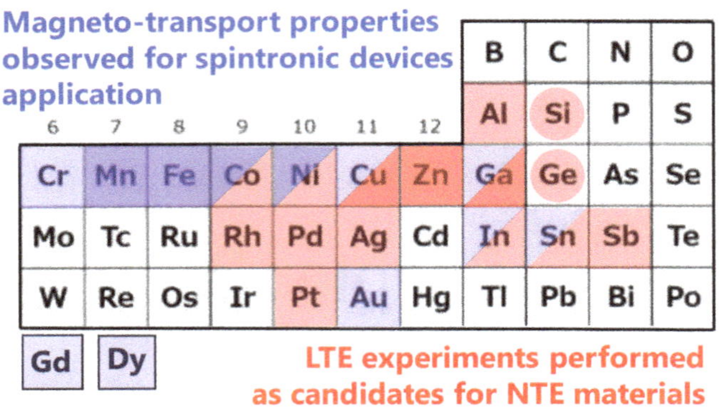

Figure 2.1. Periodic table divided into two different research regions: application oriented (blue) and curiosity-driven (red).

2.1 Non-collinear antiferromagnetic structure of Mn_3AN and Kagome lattice

Many Mn_4N-based mixed crystals doped with group 11 or higher elements have been confirmed to have noncollinear magnetic structures by synchrotron radiation measurements, such as neutron diffraction. Among them, materials with noncollinear antiferromagnetic structures, such as Mn_3NiN and Mn_3GaN, have been actively studied with physical interest because their magnetic ordering in real space has a geometric magnetic frustration structure and triangular lattice. In particular, Mn_3AN antiferromagnets often have triangular lattices, such as the Γ_{4g} and Γ_{5g} structures, and interesting phenomena such as NTE, which originates from this magnetic ordering, and L-TCR, in which the longitudinal resistivity almost does not change in a certain temperature range, have been reported. In this chapter, we first systematically introduce the studies on the magnetic ordering in Mn_3AN, noting that (1) the magnetic ordering has been investigated by neutron powder diffraction (NPD), electron spin resonance (ESR), and nuclear magnetic resonance (NMR); (2) the geometrically symmetric magnetic structure of Mn_3AN has long been the subject of *ab initio* calculations; and (3) studies on the NTE phenomenon and the L-TCR mechanism have further deepened our understanding of the magnetic ordering in the Mn_3AN system (figure 2.2).

2.1.1 Observation of magnetic order by neutron powder diffraction (NPD), electron spin resonance (ESR), and nuclear magnetic resonance (NMR) methods

NPD, ESR, and NMR have been used to investigate real-space magnetic ordering in many $Mn_{4-x}A_xN$-based polycrystalline powder samples. In this section, we will review how the magnetic structure of Mn_3AN-based mixed crystals has been clarified by these methods, using the following three types of studies: (1) phase separation of Mn_3ZnN observed by NPD method [4], (2) report on antiferromagnetic magnetic ordering in Mn_3NiN using NMR and NPD methods [6], and

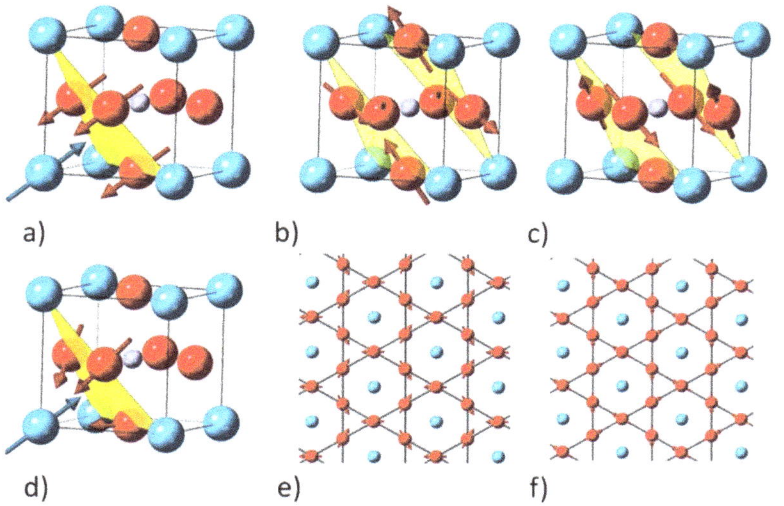

Figure 2.2. Magnetic structure of Mn_3AN: ferromagnetic (FM), ferrimagnetic (FIM), Γ_{4g}- and Γ_{5g}-antiferromagnetic (AFM) phases. Reproduced from [5].

(3) discovery of short-range order just above the Néel temperature in $Mn_{4-x}Cu_xN$ using ESR method [7].

(1) Phase separation in Mn_3 ZnN observed by NPD [4]
Figure 2.3 shows the NPD patterns for antiperovskite Mn_3ZN at 90–190 K for $2\theta = 10°–30°$ and $56°–59°$. It is noteworthy that only a single C2 (Γ_{5g}-AFM)-derived peak was observed at 190 K, whereas peaks not attributed to C2 appeared below 148 K. The C1 structure is the main reason for these results, whose magnetic structure is shown in figure 2.4. In addition to the temperature dependence of the respective magnetic moments in C1 and C2, the temperature dependence of the lattice constants of the two phases was also measured from the NPD patterns (figure 2.5). Although a clear phase separation has been observed for Mn_3ZnN by the NPD method, not all materials are like this. In the next section, we will present a few reports on Mn_3NiN, which has been reported to coexist with Γ_{4g} AFM and Γ_{5g} AFM structures, and introduce common views and differences.

(2) Observation of magnetic order of Mn_3 NiN by NMR and NPD
Regarding the magnetic ordering of Mn_3NiN [8–10]: (1) it has a Néel temperature (T_N), which is close to room temperature; (2) below T_N, the Γ_{4g} and Γ_{5g} phases are mixed; and (3) in the cryogenic region, only the Γ_{5g} phase is present. Here, we will discuss three different studies on Mn_3NiN and examine the details of magnetic ordering in Mn_3NiN.

Fruchart *et al* [8] reported that although the antiferromagnetic phase of Γ_{5g} is stable in the temperature range below 163 K, the Γ_{4g} and Γ_{5g} phases coexist in the temperature range at 163–T_N (~266 K). However, Kodama *et al* reported differently [6]: by NPD

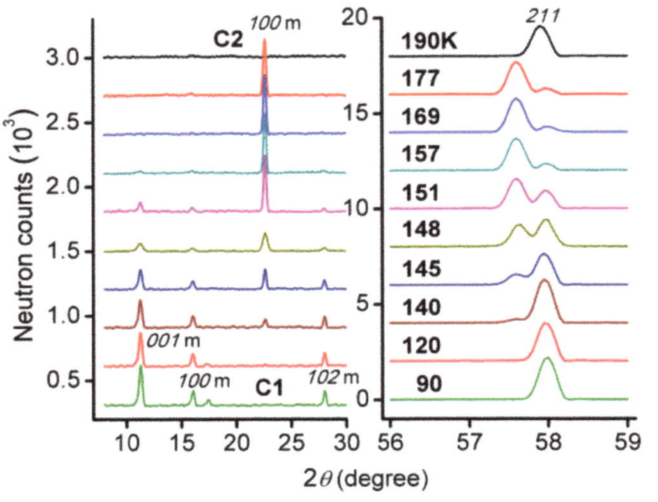

Figure 2.3. Thermal evolution of the NPD pattern of Mn_3ZnN: the 211 nuclear peak (right-hand) and typical magnetic peaks (left-hand) are shown. Note that the 100 magnetic peak at 177 K and below is from the C2 phase, while the 001, 100, and 102 peaks in the vicinity of 120 K are from the C1 phase. Reprinted with permission from [4]. Copyright (2012) American Chemical Society.

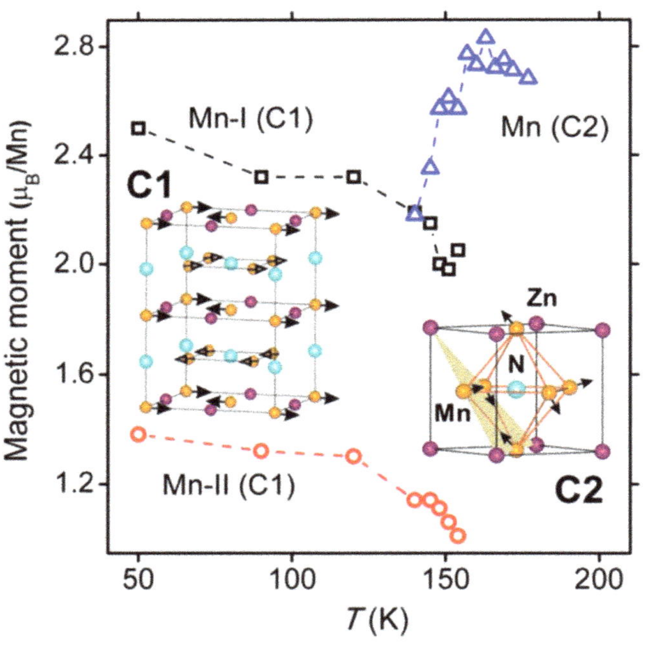

Figure 2.4. Ordered magnetic moments measured by neutron diffraction. (Insets) Magnetic structure models of the C1 (left-hand) and C2 phases (right-hand). Arrows indicate the approximate directions of the magnetic moments at the Mn atoms. Reprinted with permission from [4]. Copyright (2012) American Chemical Society.

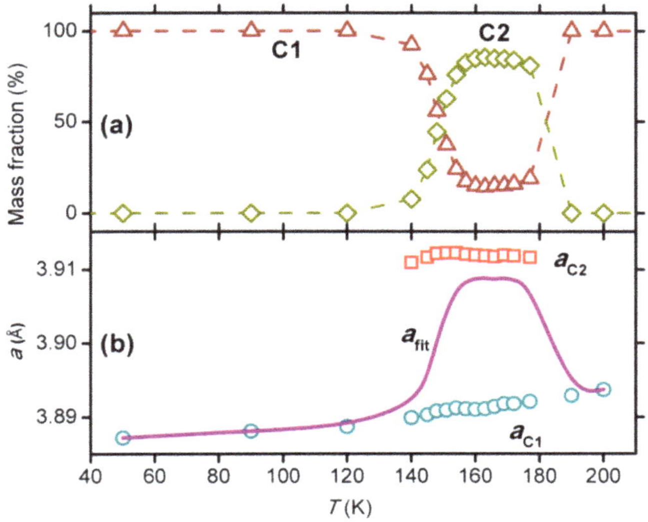

Figure 2.5. Temperature dependence of the (a) mass fraction and (b) cubic lattice parameter of Mn_3ZnN. Lattice parameter a_{fit} was calculated from the relation $a_{C1}{\times}f_{C1} + a_{C2}{\times}f_{C2}$, where f_{C1} and f_{C2} are the mass fractions of each phase. Reprinted with permission from [4]. Copyright (2012) American Chemical Society.

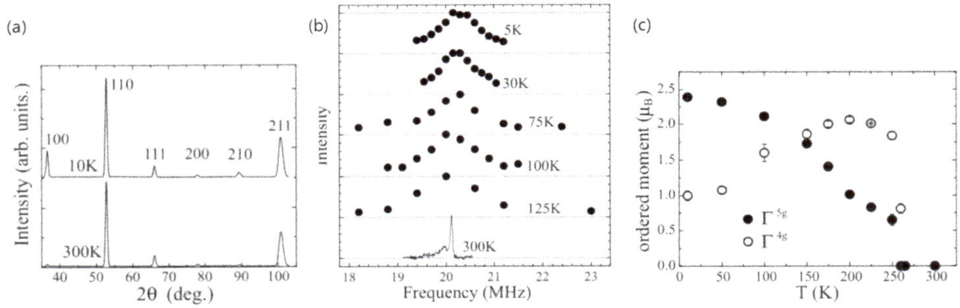

Figure 2.6. (a) Neutron-scattering profile in Mn_3NiN at 10 and 300 K. (b) NMR spectra of ^{14}N nuclei in Mn_3NiN at various temperatures. The data were collected at fixed magnetic field of 6.615 T. (c) Temperature dependence of G_{5g} and G_{4g} antiferromagnetic ordered moments with Mn_3NiN. Reprinted figure with permission from [6]. Copyright (2010) by the American Physical Society.

and nuclear magnetic resonance (NMR) spectroscopy using ^{14}N, they found that the Γ_{4g} and Γ_{5g} phases coexist in Mn_3NiN in the temperature range 10–T_N (~250 K), as shown in figure 2.6. In the low-temperature region, the Γ_{5g} phase was found to be mixed with the Γ_{4g} phase. Although the Γ_{5g} phase has a larger magnetic moment than the Γ_{4g} phase at low temperatures, the magnetic moment of the Γ_{4g} phase exceeds that of the Γ_{5g} phase around 150 K, and this magnitude relation lasts until T_N. However, the temperature dependence of the magnetization differs from those by the other groups, and thereby there is a possibility that the lack of nitrogen in the sample may have affected the results.

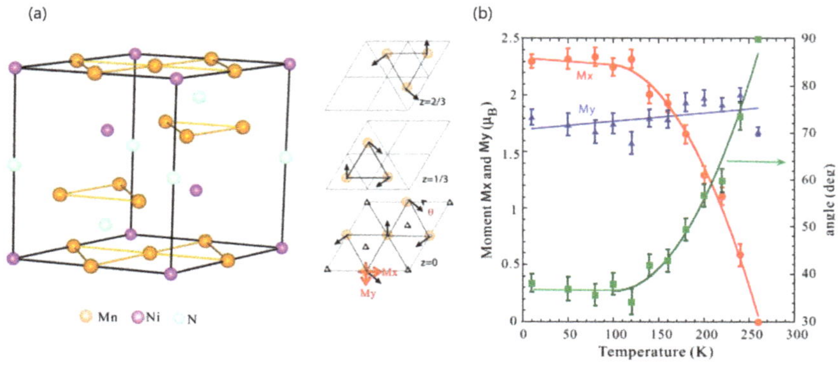

Figure 2.7. (a) Rhombohedral magnetic structure with the magnetic moments aligned in the *a-b* plane. (b) Thermal variations of M_x, M_y and the angle *q*. Reprinted from [9]. Copyright (2010), with permission from Elsevier.

Wu *et al* also analyzed the magnetic ordering of Mn_3NiN using the NPD method and reported that the magnetic moment in the (111) plane is inverted from 120 K [9]. In this study, the magnetic moment in the (111) plane is inverted in a rhombohedral unit lattice with dimension of $\sqrt{2}\ a_c \times \sqrt{2}\ a_c \times \ \sqrt{3}\ a_c$. This study suggests that the magnetic moment rotates in the (*a–b* plane). Below 120 K, the rotation angle is $\theta = 38°$, but in the temperature range of 120 –260 K it is reported to rotate toward $\theta = 90°$. The temperature characteristics obtained from the NPD are shown in figure 2.7(b).

(3) Discovery of short-range AFM order above T_C *in* $Mn_{4-x}Cu_xN$ *(*$x = 0.6–1.0$*) by ESR*

An interesting phenomenon has also been observed for $Mn_{4-x}Cu_xN$: in the case of Mn_3CuN, the crystal structure is known to change from cubic to tetragonal simultaneously with the PM-FM transition. However, for low composition samples ($0.5 < x < 1.0$), it has been reported that the temperatures of the magnetic and structural phase transitions are different. Specifically, it has been reported that cubic-FM appears just below the Curie temperature (T_C), while tetragonal-FM occurs due to the structural phase transition when the temperature is lowered further [11].

The same group has revealed by ESR that $Mn_{4-x}Cu_xN$ behaves in a special way just above T_C. Specifically, they found that at low compositions ($x < 1.0$), the mean-field ferromagnetic model is not followed and short-range AFM (SR-AFM) occurs with PM just above T_C. The details of the paper are given below.

The first measurement was made for $1/\chi$ using the Curie–Weiss rule, which reveals the SR-AFM disappearance temperature T^*. The critical exponents β, γ, and δ are also clarified. ESR measurements were also performed for $Mn_{3.1}Cu_{0.9}N$ and $Mn_{3.3}Cu_{0.7}N$. The results are shown in figure 2.8. Below T_C, two ESR spectral peaks (P_1 and P_2) with different resonance fields are observed. P_2 increases rapidly and then decreases rapidly after T^* is exceeded. The analysis of ESR spectra

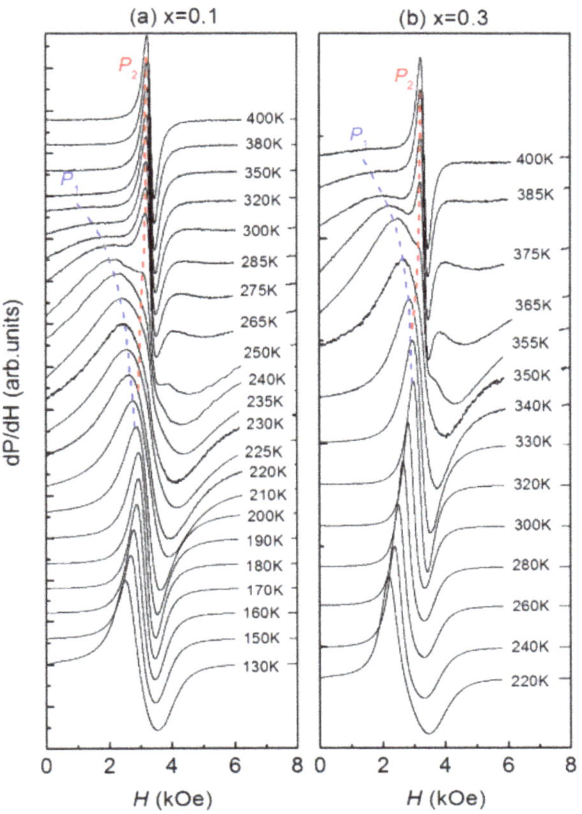

Figure 2.8. ESR spectra, *dP/dH*, as a function of temperature for $Mn_{4-x}Cu_xN$. (a) *dP/dH* for $x = 0.9$. (b) *dP/dH* for $x = 0.7$. Note that the spectra plotted here were normalized and shifted (see the text for details). The dashed lines indicate the evolutions of resonant peaks (P_1 and P_2) with temperature above T_C. Reprinted by permission from Springer Nature [7]. Copyright (2015).

confirms that P_1 corresponds to PM and P_2 to SR-AFM. However, the peak intensity of the ESR spectrum (P_2) for SR-AFM is weaker for $Mn_{3.3}Cu_{0.7}N$ than for $Mn_{3.1}Cu_{0.9}N$, indicating that SR-AFM is less likely to occur when the amount of Cu doping decreases, and that T_C and T^* are almost identical for $Mn_{3.5}Cu_{0.5}N$. This suggests that the PM + SR AFM phase does not occur in $Mn_{3.5}Cu_{0.5}N$ (figure 2.9).

2.1.2 Temperature dependence of magnetic properties and magnetic phase

The temperature dependent magnetic properties of Mn_3AN system have been reported in many studies. It is particularly interesting to note that different doping elements A give rise to completely different magnetic ordering. Some reports on the crystal structure and magnetic ordering of Mn_3AN in each temperature range are summarized in table 2.1.

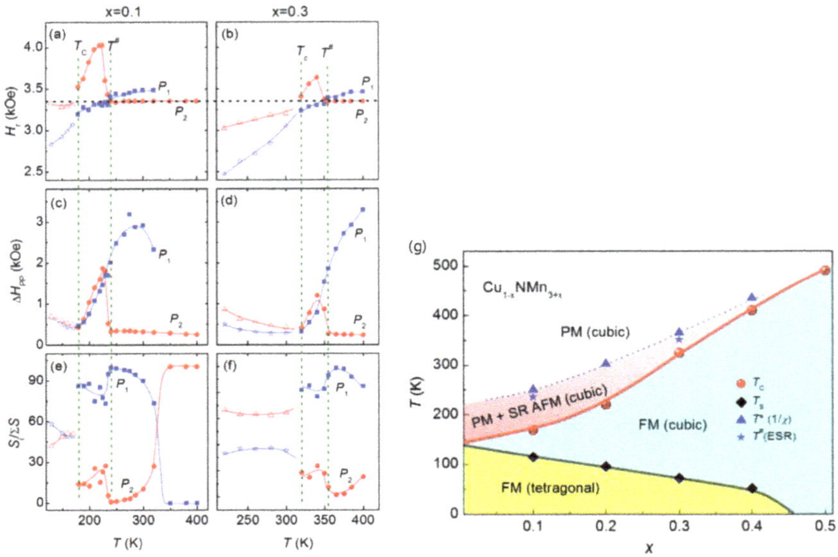

Figure 2.9. Temperature dependence of the fit parameters, i.e., the resonant field (H_r), peak-to-peak distance (ΔH_{PP}), and intensity ratio of each peak to the total intensity ($S_i/\Sigma S$) of the ESR spectra for $Mn_{4-x}Cu_xN$ (a) H_r for $x = 0.9$, (b) H_r for $x = 0.7$, (c) ΔH_{PP} for $x = 0.9$, (d) ΔH_{PP} for $x = 0.7$, (e) $S_i/\Sigma S$ for $x = 0.9$, and (f) $S_i/\Sigma S$ for $x = 0.7$. The temperatures T_C and $T^{\#}$ are indicated by the vertical dotted lines. The solid lines are guides to the eyes. P_1 and P_2 refer to the resonant peaks above T_C. (g) Phase diagram for $Mn_{4-x}Cu_xN$ with $0.6 \leqq x \leqq 1.0$. T_C represents the ferromagnetic (FM) Curie temperature and T_S represents the tetragonal-cubic structural transition temperature. T^* denotes the temperature below which $1/c(T)$ deviates from the Curie–Weiss law, and $T^{\#}$ is the onset temperature below which short-range antiferromagnetic ordering (SR AFM) occurs and coexists with the paramagnetic (PM) matrix. Reprinted by permission from Springer Nature [7]. Copyright (2015).

Table 2.1. Magnetic and crystal phases of bulk-Mn_3AN at various temperature range.

A	Phase	T_C/T_N [K]	Magnetic phase	References
Al	FM	818	–	[12]
Ni	AFM	266	Γ_{5g}(<163 K), $\Gamma_{5g}+\Gamma_{4g}$ (163–266 K)	[8]
Cu	FM	143	Tetragonal FM, SR-AFM + PM (above T_C)	[7]
Zn	AFM	183	C1 Phase (<140 K), C1+C2 Phase (140–177 K)	[4]
Ga	AFM	298	Γ_{5g}, Spin-glass behavior(<160 K)	[8]
Pd	AFM	283	–	[13,14]
Ag	AFM	290	$\Gamma_{4g}+\Gamma_{5g}$(<55 K), Γ_{5g}(55–290 K)	[8]
In	AFM	~300	AFM + enhanced FM(<50 K), AFM + weak FM(50–175 K), AFM(175–300 K)	[15]
Sn	AFM	357	Complex(<273 K), $\Gamma_{5g}+\Gamma_{4g}$ (273–357 K)	[8]
Sb	FM	360	Tetragonal FM	[16]
Pt	AFM	566	–	[17]

2.2 Negative thermal expansion (NTE) and magnetovolume effects (MVE)

The phenomenon of negative thermal expansion (NTE) has been studied in many elements, especially in materials such as Mn_3CuN, Mn_3ZnN, and Mn_3GaN. Nevertheless, being an NTE material does not mean that the volume will continue to shrink over the entire temperature range. In particular, the magnetovolume effect (MVE) associated with magnetic phase transitions in Mn_3AN-based materials causes a huge volume contraction due to temperature rise and a huge volume expansion due to temperature decrease. As a result, some researchers have attempted to explore highly efficient NTE materials by focusing on MVEs and spontaneous magnetostriction (ω_S), in addition to searching for materials with small linear or volumetric expansion coefficients. In particular, the extensive work by Takenaka *et al* [18] on Mn_3CuN based materials doped with Co, Ni, Zn, Ga, Ge, Rh, Pd, Ag, In, Sn, and Sb, and linear thermal expansion coefficient measurements will provide a guide to understanding NTE in Mn_3AN-based materials. Therefore, in this section, we will only introduce the study by Takenaka *et al* [18] (figures 2.10 and 2.11).

Coffee break: where are NTE materials?

As temperature rises, the vibrational and translational motions of atoms and molecules increase, and it is common for most materials, whether gaseous, liquid, or solid, to expand in volume as the temperature rises. However, there are some materials whose volume shrinks when the temperature rises (and expands when the temperature goes down) in a specific temperature range. A familiar example is ice.

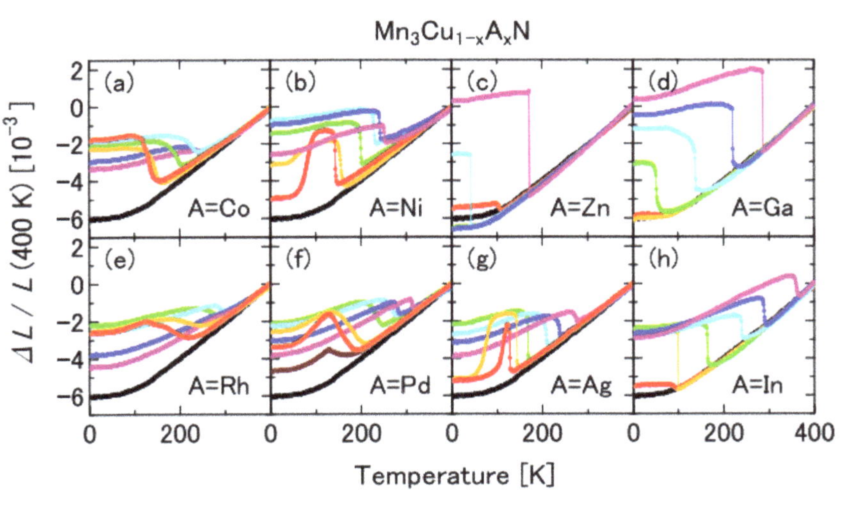

Figure 2.10. Linear thermal expansion $\Delta L(T)/L$ of $Mn_3Cu_{1-x}A_xN$ for $x = 0$ (black), 0.1 (red), 0.15 (orange), 0.3 (green), 0.5 (light blue), 0.7 (blue), and 1 (magenta): A=Co (a), Ni (b), Zn (c), Ga (d), Rh (e), Pd (f), Ag (g) and In (h). Reproduced from [18].

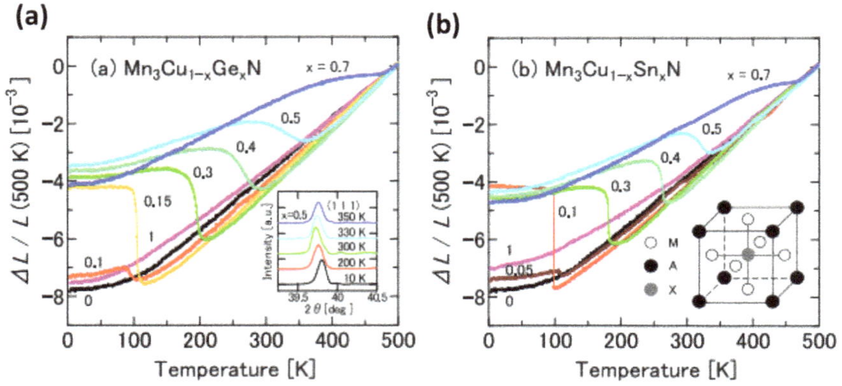

Figure 2.11. Linear thermal expansion $\Delta L(T)/L$ of $Mn_3Cu_{1-x}A_xN$: A = Ge (a) and Sn (b). Reproduced from [18].

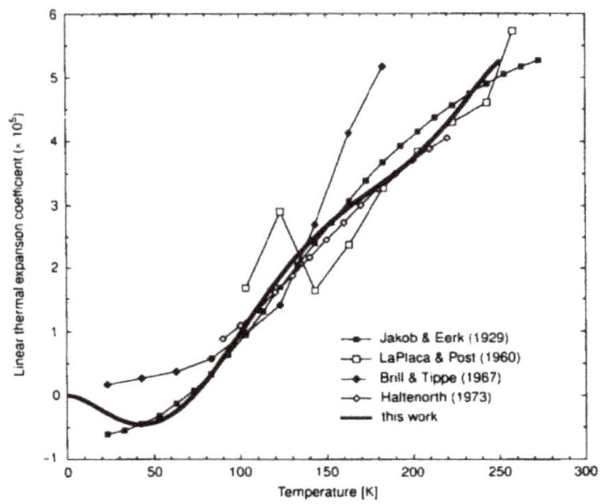

Figure 2.12. Comparison of literature data on the linear thermal expansivity coefficient for H_2O ice. Note that the data by Haltenorth were obtained on a HF-doped crystal. Although doping is known to have a very small effect on the lattice constants at 227 K (Truby, 1955) [19], its effect on the expansivity is not known exactly [20] John Wiley & Sons [1994].

It has been reported that ice has a negative coefficient of linear thermal expansion in the temperature range of approximately 70 K or lower. Figure 2.12 shows that the coefficient of linear thermal expansion decreases at low temperatures below 70 K, while it increases again in the temperature range above 70 K [20]. Next, we will discuss some engineering applications. In micro-devices, changes in position or misalignment due to thermal expansion can drastically change the properties of semiconductor devices, making it necessary to suppress thermal expansion. For this reason, development of materials that exhibit a large negative thermal expansion effect over a wide temperature range has been pursued.

2.2.1 Origin of NTE phenomena in Mn_3AN

Although there have been various discussions on the origin of NTE in the Mn_3AN system, the detailed origin has been unknown for nearly 40 years. In 2018–19, studies by Mochizuki *et al*, and Kobayashi and Mochizuki clarified the origin of NTEs in the Mn_3AN system [21, 22]. Although the focus of their work is on the elucidation of the NTEs, the considerations for magnetic anisotropy and the consideration of the exchange interaction of Mn at the face-center position that were devised in the process will be important for understanding other properties of Mn_4N-based materials. Therefore, the next subsection will describe the model for the origin of NTEs. It will also introduce the qualitative model for magnetic anisotropy that was constructed in the process of clarification.

2.2.2 Mochizuki–Kobayashi model: explanation for NTE phenomena

The physical properties of the Mn_3AN system are described by three types of electrons: itinerant Mn-4*s* electrons, localized Mn-3*d* electrons, and N-2*p* electrons that are strongly bound to the Mn atoms in the face-centered position. Mochizuki *et al*, and Kobayashi and Mochizuki described the spin of the Mn-3*d* atom in terms of the classical Heisenberg model as expressed in equation (2.1). With the nearest-neighbor antiferromagnetic interaction term (first), the next-nearest-neighbor ferromagnetic exchange interaction terms (second and third terms), and the magnetic anisotropy term (fourth term), they discuss the magnetic ordering in the Mn_3AN system.

$$
\begin{aligned}
H = &\sum_{\langle i,\mu,j,\nu\rangle} J_1 \boldsymbol{S}_{i,\mu}\cdot\boldsymbol{S}_{j,\nu} \\
&+ J_2\sum_{\langle i,\mu,j,\nu\rangle} \boldsymbol{S}_{i,\mu}\cdot\boldsymbol{S}_{j,\nu} + J_3\sum_{\langle i,\mu,j,\nu\rangle} \boldsymbol{S}_{i,\mu}\cdot\boldsymbol{S}_{j,\nu} + \kappa\sum_{i,\mu}\left(\boldsymbol{S}_{i,\mu}\cdot \boldsymbol{e}_\mu\right)^2.
\end{aligned}
\tag{2.1}
$$

First, the energy of antiferromagnetic interactions, E_{MN}, is reorganized as,

$$
E_{\mathrm{MN}} = 4NJ_1(\boldsymbol{S}_1\cdot\boldsymbol{S}_2 + \boldsymbol{S}_2\cdot\boldsymbol{S}_3 + \boldsymbol{S}_3\cdot\boldsymbol{S}_1) = 2NJ_1(\boldsymbol{S}_1+\boldsymbol{S}_2+\boldsymbol{S}_3)^2 + \text{const}, \tag{2.2}
$$

assuming that $S_1^2 + S_2^2 + S_3^2 = \text{const}$.

When the nearest-neighbor exchange interaction is antiferromagnetic ($J_1 > 0$), E_{MN} reaches a minimum at $\boldsymbol{S}_1 + \boldsymbol{S}_2 + \boldsymbol{S}_3 = 0$. Thus, the lattice spins align like a Γ_{4g} or Γ_{5g} type antiferromagnetic phase in this case. Meanwhile, when the exchange interaction is ferromagnetic ($J_1 < 0$), the sublattice spins are aligned in parallel when the norm of $\boldsymbol{S}_1 + \boldsymbol{S}_2 + \boldsymbol{S}_3$ reaches a maximum because E_{MN} is minimized. In this case, it is confirmed that the second and third terms are ferromagnetic next-nearest-neighbor interactions, which only play a role in aligning the sublattice spins in parallel, and that changing the parameters slightly has little effect on the results. In this case, there are a number of conditions for the magnetic moment to satisfy $\boldsymbol{S}_1 + \boldsymbol{S}_2 + \boldsymbol{S}_3 = 0$. In reality, however, the magnetic anisotropy contribution (fourth term in equation (2.1)) limits the number of possible magnetic ordered states. In the next section, we will discuss in detail the relationship between the magnetic anisotropy and the strength of the octahedral crystal field centered on the face-centered Mn (figure 2.13).

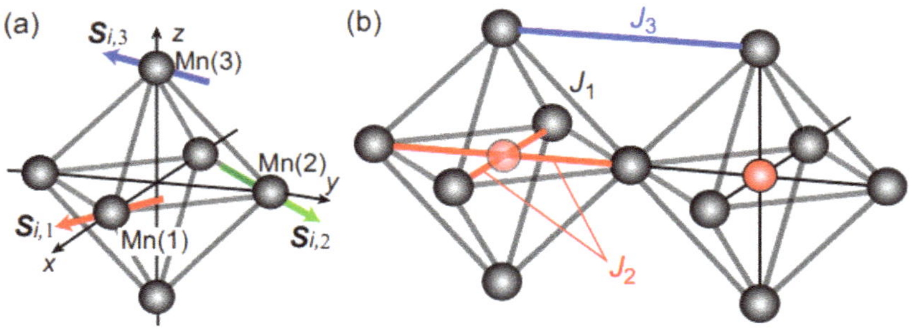

Figure 2.13. (a) Three Mn sublattices Mn(μ) with spins $S_{i,\mu}$ (μ = 1–3) on the ith octahedron. (b) Exchange interactions considered for the spin model (2.1). Reprinted figure with permission from [21]. Copyright (2018) by the American Physical Society.

2.2.3 Relationship between magnetic order and crystal field in face-centered Mn

The impact of the study about the relationship between the magnetic anisotropy and the strength of the octahedral crystal field centered on the face-centered Mn is not only the elucidation of the NTE phenomenon but also the demonstration that magnetic anisotropy in the Mn_3AN system can be considered qualitatively. In particular, the fact that the number of valence electrons of the element at position A in Mn_3AN determines the magnitude and sign of the magnetic anisotropy and nearest-neighbor exchange interaction of the material is a finding that should be applied to the development of materials in the spintronics field.

In this section, the following six points will be mentioned:

(1) Magnetic anisotropy is determined by the strength of the crystal field to which face-centered Mn is subjected.
(2) The strength of the crystal field is determined by the amount of influence exerted by the corner atom A and the body-centered nitrogen atom.
(3) The magnetic anisotropy of the crystal differs depending on the strength of the influence.
(4) If the nitrogen crystal field is greater than the A-site crystal field, then the easy direction is uniaxial (Γ_{4g}+Γ_{5g} for antiferromagnets).
(5) If the A-site crystal field is greater than the nitrogen crystal field, then the easy axis of magnetization is in one plane (Γ_{5g} for antiferromagnet).
(6) The strength of the crystal field and the sign of the nearest-neighbor exchange interaction J_1 are determined by the valence electron number of the A atom.

The octahedral structure is formed by four corner atoms A and two body-centered nitrogen atoms centered on the face-centered Mn atom. Although the energy splitting of the Mn-3d orbital is complicated, it can be classified into the following two patterns depending on the influence of the A-site crystal field and the amount of influence of the nitrogen crystal field. Assuming that the valence of the Mn atom at the face-centered position is 3+, the number of d-electron is 4. Therefore, the electrons do not enter the highest energy level when the Hund's rule prevents the upper and lower spins from being aligned. The magnetic anisotropy of the highest unoccupied orbital varies depending on the strength of the crystalline field. For example, if the nitrogen crystal field is stronger than the A-site crystal field, then there are three kinds of Mn atoms at the face-centered position, and the $3x^2–r^2$, $3y^2–r^2$, and $3z^2–r^2$ orbitals are the highest unoccupied orbitals, respectively. Therefore, the Coulomb repulsion does not work in that direction and the magnetic moment tends to be oriented in that direction. As a result, the uniaxial magnetic anisotropy (Γ_{4g} antiferromagnetic phase in NC-AFM) is found to be the more stable phase. Meanwhile, when the A-site crystal field is stronger than the nitrogen crystal field, the *yz*, *zx*, and *xy* orbitals of the Mn atoms at the face-centered position, respectively, are the highest unoccupied orbitals, which means that the easy axis of magnetization is in-plane. In addition, the Γ_{5g} phase is stable as the antiferromagnetic phase. In both cases, however, J_1 is positive when there is a difference in the strength of the crystal field, indicating that the nearest-neighbor exchange interaction is antiferromagnetic in both situations. However, if both crystal fields are in close competition, then no magnetic anisotropy is observed and the $3x^2–r^2$ and *yz* orbitals, $3y^2–r^2$ and *zx* orbitals, and $3z^2–r^2$ and *xy* orbitals are degenerated. In such a system, the ferromagnetic state with antiferromagnetic ordering, in which two orthogonal electron orbitals are alternately arranged, is known to be stable, and thus $J_1 > 0$. We have mentioned that the number of valence electrons of A-site atoms determines the strength of the A-site crystal field, that the comparison of the A-site crystal field with the nitrogen crystal field yields the pattern of the highest unoccupied orbital, and that the pattern of the highest unoccupied orbital gives rise to the choice of magnetic anisotropy, easy axis/easy plane. It is also mentioned that when there is a difference in the magnitude of both crystal fields, the nearest-neighbor exchange interaction J_1 is positive and antiferromagnetic. Meanwhile, when the strength of the A-site and nitrogen crystal fields are antagonistic, J_1 is negative and shows ferromagnetic behavior, as shown in figures 2.14 and 2.15.

2.2.4 Relationship between nearest exchange coupling J_1 and bond length θ

In the previous sections, we have described the spin moments of Mn-3*d* orbitals in terms of the classical Heisenberg model and discussed their magnetic anisotropy. In this section, we would like to focus again on the nearest-neighbor exchange interaction J_1, which appeared in section 2.2.2. In this case, J_1 is the sum of the antiferromagnetic interaction (J^{AF}) between Mn-Mn in the Mn_6N octahedron and the ferromagnetic interaction (J^{FM}) originating from the Kanamori–Goodenough rule of 90° Mn–N–Mn via a nitrogen atom at the body-centered position. When the Hamiltonian is written by the classical Heisenberg model, which incorporates the bond length expansion and contraction by,

$$\begin{aligned} H = \sum_{\langle i,\mu,j,\nu\rangle} J_1(\delta)\boldsymbol{S}_{i,\mu}\cdot\boldsymbol{S}_{j,\nu} + J_2\sum_{\langle i,\mu,j,\nu\rangle}\boldsymbol{S}_{i,\mu}\cdot\boldsymbol{S}_{j,\nu} + J_3\sum_{\langle i,\mu,j,\nu\rangle}\boldsymbol{S}_{i,\mu}\cdot\boldsymbol{S}_{j,\nu} \\ +\kappa\sum_{i,\mu}\left(\boldsymbol{S}_{i,\mu}\cdot\boldsymbol{e}_\mu\right)^2 + K_1\sum_k\delta_k^2 - K_2\sum_k\delta_k^3. \end{aligned} \tag{2.3}$$

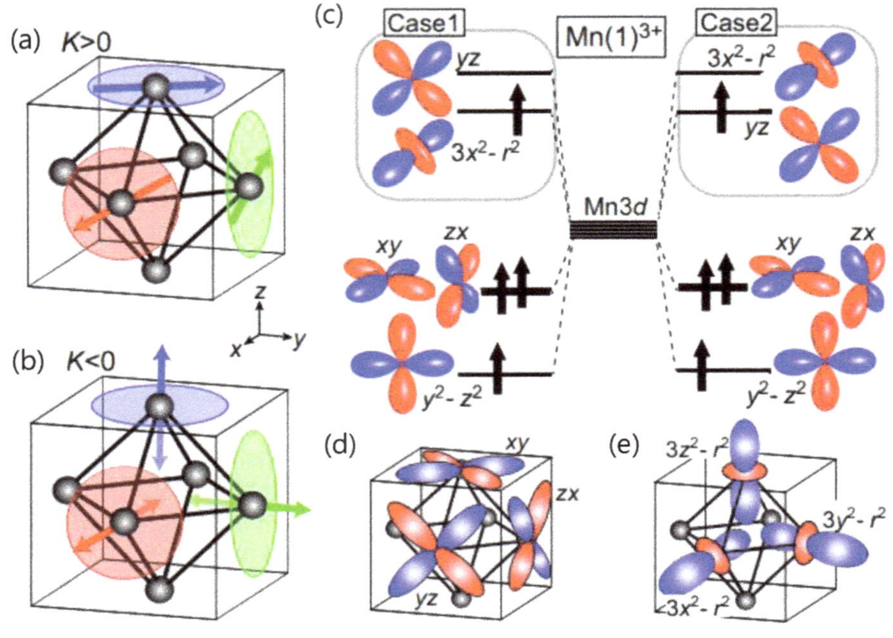

Figure 2.14. (a) Easy-plane magnetic anisotropy with $K > 0$. (b) Easy-axis magnetic anisotropy with $K < 0$. (c) Orbital-level schemes of the Mn^{3+} ion on the Mn(1) sublattice. Two cases, labeled 1 and 2, are possible depending on the ratio of the crystal-field strengths for A ions and for N ions acting on the Mn-3*d* orbitals; Case 1 (Case 2) obtains when the crystal field for A (N) ions is stronger. (d) and (e) Unoccupied orbitals of the highest levels on the three Mn sublattices for Case 1 [Case 2]. Reprinted figure with permission from [21]. Copyright (2018) by the American Physical Society.

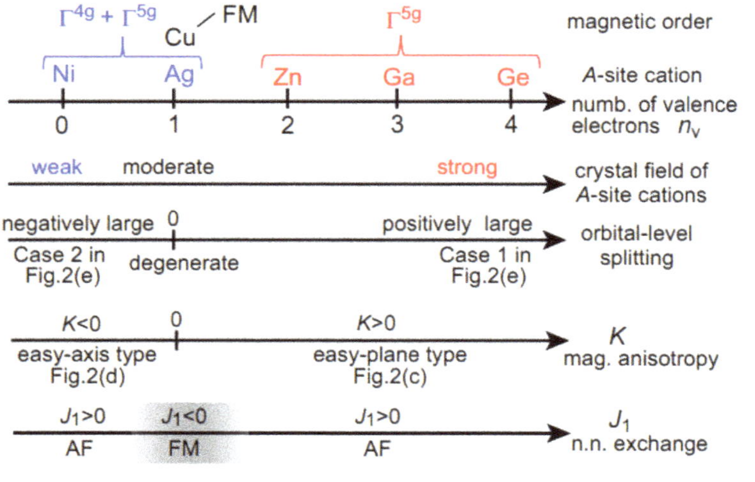

Figure 2.15. Systematic evolution of the magnetic orders upon variation of the A-site species or the number of valence electrons. The A-site ions govern the signs of the magnetic anisotropy K and the nearest-neighbor-coupling J_1 by modulating the orbital-level schemes via generating the crystal field, which competes with the crystal field of the N ions. Reprinted figure with permission from [21]. Copyright (2018) by the American Physical Society.

It is noted that J_1 depends on the bond elongation length δ, and that a term for the harmonic elastic energy (5th term) and a term for the anharmonic elastic energy (6th term) have been added.

In general, the direct antiferromagnetic exchange interaction is expressed as,

$$J^{\mathrm{AF}} = \frac{4t_{\mathrm{dd}}^2}{U}. \tag{2.4}$$

Here, t_{dd} is the hopping integral between the nearest-neighbor transition metal d orbitals, and U is the energy of the Coulomb interaction. Ferromagnetic exchange interactions via 90°-bonds are given by,

$$J^{\mathrm{FM}} = -\frac{4t_{\mathrm{dp}}^4 J_H}{\Delta^2 U^2}. \tag{2.5}$$

The t_{dp} is the hopping integral in the interaction between the d orbital of the transition metal and the p orbital of the ligand, J_{H} is the strength of the bond due to the Hund rule, and Δ is the magnitude of the charge transfer energy. The point is how the t_{dd} and t_{dp} change with (small) changes in the bond stretch length δ. The bond stretch length is given by,

$$1 + \delta = \frac{l_0 + \Delta l}{l_0}. \ (2.6) \tag{2.6}$$

Here, l is the bond length, l_0 is the original bond length, Δl is the change in bond length. Using the relationships such as that $t_{\mathrm{dd}}^2 \propto l^{-10}$and $t_{\mathrm{dp}}^4 \propto l^{-14}$, equations (2.4) and (2.5) are changed into equations (2.7) and (2.8), respectively.

$$J^{\mathrm{AF}} \propto t_{\mathrm{dd}}^2 \propto (1 + \delta)^{-10} \sim 1 - 10\delta. \tag{2.7}$$

$$J^{\mathrm{FM}} \propto t_{\mathrm{dp}}^4 \propto (1 + \delta)^{-14} \sim 1 - 14\delta. \tag{2.8}$$

Since $J_1 = J^{\mathrm{AF}} - | J^{\mathrm{FM}} |$,

$$\begin{aligned} J_1(\delta) &\sim J_0^{\mathrm{AF}}(1 - 10\delta) - | J_0^{\mathrm{FM}} | (1 - 14\delta). \\ &= (14 | J_0^{\mathrm{FM}} | - 10J_0^{\mathrm{AF}})\delta + (J_0^{\mathrm{AF}} - J_0^{\mathrm{FM}}). \end{aligned} \tag{2.9}$$

The intensity ratio of J_0^{AF} to J_0^{FM} in equations (2.9) is a major factor in determining the magnetic ordering, and calculations show that a ferromagnetic order is stable if $| J_0^{\mathrm{FM}}/J_0^{\mathrm{AF}} | > 1$, while a Γ_{5g}-type antiferromagnetic order is stable if it is below 10/14 $\sim$ 0.714 The calculation shows that the ferromagnetic order is stable when the value of J_0^{AF} is greater than 1. Furthermore, in the following conditions,

$$0.714 < \frac{| J_0^{\mathrm{FM}} |}{J_0^{\mathrm{AF}}} < 1, \tag{2.10}$$

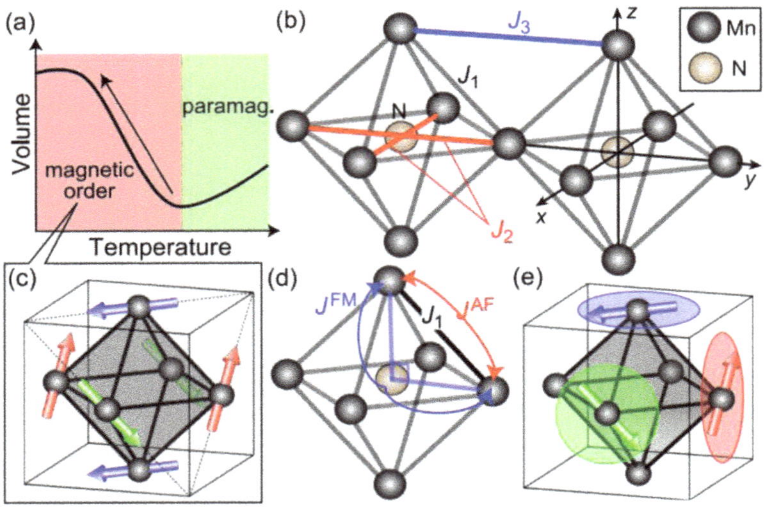

Figure 2.16. (a) Schematic figure of the negative thermal expansion where the crystal volume shows a pronounced expansion upon cooling at the magnetic transition. (b) Exchange interactions considered for the classical Heisenberg model in equations (2.3). (c) Γ_{5g}-type antiferromagnetic order in Mn_3AN. (d) Two opposite contributions to the nearest-neighbor exchange interaction J_1, i.e., the antiferromagnetic contribution J_{AF} and the ferromagnetic contribution J_{FM}. (e) Easy magnetization plane for each Mn sublattice. Reprinted figure with permission from [22]. Copyright (2019) by the American Physical Society.

the energy gain due to J_1 is large, a situation arises in which δ becomes smaller as the temperature increases in a particular temperature range.

Although NTE phenomena can occur in both Γ_{4g}-AFM and Γ_{5g}-AFM phases, it is confirmed that materials such as Mn_3NiN and Mn_3AgN, where the value of $| J_0^{FM}/J_0^{AF} |$ is less than 0.714, do not exhibit large NTE. It has also been confirmed that the NTE phenomenon does not occur for materials such as Mn_3GaN and Mn_3ZnN with values between 0.714 and 1, while it does not occur for Mn_3GeN, which may happen because the value of $| J_0^{FM}/J_0^{AF} |$ is greater than 1. Although Mn_3CuN itself is not an NTE material because of its ferromagnetic behavior, it is expected that NTE materials that satisfy equations (2.10) may be obtained by doping various elements into Mn_3CuN. In fact, Takenaka *et al* observed NTE in many materials by introducing various elements [18] (figures 2.16 and 2.17).

2.3 Measurement of thermal coefficient of resistivity (TCR) and magnetic order

The temperature dependence of the longitudinal resistivity ρ_{xx} has been measured in the Mn_3AN system and related materials. Among them, Mn_3NiN [10, 23], $Mn_{4-x}Ag_xN$ [24], and $Mn_3Ag_{1-x}Cu_xN$ [25–27] have been reported to show no change in ρ_{xx} with temperature in a specific temperature region, including room temperature, and they have been actively developed as low temperature coefficient of resistivity (L-TCR) materials. In this case, the following parameters are

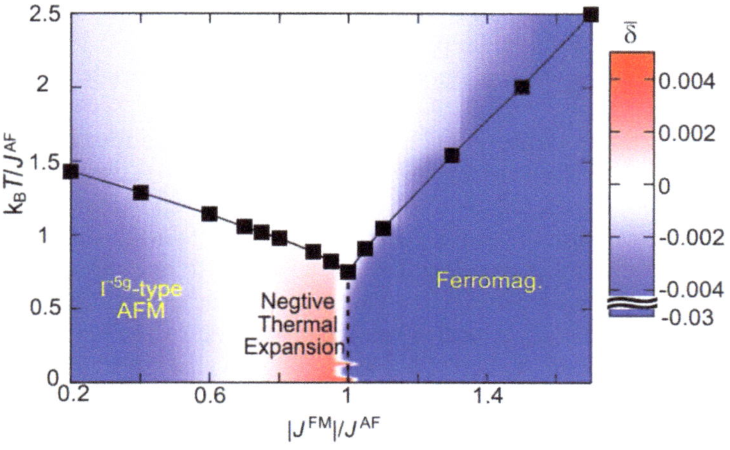

Figure 2.17. Phase diagram of the spin model in equations (2.3) and color map of the averaged difference of normalized bond length $\bar{\delta}$ in plane of temperature $k_{\mathrm{B}T}/J_{\mathrm{AF}}$ and the ratio $|J_0^{\mathrm{FM}}/J_0^{\mathrm{AF}}|$, which indicates the occurrence of NTE in the limited area of the Γ_{5g} -type antiferromagnetic phase with $0.7 <| J_0^{\mathrm{FM}}/J_0^{\mathrm{AF}} | < 1$. Reprinted figure with permission from [22]. Copyright (2019) by the American Physical Society.

optimized. It is noted that ρ_0 is often used as the electrical resistivity at room temperature.

$$\mathrm{TCR} = \frac{1}{\rho_0}\frac{d\rho}{dT}. \tag{2.11}$$

In some cases, electrical properties have been investigated without the aim of developing TCR materials. It has been reported that for some materials such as $Mn_{4-x}Ag_xN$ [28] and $Mn_{4-x}Pd_xN$ [29], because the composition of Ag and Pd is increased, the temperature dependence of resistance changes from metal to semiconductor type (metal-semiconductor transition; MST).

In the first subsection we will discuss TCR materials. In the second subsection we will discuss in detail the materials for which MST has been confirmed.

2.3.1 L-TCR materials: Mn_3NiN, Mn_3CuN, Mn_3AgN

Among L-TCR materials, those based on Mn_3NiN, Mn_3CuN, and Mn_3AgN have been studied extensively. Materials whose electrical resistivities are insensitive to temperature change are used in various devices, such as electrical resistors, and the search for such materials continues in antiperovskite manganese nitrides. Recently, a study on $Mn_3Ag_{1-x}Cu_xN$ with TCR measurements has been reported. The origin of the L-TCR effect is proposed as summarized below.

The origin of the L-TCR effect or resistivity anomaly is often discussed with the following three types of factors:

(1) Scattering phenomena originating from changes in real-space magnetic order and short-range order due to magnetic phase transitions.

(2) Changes in electronic structure and band structure due to structural changes associated with magnetic phase transitions.
(3) Resistivity saturation phenomena in bad metal materials.

As an example of the discussion of (1), we would like to introduce the work by Deng *et al* who reported nearly zero temperature coefficient of resistivity (NZ-TCR) of Mn_3NiN [10] and $Mn_{3+y}Ag_xN$ [24]. In their paper reporting the NZ-TCR of Mn_3NiN, Deng *et al* simultaneously performed NPD measurements and reported that in the temperature region where the electrical resistivity changes from metallic to NZ-TCR type, the long-range magnetic order disappears and a short-range antiferromagnetic order is present. The scattering due to the short-range order is negative. There are also reports that scattering by the short-range order gives negative TCR [30]. It is considered that the L-TCR may result from the offsetting effects on the electrical resistivity that result from electron-phonon scattering and scattering by the short-range order.

As an example of the discussion of (2), we would like to cite the report by Sun *et al* [10], wherein the TCR of Mn_3NiN was reported. First, we list three features of the band structure of the Mn_3AN system.

(i) According to calculations using Labbé–Jardin's tight-binding approximation method [31], a localized electron system band with a hybridization of Mn-3*d* and N-2*p* orbitals is observed. In addition, the hybridized band has a singularity and the density of states is divergent.
(ii) The Fermi level is located inside or near the hybridized band mentioned in (i).
(iii) The bands in itinerant electron systems are energetically broadened.

From these factors, it is expected that Fermi surfaces also shift from one Brillouin zone to another when the lattice constant changes abruptly. In fact, it has been reported that the lattice constant changes abruptly at the temperature at which the temperature dependence of electrical resistivity switches from the metallic type to the L-TCR type. The band structure of Mn_3NiN is also expected to be semimetallic and the band gap is expected to change gradually in the temperature region where L-TCR is confirmed. As a result, we consider that L-TCR may be maintained by a delicate balance regarding carrier density and mobility. Although this was the only study that directly considered L-TCR from the electronic structure, the research group that is introduced next mentions *s-d* scattering based on the characteristics of the electronic structure of the Mn_3AN system.

Finally, as for (3), although there are many papers that discuss the resistivity-saturation phenomenon, we would like to introduce the discussion by Hadano *et al* [26]. As mentioned earlier, what is still characteristic in the band structure of the Mn_3AN system is (1) a narrow band of localized *d*-electron systems in which the Mn-3*d* and N-2*p* orbitals are hybridized and (2) an energetically broad itinerant electron system derived from Mn-4*s* electrons. In such systems, high mobility *s* electrons scatter with low mobility *d*-electron states (*s-d* scattering). It has been reported that resistivity-saturation phenomena occur in bad metal systems where

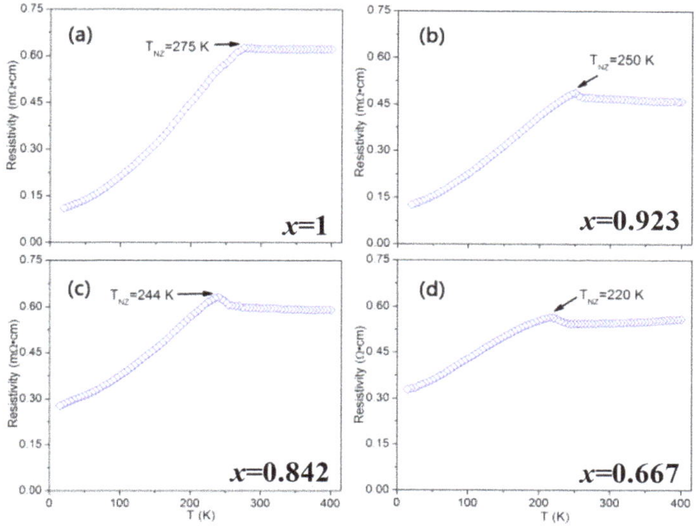

Figure 2.18. Temperature dependence of the electrical resistivity [(a) $x = 1$, (b) $x = 0.923$, (c) $x = 0.842$, and (d) $x = 0.667$] for $Mn_{4-x}Ni_xN$. Reprinted from [23], with the permission of AIP Publishing.

inelastic scattering such as *s-d* scattering occurs [32]. Based on the above, this research group considers that L-TCR may also be observed in $Mn_3Ag_{1-x}Cu_xN$.

The same group also reported the magnetoresistance effect of Mn_3AgN at 300 K. They found that the absolute value of the magnetoresistance ratio $[\rho(B)-\rho(0\ T)]/\rho(0\ T)$ of this sample was smaller than that of Au and Mn. This indicates that the cyclotron motion of electrons due to the external magnetic field is inhibited by strong magnetic scattering, supporting the small electronic relaxation time. Therefore, this small magnetoresistance ratio also supports the strength of magnetic scattering, and consequently supports the claim that *s-d* scattering, which is strong inelastic scattering, may be the origin of L-TCR (figures 2.18–2.21).

2.3.2 Electrical properties: metal-semiconductor transition in $Mn_{4-x}Ag_xN$ and $Mn_{4-x}Pd_xN$

In addition to the L-TCR effect, we would like to mention examples of $Mn_{4-x}Ag_xN$ and $Mn_{4-x}Pd_xN$ concerning MST. Both samples were prepared by RF sputtering on Si substrates with a small amount of doping ($x < 1.0$) and the temperature dependence of electrical resistivity was investigated.

For $Mn_{4-x}Ag_xN$, samples with $x = 0.0$, 0.18, 0.42, 0.63, 0.81, and 0.87 were prepared on Si substrates and the temperature dependence of their electrical resistivity was reported [28]. The samples with $x = 0.81$ and 0.87 exhibit semiconductor-type behavior, in which the electrical resistivity decreases with increasing temperature, while the samples with $x = 0.0$ and 0.18 exhibit metallic-type behavior, in which the electrical resistivity increases with increasing temperature. For the intermediate compositions $x = 0.42$ and 0.63, the electrical resistivity increases at low temperatures, while it shows a maximum value in the intermediate region and

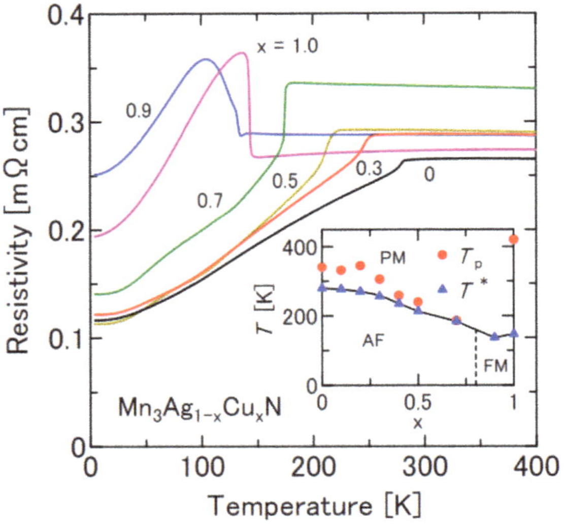

Figure 2.19. Temperature dependence of the dc resistivity ρ(T) measured on $Mn_3Ag_{1-x}Cu_xN$. Inset shows the relation between magnetic transition temperature T^* and the resistivity-peak temperature T_p. The magnetically ordered state is AF and FM for $x \leqslant 0.7$ and $x \geqslant 0.9$, respectively. Reprinted from [25], with the permission of AIP Publishing.

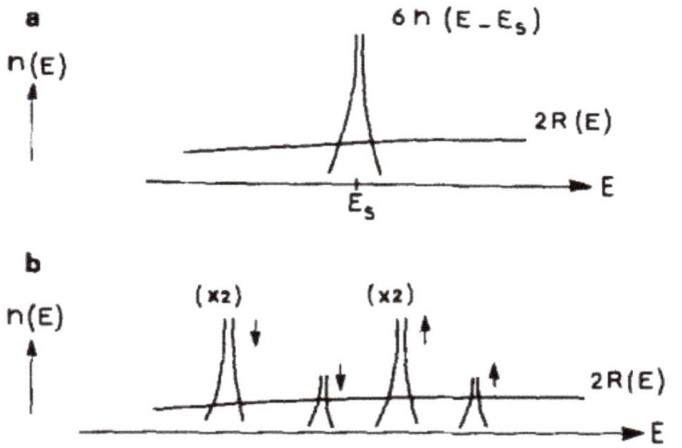

Figure 2.20. Electronic density of states of the p-d_{eg} subband as calculated in Labbé–Jardin model. Reprinted from [32], Copyright (1983), with permission from Elsevier.

decreases at high temperatures. However, the L-TCR effect was reported in the same group that prepared samples by the solid phase reaction method, resulting in completely different electrical resistivity characteristics [23]. In all of the samples, it was confirmed that the metallic behavior of the electrical resistivity increases as the temperature rises in the low-temperature region. Above a certain temperature, the electrical resistivity remains almost unchanged (figures 2.22 and 2.23).

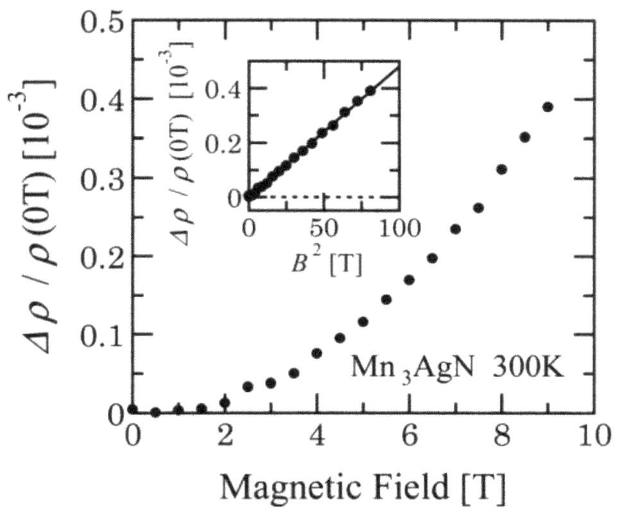

Figure 2.21. Magnetoresistance of Mn_3AgN at 300 K against magnetic field B up to 9 T. Magnetoresistance is defined as $\Delta\rho/\rho(0\ \mathrm{T})=[\rho(B)-\rho(0\ \mathrm{T})]/\rho(0\ \mathrm{T})$. The inset shows $\Delta\rho/\rho(0\ \mathrm{T})$ as a function of B^2. Reprinted from [26], with the permission of AIP Publishing.

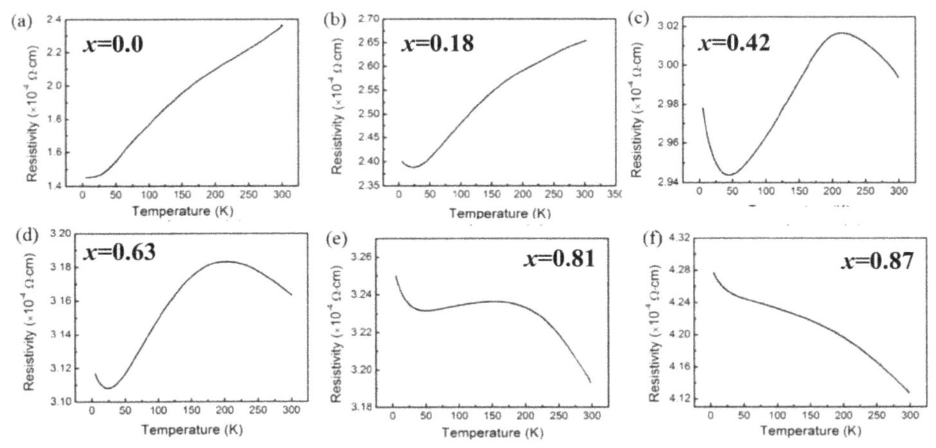

Figure 2.22. Temperature dependence of electrical resistivity for the $Mn_{4-x}Ag_xN$/Si with (a) $x = 0.0$, (b) 0.18, (c) 0.42, (d) 0.63, (e) 0.81, and (f) 0.87. Reproduced with permission from reference. Reproduced from [28].

In $Mn_{4-x}Pd_xN$, samples with x = 0.0, 0.12, 0.17, 0.27, and 0.36 have been prepared on Si substrates and the temperature dependence of electrical resistivity was also reported [28]. However, the electrical resistivity decreases with increasing temperature in the temperature range above $x = 0.12$. The same behavior is observed for $x = 0.17$ and 0.27. However, at $x = 0.36$, the maximum electrical resistivity is no longer observed in the intermediate temperature region, and the temperature

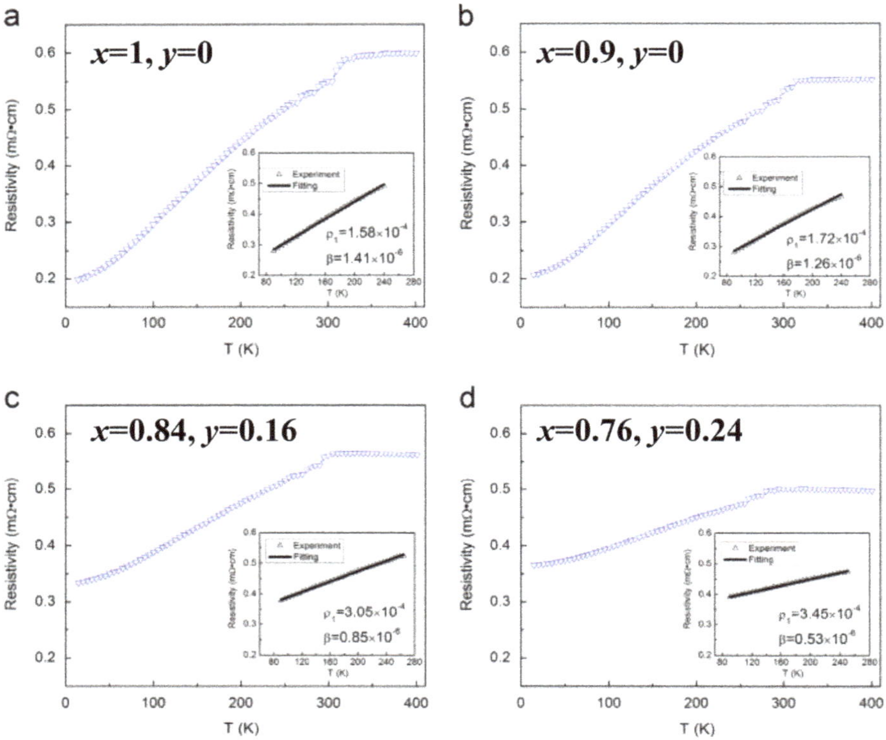

Figure 2.23. Temperature dependence of the electrical resistivity for $Mn_3Ag_xMn_yN$ compounds: (a) $x = 1$ ($y = 0$), (b) $x = 0.9$ ($y = 0$), (c) $x = 0.84$ ($y = 0.16$), and (d) $x = 0.76$ ($y = 0.24$). Reprinted from [24], Copyright (2015), with permission from Elsevier.

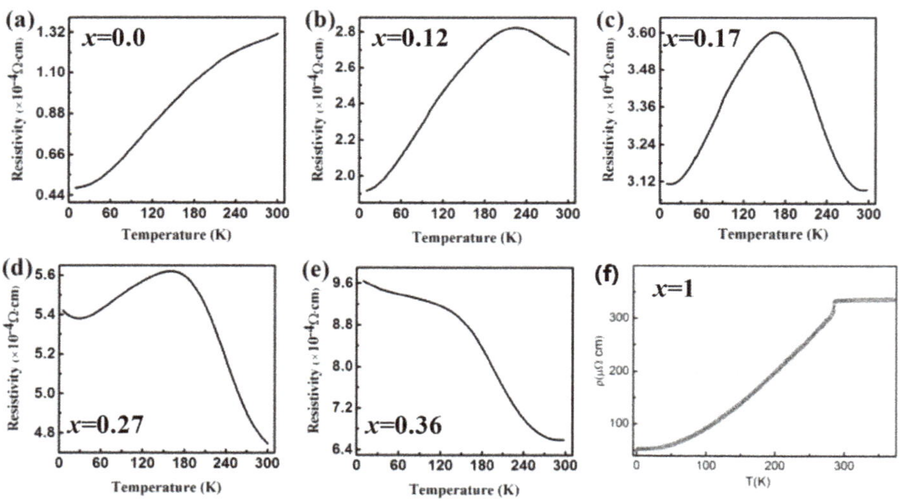

Figure 2.24. Temperature dependence of the electrical resistivity measured in the temperature range from 10 to 300 K in the antiperovskite $Mn_{4-x}Pd_xN$ deposits with (a) $x = 0.0$, (b) 0.12, (c) 0.17, (d) 0.27, (e) 0.36, respectively. Reproduced from [29]. © IOP Publishing Ltd. All rights reserved. (f) The temperature dependent resistivity of Mn_3PdN. Reprinted from [14]. Copyright (2017), with permission from Elsevier.

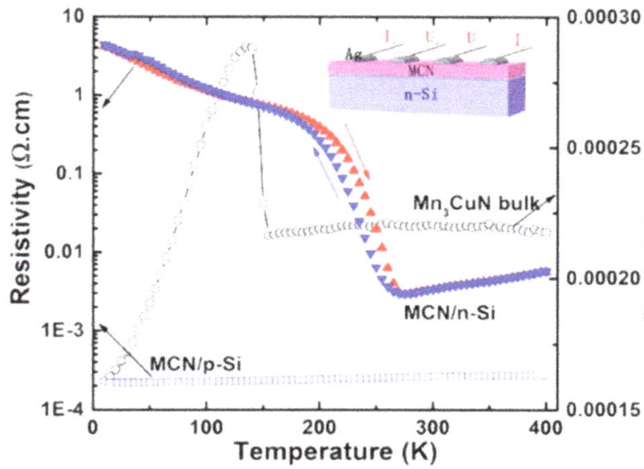

Figure 2.25. Temperature dependence of resistivity of the Mn_3CuN thin film on n-Si and p-Si(001) substrates (left axes) and the bulk (right axes). The inset shows the model for resistivity measurement. Reprinted with permission from [33]. Copyright (2017) American Chemical Society.

characteristics are completely semiconductor type (figure 2.24.(a)–(e)). However, a group that prepared Mn_3PdN by the solid-state reaction method reported that it exhibits metallic-type behavior (figure 2.24.(f))[14].

MST has also been reported in Mn_3CuN/n-Si heterojunction structures [33]. Interestingly, the behavior of Mn_3CuN/n-Si is quite different from that of Mn_3CuN (bulk) and Mn_3CuN/p-Si: the electrical resistivity of the Mn_3CuN/n-Si structure decreases with increasing temperature, and MST was reported to occur at $T = 240$ K. Meanwhile, it was also reported that the electrical resistivity of Mn_3CuN/p-Si is constant with temperature, while Mn_3CuN (bulk) shows a singularity in the temperature region around $T = 120$ K. Interestingly, Si substrates are used in all the papers reporting MST (figure 2.25).

Bibliography

[1] Wijn H P J 1988 *Alloys and Compounds of d-Elements with Main Group Elements. Part 2* (Berlin: Springer)

[2] Song B, Jian J, Bao H, Lei M, Wang G, Xu Y and Chen X 2008 *Appl. Phys. Lett.* **92** 192511

[3] Asano K, Koyama K and Takenaka K 2008 *Appl. Phys. Lett.* **92** 161909

[4] Sun Y, Wang C, Huang Q, Guo Y, Chu L, Arai M and Yamaura K 2012 *Inorg. Chem.* **51** 7232

[5] Coey J M D, Givord D and Fruchart D 2022 *ECS J. Solid State Sci. Techol* **11** 055002

[6] Kodama K, Iikubo S, Takenaka K, Takigawa M, Takagi H and Shamoto S 2010 *Phys. Rev.* **B81** 224419

[7] Lin J *et al* 2015 *Sci Rep.* **5** 7933

[8] Fruchart D and Bertaut E F 1978 *J. Phys. Soc. Jpn.* **44** 781

[9] Wu M, Wang C, Sun Y, Chu L, Yan J, Chen D, Huang Q and Lynn J W 2013 *J. Appl. Phys.* **114** 123902

[10] Sun Y, Wang C, Chu L, Wen Y, Niea M and Liu F 2010 *Scr. Mater.* **62** 686

[11] Lin J C, Wang B S, Lin S, Tong P, Lu W J, Zhang L, Zhao B C, Song W H and Sun Y P 2012 *J. Appl. Phys.* **111** 113914
[12] Lin J C, Wang B S, Tong P, Lu W J, Zhang L, Zhu X B, Yang Z R, Song W H, Dai J M and Sun Y P 2011 *Appl. Phys. Lett.* **98** 092507
[13] Takenaka K, Hamada T, Shibayama T and Asano K 2013 *J Alloy Compd.* **577** S291
[14] Tan S, Gao C, Yuan H, Wu J, Wang C, Cao R, Sun Y and Solid J 2021 *State Chem* **302** 122389
[15] Sun Y S, Guo Y F, Wang X X, Yi W, Li J J, Zhang S B, Sathish C I, Belik A A and Yamaura K 2012 *J. Phys.: Conf. Ser.* **400** 032094
[16] Takenaka K, Shibayama T, Kasugai D and Shimizu T 2012 *Jpn. J. Appl. Phys.* **51** 043001
[17] Chou R, Sun Y, Lu H and Lu G H 2018 *Int. J. Mod. Phys.* **32** 1850314
[18] Takenaka K, Ichigo M, Hamada T, Ozawa A, Shibayama T, Inagaki T and Asano K 2014 *Sci. Technol. Adv. Mater.* **15** 015009
[19] Truby F K 1955 *Science* **121** 404
[20] Rottger K, Endriss A and Ihringer J 1994 *Acta Crysta* B **50** 644
[21] Mochizuki M, Kobayashi M, Okabe R and Yamamoto D 2018 *Phys. Rev.* **B97** 060401
[22] Kobayashi M and Mochizuki M 2019 *Phys. Rev. Mater.* **3** 024407
[23] Deng S, Sun Y, Wang L, Wu H, Shi K, Hu P, Huang Q and Wang C 2016 *Appl. Phys. Lett.* **108** 041908
[24] Deng S, Sun Y, Yan J, Shi Z, Shi K, Wang L, Hu P, Malik M I and Wang C 2015 *Solid State Commun.* **222** 37
[25] Takenaka K, Ozawa A, Shibayama T, Kaneko N, Oe T and Urano C 2011 *Appl. Phys. Lett.* **98** 022103
[26] Hadano M, Ozawa A, Takenaka K, Kaneko N, Oe T and Urano C 2012 *J. Appl. Phys.* **111** 07E120
[27] Lukose C C, Zoppi G and Birkett M 2022 *J. Mater. Sci. Technol.* **99** 138
[28] Lu N P, Xu T, Cao Z X and Ji A L 2016 *J. Phys. D: Appl. Phys.* **49** 045308
[29] Xu T, Ji G P, Cao Z X and Ji A L 2018 *J. Phys. D: Appl. Phys.* **51** 055303
[30] Jen S U and Liou S S 1999 *J. Appl. Phys.* **85** 8217
[31] Jardin J P and Labbe J 1983 *J. Solid State Chem.* **46** 275
[32] Gunnarsson O, Calandra M and Han J E 2003 *Rev. Mod. Phys.* **75** 1085
[33] Shi K, Wang C, Ying Sun, Wang L, Deng S, Hu P, Lu H, Hao W, Wang T and Tang W 2017 *ACS Appl. Mater. Interfaces* **9** 12592

Chapter 3

Crystal and magnetic structure of Mn_4N

3.1 Crystal structure of antiperovskite transition metal nitrides

The crystal structure of Mn_4N is shown in figure 3.1, which is an antiperovskite structure (space group; $Pm\bar{3}m$) consisting of a body-centered N atom and face-centered Mn atoms. The Mn atoms are grouped into two: those at the corner sites, Mn(I); and the others at the face-centered sites, Mn(II). This face-centered cubic phase ideally contains one N atom in the body center of each cube for stoichiometric Mn_4N. Therefore, Mn_4N can be written as $Mn(II)_3Mn(I)_1N$. The N atoms consequently form a superlattice. The lattice constant for 20 at.% N is 0.3864 nm, which is the average of the very closely agreeing values of references [1–4]. In the presence of magnetization, the latter Mn sites can be further divided into two sites: Mn(II)A and Mn(II)B. Since the interatomic distance between the N and Mn(I) atoms is different from that between the N and Mn(II) atoms, the magnetic moment of Mn(I) atoms is different from that of Mn(II) atoms, even though the elements are the same. According to first-principles calculations on Mn_4N [5, 6], Mn(II) atoms are more itinerant than Mn(I) atoms because of the hybridization of Mn(II) 3*d* and N *sp* states. In contrast, Mn(I) atoms are more localized than Mn(II) atoms because of the larger interatomic distance between N and Mn(I) than that between N and Mn(II). The magnetic properties of Mn_4N can be modulated by substituting Mn(I) and/or Mn(II) atoms with various elements, such as magnetic (Fe, Co, and Ni) and non-magnetic elements. Since other 3*d* nitrides such as Fe_4N, Co_4N, and Ni_4N have the same crystal structure as Mn_4N, a variety of compounds are available based on this crystal structure, as shown in figure 3.2.

3.2 Mn_4N bulk crystals

3.2.1 Magnetic order by neutron diffraction

Neutron diffraction measurement on Mn_4N powder samples shows that Mn_4N is ferrimagnetic [7], and the magnetic moment of Mn at the I site is 3.53 μ_B at 300 K,

doi:10.1088/978-0-7503-5477-6ch3

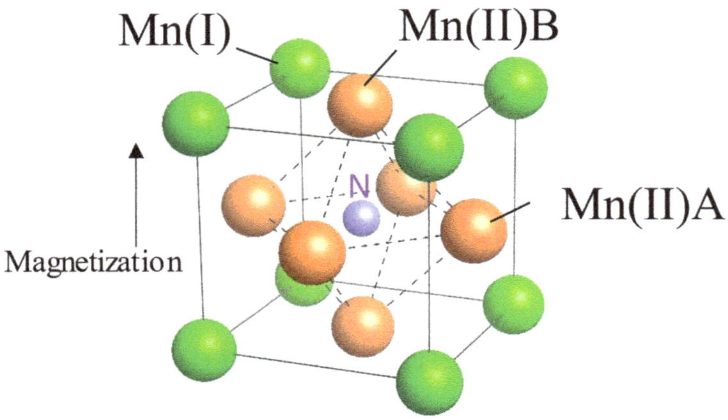

Figure 3.1. Crystal structure of Mn_4N. An N atom is at the body center and transition metal atoms are at corner (I)- and face-centered (II) sites.

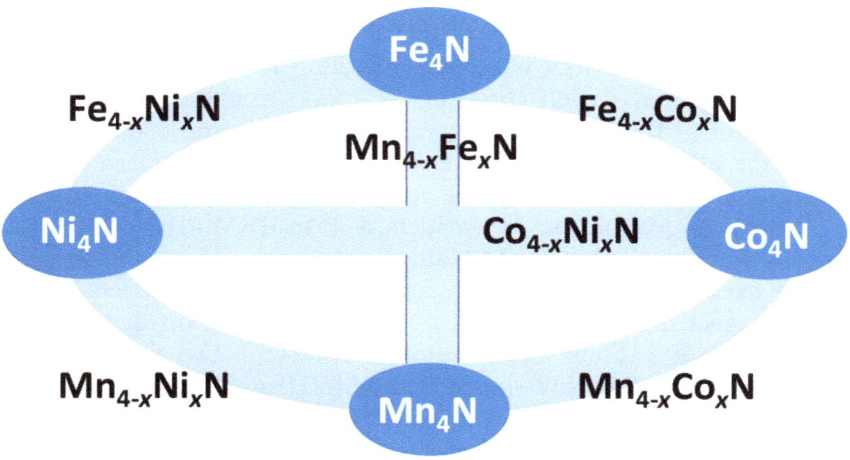

Figure 3.2. Schematic of 3*d* nitride-based compounds.

and that at the II site is -0.89 μ_B. The Néel temperature T_N is measured to be 745 K, which is much higher than room temperature. The easy axis of magnetization in Mn_4N bulk is in the <111> direction, with a non-collinear magnetization component at II-sites [8, 9], as shown in figure 3.3 [10]. Mn_3AN (A= Ga, Zn, etc) alloys have been actively investigated [11] due to large negative thermal expansion [12], which is caused by such a non-collinear magnetic structure.

3.2.2 Magnetic compensation of impurity-doped Mn_4N bulk

The occurrence of magnetic compensation in Mn_4N bulk crystals partially substituted by In and Sn was first reported in reference [13] based on experiment. Samples were prepared by nitriding of Mn–In and Mn–Sn alloys. Figure 3.4 shows

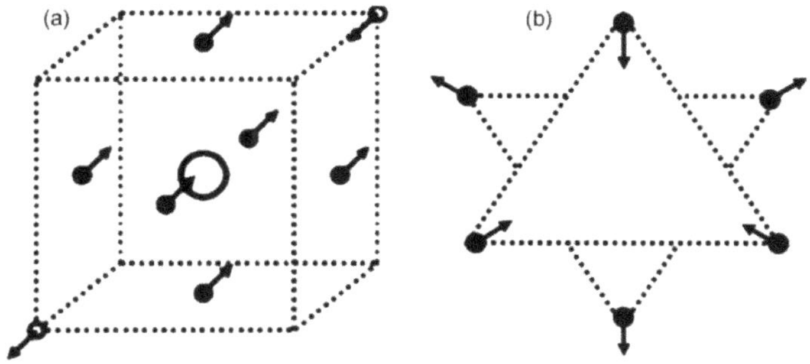

Figure 3.3. Schematics of (a) magnetic structure of Mn_4N bulk along the [111] direction and (b) that of Mn(II) on the (111) plane. Reprinted from [10]. Copyright (2008), with permission from Elsevier.

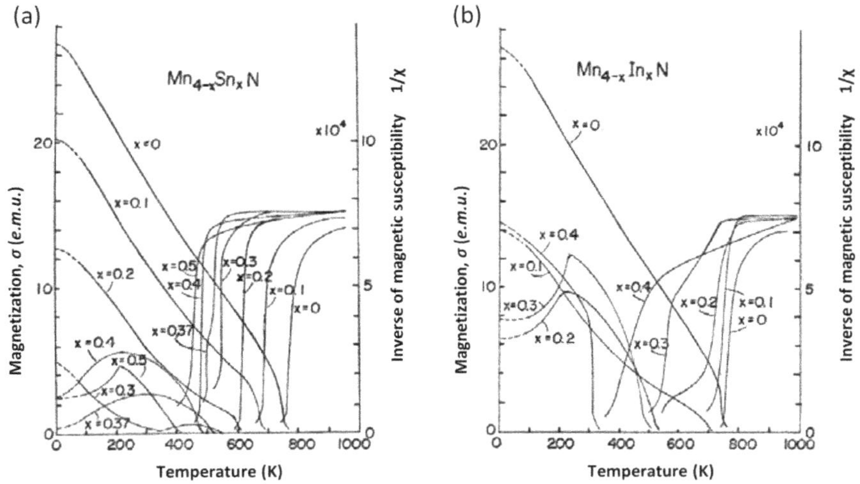

Figure 3.4. Temperature dependences of magnetization and the inverse of magnetic susceptibility for (a) $Mn_{4-x}Sn_xN$ and (b) $Mn_{4-x}In_xN$. Reprinted from [13], with the permission of The Physical Society of Japan.

the temperature dependences of magnetization and the inverse of magnetic susceptibility for (a) $Mn_{4-x}Sn_xN$ and (b) $Mn_{4-x}In_xN$. It is noted that the magnetization versus temperature (M–T) curves that have a maximum value at a certain temperature were obtained in both samples. This is attributed to the difference in the each sublattice's temperature dependence of the magnetization, as shown in figure 3.5.

Figure 3.6 shows the magnetic moments per formula unit determined by the extrapolation of the saturation magnetization to 0 K, wherein the magnetization above the compensation point is plotted in the negative region. The origin of the magnetic compensation is explained by the position of the Fermi level, leading to the change in the magnetic moment of Mn atoms.

Recently, such works have been expanded to polycrystalline Mn_4N bulk crystals doped with various impurity elements [14]. Samples were formed by arc-melt to prepare homogeneous polycrystalline ingots. As shown in figure 3.7, it was reported

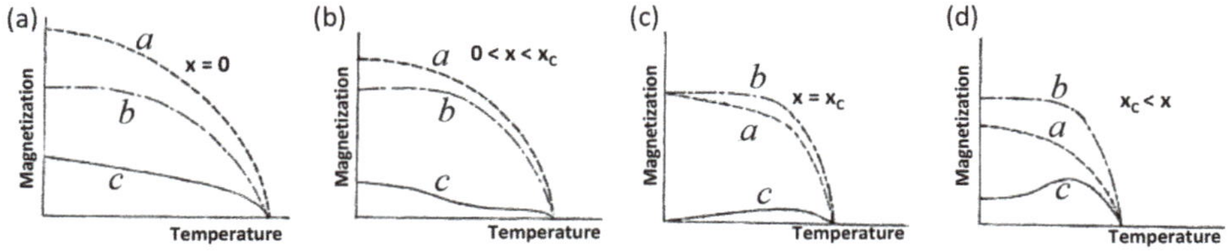

Figure 3.5. Schematic M–T curves at various impurity compositions x with respect to the magnetic compensation composition x_C. Marks a, b, and c denote the magnetizations of the Mn(I) sublattice, of the Mn(II) sublattice, and of the resultant, respectively. Reprinted from [13], with the permission of The Physical Society of Japan.

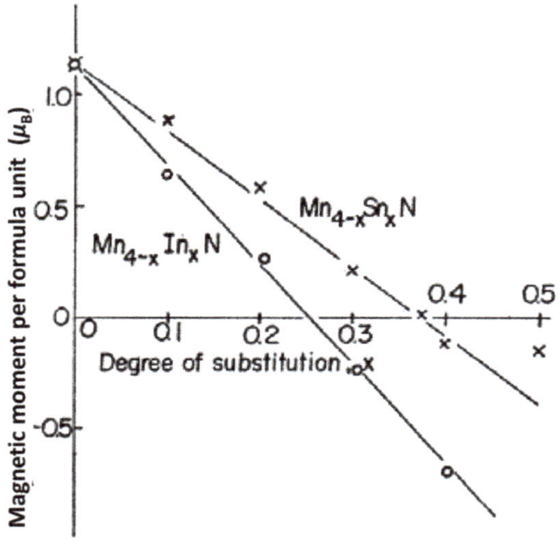

Figure 3.6. Degrees of substitution dependent magnetic moments of Mn_4N partially substituted by In and Sn at 0 K. Reprinted from [13], with the permission of The Physical Society of Japan.

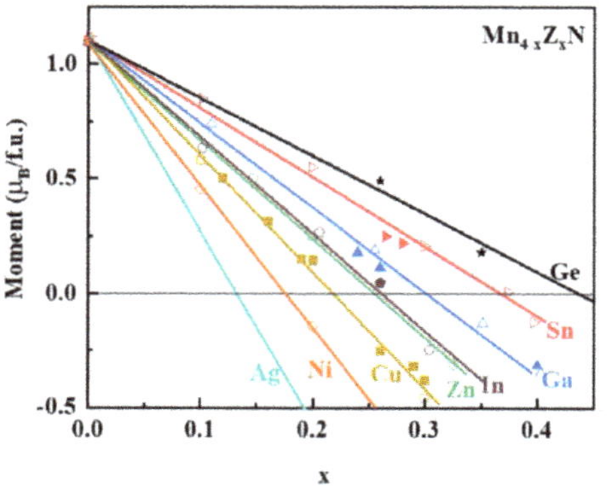

Figure 3.7. Summary of the net magnetic moment of $Mn_{4-x}Z_xN$ as a function of composition x for various impurities. Solid and open data points are obtained by experiment. Reproduced from [14]. CC BY 4.0.

that not only Sn and In but also Ag, Ni, Cu, Zn, Ga, and Ge are candidates that induce the magnetic compensation in Mn_4N at room temperature.

3.3 Mn_4N thin films

3.3.1 History of thin-film growth

Mn_4N thin-film growth started around 1994, while the research on bulk Mn_4N has started by more than 50 years ago [15]. Extensive studies have since been conducted on thin-film growth of Mn_4N by various methods, such as sputtering, molecular beam epitaxy, and pulsed laser deposition on a variety of substrates, such as Si, glass, SiC, GaN, MgO, and $SrTiO_3$(STO). Table 3.1 summaries several reports on Mn_4N thin-film growth. It is noted that a lattice mismatch $\Delta a/a$ given by equation (3.1) influences the crystalline quality of grown films.

$$\frac{\Delta a}{a} = \frac{a_{\mathrm{film}} - a_{\mathrm{sub}}}{a_{\mathrm{sub}}}. \tag{3.1}$$

Here, a_{film} and a_{sub} are in-plane lattice constant of Mn_4N and substrate used, respectively. The $\Delta a/a$ is –7.6% for Mn_4N/MgO(001) and –0.4% for Mn_4N/STO (001).

3.3.2 Characterizations of Mn_4N epitaxial films on MgO(001) and $SrTiO_3$(001)

Here, specific examples of Mn_4N thin-film growth and their properties are presented, focusing on results obtained in the author's laboratory [11]. Although MgO(001) is the substrate that is most often used to form Mn_4N epitaxial thin films [6, 20, 21], Mn_4N epitaxial films on STO(001) exhibit more remarkable properties, in

Table 3.1. Several reports on Mn_4N thin-film growth. Substrate, growth method, uniaxial isotropy constant (K_u), ratio of out-of-plane lattice constant (c) to in-plane lattice constant (a), c/a, saturation magnetization (M_S), and crystal orientation of grown films are specified.

Substrate	Method	K_u [MJ m^{-3}]	c/a	M_S [kA m^{-1}]	Orientation	References
Si(100)	Sputter	—	—	49	<100>, <111>	[16]
Glass	Sputter	—	—	40	<100>	[17]
6H-SiC, GaN(0001)	MBE	—	—	80	<111>	[18]
MgO(001), STO (001)	MBE	0.22	0.99	145	<100>	[19]
MgO(001)	PLD	0.16	0.99	157	<100>	[20]
MgO(001)	Sputter	0.088	0.99	110	<100>	[21]
MgO(001)	Sputter	0.01–0.1	0.99	50–100	<100>	[6]

particular a few mm-sized huge magnetic domain structure, with perfect remanent magnetization, poor nucleation, and a sharp magnetization switching [22], which has never been seen before. These properties together with a very small magnetization and large PMA, demonstrate the potential of Mn_4N films in spintronics applications.

First, details of the growth conditions of the Mn_4N thin films and the results of x-ray diffraction properties of the thin films are presented. Next, we discuss the basic magnetic properties of the films, such as the magnetization hysteresis curves and the magneto-transport properties. In addition, the magnetic domain texture is investigated by magneto-optical Kerr effect microscopy and magnetic force microscopy in the as-deposited state, as well as partially reversed state by the magnetic field. Finally, a comparison of experimental results between Mn_4N films on MgO(001) and those on STO(001) is made, and the relationship between crystallinity of Mn_4N films and the DW properties is discussed.

(1) MBE growth of Mn_4N films

A schematic of the MBE system is shown in figure 3.8, which is equipped with reflection high-energy electron diffraction (RHEED) system, three Knudsen cells (K-Cells) for Mn and other metal sources, and a radio-frequency (RF) sputtering gun for capping materials to prevent oxidation of grown layers. An RF-plasma gun for nitrogen source for Mn_4N is also equipped. The base pressure of the growth chamber is in the order of 10^{-7} Pa. It is monitored by a nude ion gauge. The actual thin-film growth procedure is shown below.

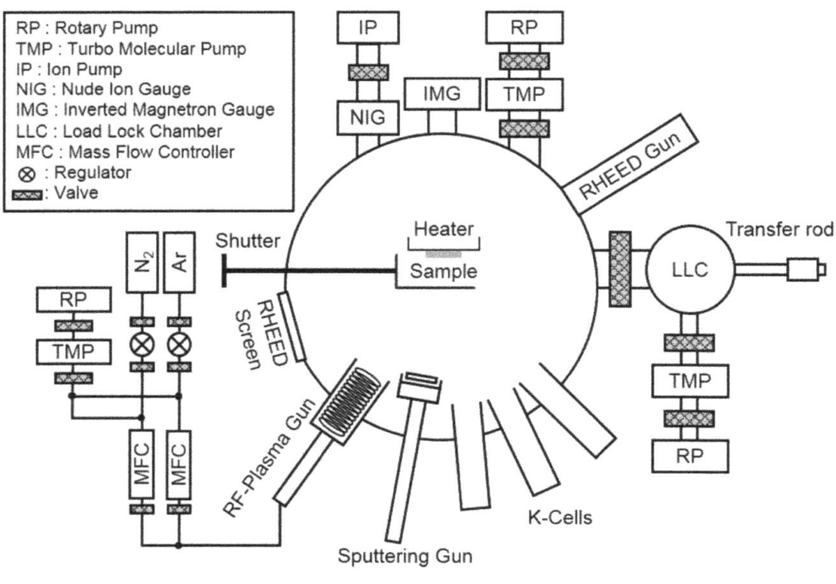

Figure 3.8. Set-up of the MBE chamber.

a. Substrate preparation

STO(001) and MgO(001) substrates are washed with organic solvents in an ultrasonic bath. It is noted that they are not immersed in deionized water to avoid dissolution of MgO. Prior to growth, the MgO substrate is annealed at 600 °C in the growth chamber to remove organic contamination. Meanwhile, the STO wafers are etched in a buffered HF solution with a composition of 35%-NH_4F and 5%-HF by weight to obtain a Ti–O_2 terminated step surface [23]. To avoid the formation of oxygen vacancies, the STO is not annealed prior to the deposition.

b. Deposition technique

There are three key parameters that should be optimized: the substrate temperature T_S during deposition, the deposition rate of Mn, and the N-supplying condition. As for the N_2 gas, the N is supplied as an electrically neutral atom through the gas line and plasma gun to excite N-plasma, while N-related ions are prevented from reaching the substrate by applying a bias voltage of 300 V. The flow rate and the pressure during the deposition is controlled to be around 1 sccm and 5 mPa, respectively. The RF input power is usually set to 105 W and the emission spectrum of the nitrogen plasma is monitored by an optical spectrometer. The deposition rate of Mn can be controlled by the crucible temperature of the Mn K-cell. When Mn_4N films are formed, the deposition rate of Mn is kept constant to be at around 1.3 nm min^{-1} by setting the crucible temperature at 835 °C. The substrate temperature is kept at 450 °C to react Mn with N atoms during the deposition.

The surface of Mn_4N is covered by SiO_2 or Ta capping layers by sputtering after the growth of Mn_4N to prevent oxidization. The Ar pressure is 0.1 Pa during the deposition of the capping layers. The sputtering gun generates an Ar^+ plasma and sputters the target source to deposit Ta or SiO_2 on grown layers.

(2) Properties of Mn_4 N films

The crystalline quality of grown films is characterized by RHEED and XRD.

a. **RHEED**

The MBE chamber used is equipped with a RHEED system (figure 3.9) to characterize the surface of grown layers. The electron beam is accelerated at 20 keV. This electron beam is irradiated onto the sample surface at an angle of

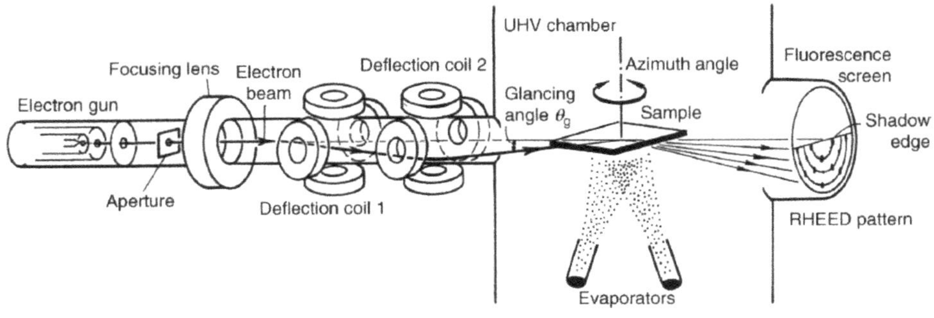

Figure 3.9. Schematic RHEED diagrams. Taken from T Gushi's PhD thesis [11].

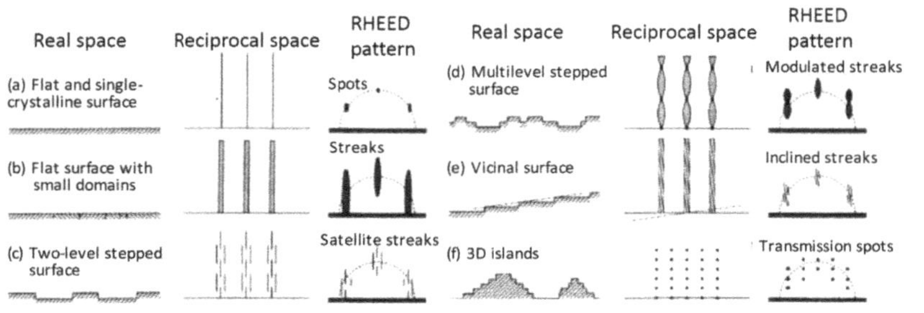

Figure 3.10. Typical RHEED patterns for various surface states. Taken from Gushi's PhD thesis [11].

incidence of around 0.2° to obtain a forward scattered diffraction pattern on the screen coated with fluorescent materials [24]. Figure 3.9 shows a schematic of a typical RHEED system. The brightness, the focus, the contrast of the diffraction pattern can be controlled. The RHEED pattern is quite sensitive to the crystal orientation, crystalline quality, and the surface flatness of grown layers, as shown in figure 3.10. Figure 3.11(a) presents a simulated RHEED pattern of *c*-axis-oriented Mn_4N epitaxial films observed along the MgO or STO[100] azimuth [25]. It is noted that the superlattice diffraction, such as 010, is ascribed to the long-range order of N atoms at the body center. Furthermore, the degree of N-order can be evaluated quantitatively from the superlattice diffraction intensities. Figures 3.12(b) and (c) show the RHEED streaks of Mn_4N layers grown on MgO(001) and STO(001), respectively. These RHEED streaks indicate a good crystalline quality of the grown layers. The superlattice diffraction also appears at the arrows in figures 3.11(d) and (e).

b. **X-ray diffractometry (XRD)**

X-ray diffractometry techniques have been used to characterize MBE-grown layers. Incident x-rays with a Cu Kα radiation source diffract based on Bragg's law expressed as,

$$2d_{\mathrm{hkl}} \sin\theta = n\lambda. \tag{3.2}$$

In the experiment, x-rays are made monochromatic using Ge(220) crystals. Here, d_{hkl} denotes the distance between (*hkl*) planes in the crystal, θ is the incident angle of x-ray with respect to the film plane, n is the diffraction order, and λ is the wavelength of x-ray. It is around 0.154 056 nm. Here, the nth diffraction peak from the (*hkl*) plane is denoted as *nh nk nl*. For example, the first order diffraction peak of the $Mn_4N(002)$ plane ($n = 1$) is expressed as 'Mn_4N 002'. In addition, the second order diffraction peak related to the $Mn_4N(001)$ plane ($n = 2$) is also expressed as 'Mn_4N 002'. Figure 3.12 (a) shows the XRD measurement set-up [26]. By adjusting the angle of the substrate relative to the x-ray detector, the orientation axis to focus on can be selected, such as the out-of-plane plane ($2\theta/\omega$) or the in-plane ($2\theta_\chi/\varphi$) axis. A series of diffraction peaks identify the type of crystals and provides information about the quality of the film.

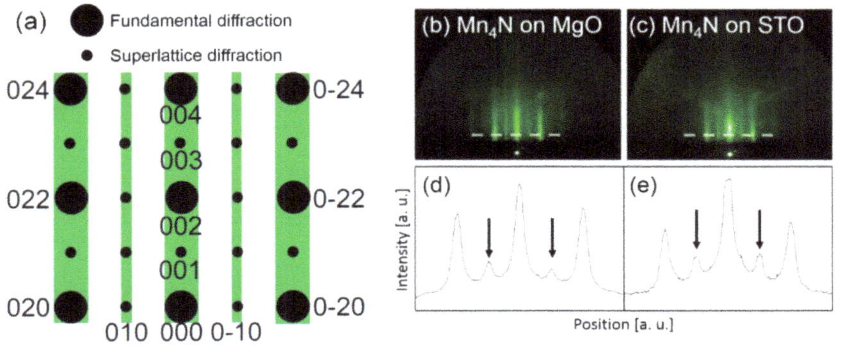

Figure 3.11. (a) Simulated RHEED pattern of antiperovskite nitrides [25]. (b), (c) RHEED patterns of Mn_4N layers grown on MgO(001) and STO(001) taken along the [100] azimuth. (d), (e) The brightness profiles of RHEED streaks in (c) and (d), respectively. Arrows indicate the superlattice diffractions.

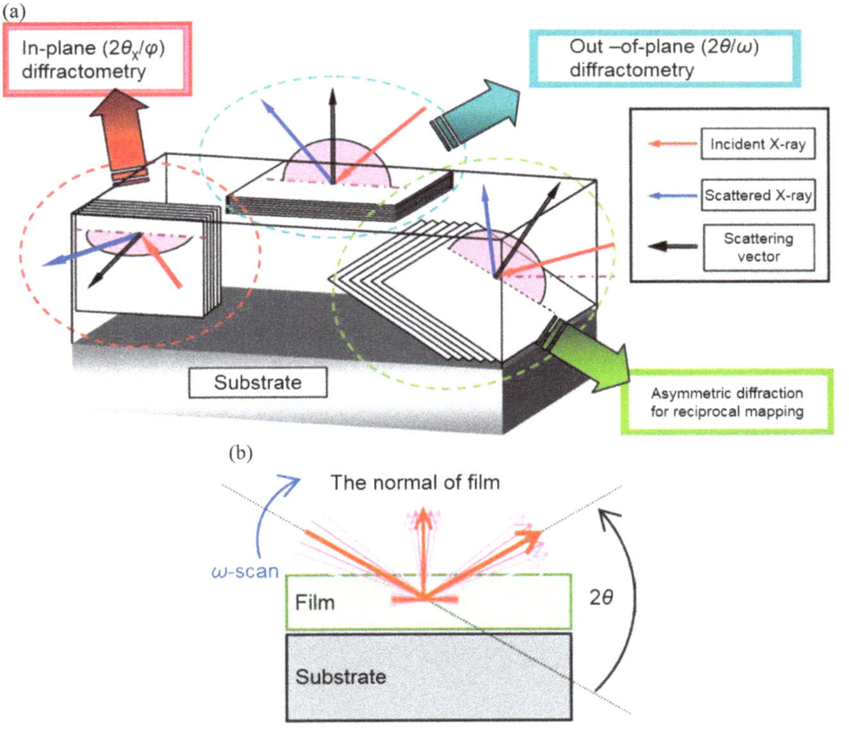

Figure 3.12. Geometries of (a) in-plane and out-of-plane x-ray diffraction measurements. Reproduced with permission from [26]. (b) ω-rocking curve measurements. Reproduced with permission from [27].

X-ray ω-scan rocking curve measurements are useful to evaluate how well the crystalline axis of the film is oriented normal to the sample plane. Figure 3.12(b) shows its measurement principle [27]. After setting 2θ at a certain diffraction angle satisfying Bragg's law, the angle of sample is tilted

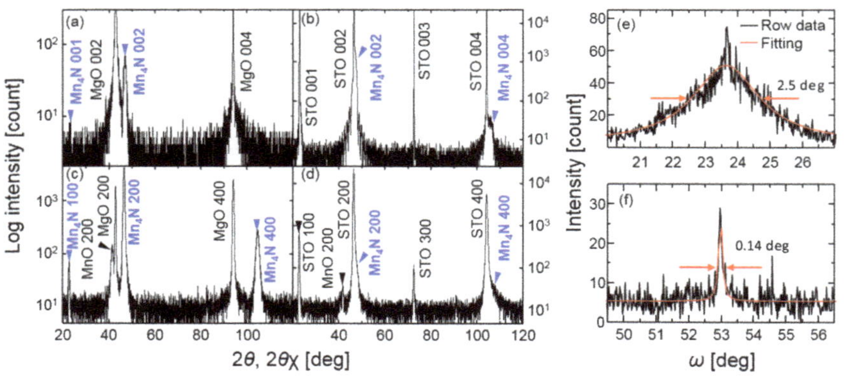

Figure 3.13. (a), (b) Out-of-plane XRD patterns. The blue marks indicate the peaks of *c*-axis-oriented-Mn_4N. (c), (d) In-plane XRD patterns. The incidence angle is set at $\omega = 0.4°$, and the scattering vector is along the [100] azimuth of MgO or STO. (e), (f) ω-scan rocking curves for (e) Mn_4N 002 on MgO, and (f) Mn_4N 004 on STO. The black and red lines represent raw data and Lorentzian curve fits, respectively. Reproduced from [22].

by ω and the tilt of the crystalline axis normal to the sample plane can be qualitatively evaluated.

Figure 3.13 presents the results of XRD measurement on the Mn_4N epitaxial films on MgO(001) and STO(001). The x-ray ω-scan rocking curves are also shown in (e) for Mn_4N/MgO(001) and (f) Mn_4N/STO(001), wherein the 2θ is set at 002 and 004 of Mn_4N, respectively. For Mn_4N/MgO, the diffraction peaks of the film and those of the substrate are well distinguished, allowing us to extract enough information on the Mn_4N grown layers. In contrast, the diffraction peak of Mn_4N 002 is close to that of STO 002. We therefore have to use the higher-order Mn_4N 004 peak, which is well distinguished from the STO 004 peak. The obtained rocking curves are fitted well by a Lorentzian curve. These observations demonstrate the epitaxial growth of Mn_4N films on both MgO(001) and STO(001). However, there is a big difference in the full-width at half maximum between figures 3.13(e) and (f). This result indicates that Mn_4N films are much better oriented on STO than those on MgO, which is consistent with what has been previously reported on Fe_4N films [25].

c. **Magnetic properties**

The magnetization of the sample is measured by detecting the stray magnetic field emitted from the sample. Here we see two kinds of magnetometer, a vibrating sample magnetometer (VSM) and a superconducting quantum interference device (SQUID). In both techniques, the magnetic moment of the entire sample, including the capping layer and substrate, in addition to the grown Mn_4N films, is measured.

Vibrating sample magnetometer (VSM)

Stray magnetic fields caused by electromagnetic induction is detected by VSM. Figure 3.14 presents a schematic of VSM. The sample is placed between the magnetic poles to magnetize by a constant magnetic field. The inductive voltage generated in the pick-up coils is proportional to the time-derivative of magnetic flux passing through them according to Faraday's law. Therefore, the constant magnetic field by the magnetic poles does not generate any inductive voltage in the pick-up coils. To make the magnetic flux passing through the pick-up coils time-varying, the position of sample is vibrated at around 80 Hz. The pick-up coils use a lock-in technique to detect the induced oscillating voltage caused by the vibration of the sample.

The magnetization-to-voltage conversion ratio is calibrated using a standard sample such as Ni, which can therefore be measured the sample's magnetization from the oscillating voltage.

Superconducting quantum interference device (SQUID)

The Josephson effect in superconductors [28] is utilized in the SQUID, which appears in superconductor/insulator/superconductor layered structure. In 1962, B Josephson predicted that Cooper pairs can tunnel through a barrier layer from one superconductor to another [29]. The two superconductor layers are separated by an insulating barrier layer with 'weak link' at distances of about 2 nm. The point is that the distance between the two superconductors is shorter than the coherent length of the superconductor. When the current is smaller than the critical current I_c, the Cooper pairs can tunnel through the insulating barrier layer without any voltage drops. This current is called Josephson current. The magnitude of Josephson current is quite sensitive to the magnetic flux going through a ring circuit containing one or more Josephson junctions. The ring device shown in figure 3.15 consists of two such junctions, and is called SQUID, which can sense a tiny magnetic field. Once a bias current I_b is set to be slightly greater than I_c, a voltage drop depending on the current flowing in the loop I_{loop} can be detected. When a magnetic field penetrates the

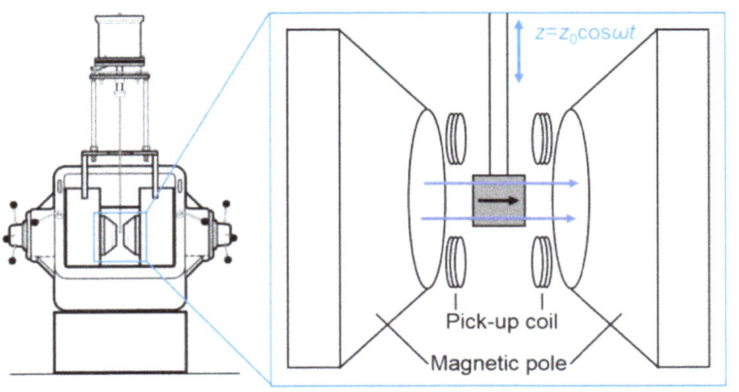

Figure 3.14. Schematic of the VSM.

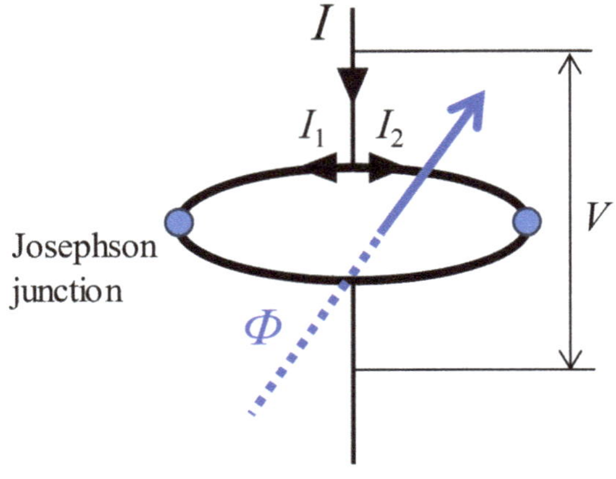

Figure 3.15. Josephson junction and SQUID principle.

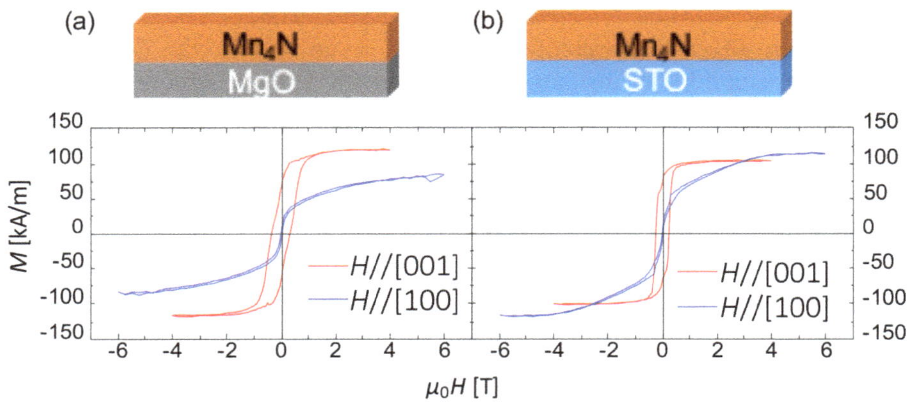

Figure 3.16. Magnetization curves measured on the (a) Mn_4N/MgO and (b) Mn_4N/STO samples at room temperature. The red and blue lines correspond to out-of-plane and in-plane magnetic field measurements, respectively. Reproduced from [22].

SQUID ring, a screening current due to the Meissner effect tries to cancel the magnetic flux and I_{loop} ($= \Phi_{loop}/L_{loop}$) changes. Therefore, we can measure the magnitude of the magnetic field. In contrast to VSM, SQUID does not measure the stray magnetic field but the magnetic flux itself. Therefore, there is no need to vibrate samples.

Figure 3.16 presents the out-of-plane and in-plane magnetic field ($\mu_0 H$) versus magnetization (M) curves, obtained on (a) Mn_4N films on MgO and (b) Mn_4N films on STO by the magnetic properties measurement system (MPMS; Quantum Design) with an external magnetic field up to 4–6 T. This MPMS is equipped with SQUID and VSM options. The samples are cut into pieces, and inserted into a straw to conduct the measurement. From the out-of-plane hysteresis curves, the M_S of Mn_4N films on MgO and those on STO can be estimated to be 119 and 103 kA m^{-1}, respectively. These values are in good agreement with those reported [21]. From the in-plane magnetization curve of the Mn_4N films on STO, its anisotropy field H_K is estimated to be around 4 T. The uniaxial anisotropy K_u is thus calculated to be 1.1×10^5 J m^{-3} for both samples. The demagnetization field energy is also taken into account.

Magneto-transport properties

Anomalous Hall effect (AHE) measurements are often conducted to detect the magnetization switching precisely. This is because the characteristics of only Mn_4N films with current flow can be evaluated since the substrate (MgO, STO) and the capping layers (SiO_2) are insulating. Both the Hall coefficient and resistivity are measured using the van der Pauw method [30, 31]. As shown in figure 3.17, the four corners of the film sample are connected by wire bonding. By using this van der Pauw method, we will be able to measure the resistivity and Hall coefficient of arbitrary shaped films without patterning them into a Hall bar as long as the film is flat and homogeneous.

To measure the resistivity of the Mn_4N films, we pass the current from point **a** to point **b**, I_{ab} in figure 3.17 and measure the voltage drop between **c** and **d**, V_{cd}. The direction of the current is then reversed and the voltage drop is measured in the same manner to take the average. The ratio of V_{cd} to I_{ab} is defined as $R_{ab,cd} = V_{cd}/I_{ab}$. The resistance obtained with the other combination $R_{ac,bd}$ is also measured to take into account eventual asymmetries of measured sample. There is the following relationship between $R_{ab,cd}$, $R_{ac,bd}$, and the resistivity ρ_{xx},

$$\exp\left(-\frac{\pi t}{\rho_{xx}} R_{ab,\,cd}\right) + \exp\left(-\frac{\pi t}{\rho_{xx}} R_{ac,\,bd}\right) = 1. \qquad (3.3)$$

Here, t is the thickness of the Mn_4N films.

The schematic for Hall measurement is shown in the right-hand part of figure 3.17. The transverse voltage V_y in magnetic layers depends mainly on the

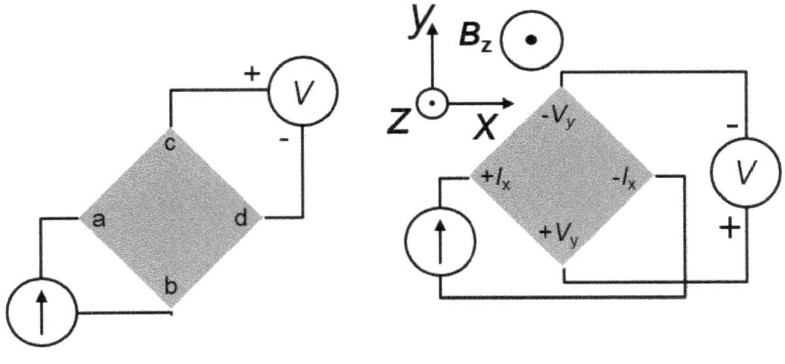

Figure 3.17. Measurement circuit used for resistivity measurement (left-hand) and Hall measurement (right-hand). The chirality of *I*–*V*–*B* is defined like an ordinary Hall measurement. Positive carriers generate positive ordinary Hall voltage.

Hall effect, the anomalous Hall effect, and the planar Hall effect. V_y can be given in the following when a magnetic flux density is in the normal direction, B_z, [32–34].

$$V_y = \left(R_0 + R_{\rm H}\frac{B_z}{t} + \rho_{\rm AHE}\frac{\cos\theta}{t} + \frac{\rho_{//} - \rho_{\perp}}{t}\sin^2\theta \sin\varphi \cos\varphi \right) I_x \tag{3.4}$$

Here, R_0 is the resistive offset, $R_{\rm H}$ is the ordinary Hall coefficient, $\rho_{\rm AHE}$ is the anomalous Hall resistivity, $\rho_{//}$ ($\rho_{\perp}$) is the resistivity for a magnetization parallel (perpendicular) to the current, $\theta(\varphi)$ is the elevation (azimuth) angle of the magnetization shown in figure 3.18, and I_x is the longitudinal current. Since the sample is not perfectly square, the transverse voltage V_y usually includes a resistive offset (R_0), which depends very slightly on the magnetic field. The second term of the equation, which is proportional to the magnetic flux density, is the ordinary Hall component. In Mn_4N, $R_{\rm H}$ is very small because Mn_4N is metal and thus the career density is so high. This component is proportional to the external magnetic field, and thus can be subtracted from the slope in the high field region. The third term is the AHE voltage, which is proportional to the magnetization normal to the sample plane. Since the magnitude of the third term is inversely proportional to the thickness of Mn_4N layer, the magneto-transport measurement method is useful to ultrathin films with particularly small $M_{\rm S}$, where stray magnetic fields are difficult to detect by VSM. The fourth term represents the planar Hall effect, which has the same origin as the anisotropic magnetoresistance (AMR) effect, depending on the angle between the magnetization and the current. The AMR ratio of Mn_4N is usually smaller than 0.1% at room temperature [21]. Thus, this term is negligible.

The origins of AHE such as skew scattering [35, 36], side jump [37], and intrinsic deflection due to Berry's phase curvature [38, 39] are the same as those of SHE. Electrons with up-spin and those with down-spins are deflected in the opposite

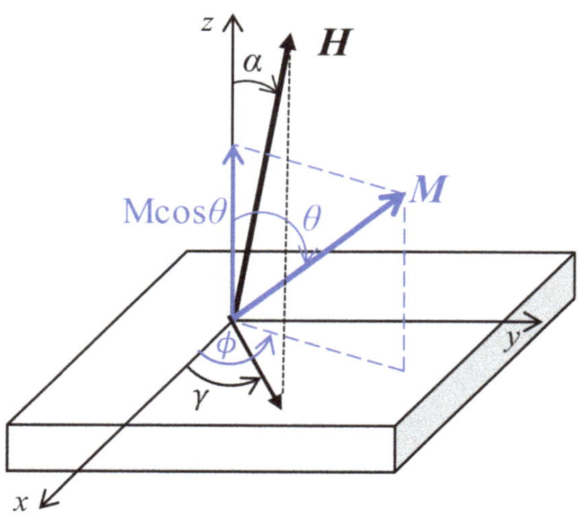

Figure 3.18. Cylindrical coordinates to describe the magnetization direction.

direction. In Pt, a non-magnetic heavy metal [40], a pure spin current normal to both the magnetization and the current is generated.

The magnitude of the AHE is expressed by the anomalous Hall angle θ_{AH}. They are given using ρ_{AHE} and the longitudinal resistivity ρ_{xx} as,

$$\theta_{AH} = \frac{\rho_{AHE}}{\rho_{xx}}, \ \sigma_{AH} = \frac{\rho_{AHE}}{\rho_{xx}^2 + \rho_{AHE}^2}. \tag{3.5}$$

Here, θ_{AH} is useful to evaluate the conversion ratio from the longitudinal voltage (current) into the transverse voltage (current). Similarly, σ_{AH} is useful to theoretically classify the origin of the AHE by comparing it with the temperature dependence of the longitudinal conductivity σ_{xx} [41]. For instance: (i) in the high conductivity regime, the skew scattering dominates the AHE and the linear σ_{AH} becomes σ_{xx}; (ii) a scattering-independent or intrinsic regime where σ_{AH} is independent on σ_{xx}; and (iii) a bad metal regime where σ_{AH} decreases with σ_{xx} faster than linearly at high temperatures.

We now see the transport measurement results. The resistivity ρ_{xx} extracted from the van der Pauw measurement is 187 μΩcm for Mn_4N films on MgO and 181 μΩcm for Mn_4N films on STO at room temperature. Figure 3.19 shows the AHE hysteresis loops with magnetic fields applied normal to the sample surface. The AHE angle θ_{AH} is negative but high (–2%), which is in line with those previously reported on Mn_4N [20, 42]. The Mn_4N films on MgO show a smooth hysteresis loop (red), whereas the magnetization of the Mn_4N films on STO switches very sharply, with a full remanence at zero magnetic field (blue). The coercivity is greater for the former (red) than the latter (blue). In spite of these differences, very similar spontaneous magnetization of the two systems indicate that the grown films of Mn_4N are essentially the same regardless of the substrate. However, the structural features

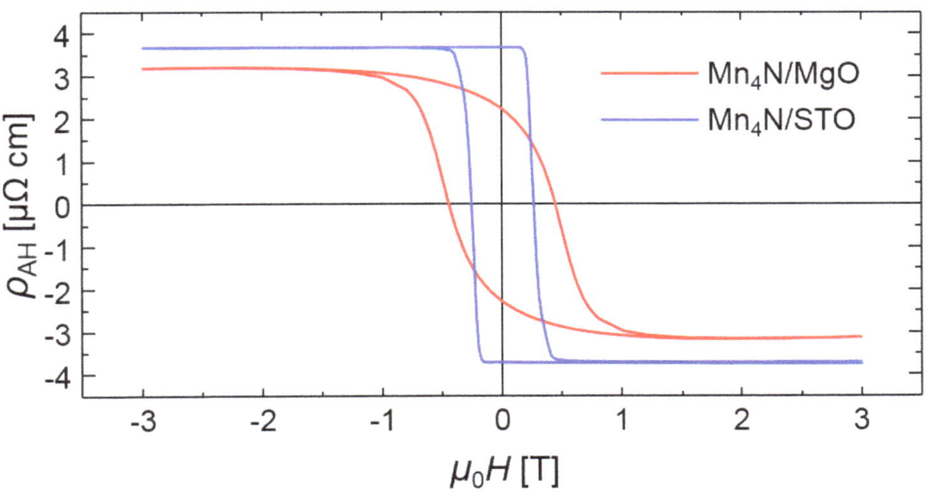

Figure 3.19. Anomalous Hall effect hysteresis loops measured on the Mn_4N/MgO(red) and Mn_4N/STO(blue) samples.

suggest that the nature and the density of structural disorder, i.e., crystal defects, are different, and thus the increased disorder is responsible for the greater coercivity and the slanted loop in Mn_4N films on MgO.

We can evaluate in-plane magnetization properties by magneto-transport measurement when magnetic fields are applied in the in-plane direction (define as y-axis) [43–45]. The experimental geometry is shown in figure 3.20. Since the anomalous Hall voltage ΔV_y is proportional to the vertical component of magnetization ($M_z = M_S \cos\theta$), the in-plane component of magnetization (M_y) is given as,

$$\frac{M_z(H)}{M_S} = \frac{\Delta V_y(H)}{\Delta V_{AH}} \tag{3.6}$$

$$M_y(H) = M_S\sqrt{1 - \left(\frac{M_z(H)}{M_S}\right)^2} \tag{3.7}$$

Here, ΔV_y is the transverse voltage without resistive offset for each H and ΔV_{AH} is the saturated anomalous Hall voltage. The magnetization is always set to be normal to current (I_x) to make the AMR contribution negligible. The magnetic anisotropy constant K_u can be derived from the area enclosed by out-of-plane and in-plane M–H loops as,

$$K_u = \frac{\mu_0 M_S^2}{2} + \mu_0 \int_0^{H_{MAX}} \left| M_z dH_z - M_y dH_y \right|. \tag{3.8}$$

Here, the first term is the demagnetizing energy and the second term is the effective anisotropy.

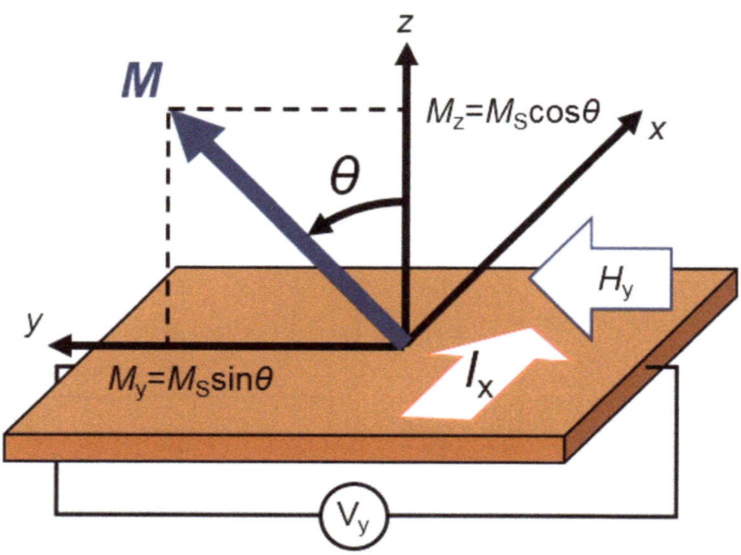

Figure 3.20. Geometry of in-plane AHE.

Figures 3.21(a) and (b) present the hysteresis loops of Mn_4N films grown on MgO (001) and STO(001), respectively, with magnetic fields applied in the in-plane direction. The red lines denote the out-of-plane magnetization component expressed by equation (3.6) and the blue lines are the in-plane magnetization component derived from equation (3.7). The M_y on MgO increases rapidly at a relatively small magnetic field and shows weak anisotropy, while M_y on STO rises linearly with increasing magnetic field and shows an anisotropic field $H_K = 4.5$ T. According to equation (3.8), the magnetic anisotropy energy K_u is calculated to be 0.12 MJ m^{-3} for MgO and 0.20 MJ m^{-3} for STO. While the results of magnetization measurements by MPMS are almost the same as those for MgO and STO substrates, magneto-transport based analysis reveals a clear difference between the two substrates in terms of magnetic anisotropy. This difference is attributed to the degree of orientation, which is confirmed by x-ray ω-rocking curve measurements.

(3) Magnetic domain structure

The domain structures of magnetic films can be observed by several methods, and the most common are magneto-optical methods using the Faraday effect for transmitted light and the Kerr effect for reflected light. Photoemission electron microscopy (PEEM), magnetic force microscopy (MFM), Lorentz microscopy, and scanning electron microscopy with polarization analysis are also used. In this session, Kerr microscopy and scanning probe microscopy (SPM) are briefly introduced.

a. Magneto-optical Kerr effect (MOKE) imaging

The magneto-optical Kerr effect (MOKE) was discovered in 1877 [46]. It uses the fact that once linearly polarized incident light is reflected on a magnetic surface, it is converted to elliptically polarized light. The sign of the Kerr rotation (*i.e.*, the rotation of the light polarization) and the magnitude depend on the magnetization direction of the film [47]. This makes it possible to detect changes in magnetization at a specific location of the sample, and thus to make a two-dimensional image reflecting the magnetic domain configurations. Depending on the magnetization direction relative to the incident light, there are three types of configurations, as shown in figure 3.22. The polar MOKE is usually used for PMA films, and longitudinal and transverse MOKEs are used for in-plane magnetization films.

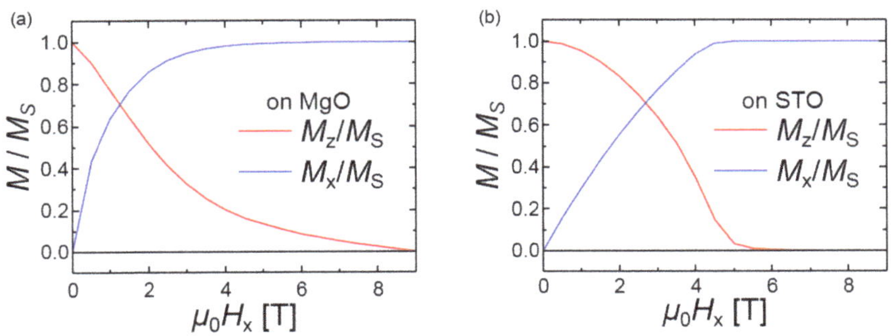

Figure 3.21. In-plane magnetization loops of Mn_4N films on (a) MgO and (b) STO.

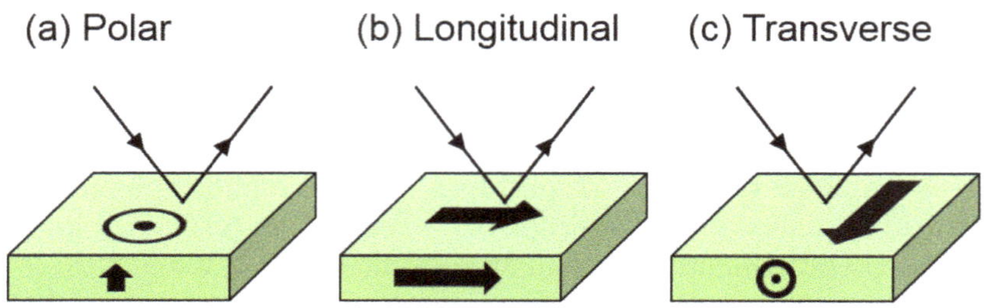

Figure 3.22. Three types of MOKE configuration.

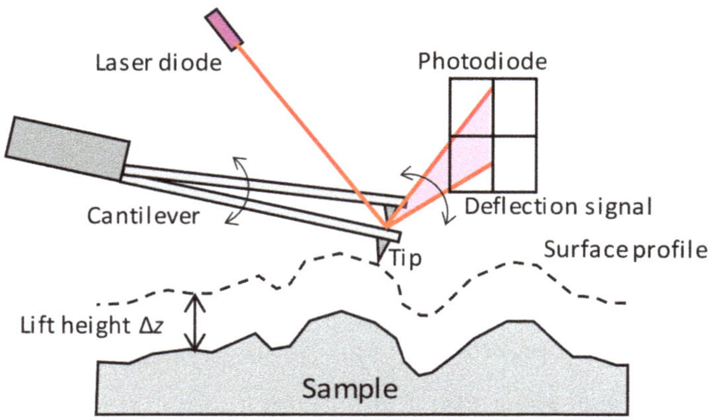

Figure 3.23. AFM set-up and operation principle.

The spatial resolution of the MOKE microscopy is limited by the wavelength of incident light. In a set-up with the highest magnification objective lens, 1 pixel corresponds to 36.5 nm. Meanwhile, the technique is relatively fast and the maximum viewing size is comparable to that of an optical microscope. Therefore, it can be applied to time-resolved measurements, such as DW motion experiments and observation of large magnetic domains.

b. Magnetic force microscopy (MFM) imaging

By using a special AFM chip coated with a magnetic thin film, magnetic signals from the sample surface can be detected and magnetic domain structures can be obtained with a higher spatial resolution than with a MOKE microscope. Figure 3.23 shows the operation principle of AFM [48]. MFM is a type of scanning probe microscope that uses a probe to scan a sample surface and detect magnetic signals. AFM operates by atomic force, which has a very short operating range between the tip and the sample surface, and can produce topographic images of surfaces. In other words, the harmonic frequency of the cantilever, monitored by laser reflection and photodiode, is modulated by atomic force to adjust the height, while keeping the frequency constant and output height data. Meanwhile, the magnetic force has a longer interaction length than an atomic force. Therefore, by lifting the tip off the sample surface and keeping it high, the AFM signal can be drastically reduced and the magnetic signal can be extracted from the sample. Since magnetic force also modulates the harmonic frequency of the tip, the phase change $\Delta\varphi$ of the oscillation can be detected, which is given by $\Delta\varphi = \frac{Q}{k}\frac{\partial F_z}{\partial z}$. Here, k, Q, z and F_z are the spring constant, the quality factor, the height of the tip, and the force acting in the z direction, respectively. The domain structure of the magnetic layer can be estimated from the two-dimensional mapped phase change distribution. The actual MFM image is quite different between PMA and IMA films. This happens because the MFM tip detects the magnetic force caused by the stray magnetic field due to the magnetization. In this book, we focus on perpendicularly magnetized films such as Mn_4N.

Figure 3.24 shows a schematic of the stray magnetic field from a perpendicularly magnetized film and a simplified MFM signal. The film with PMA emits stray magnetic field everywhere, allowing direct observation of magnetization orientation. In MFM images, there are two types of magnetic signals, called monopolar and bipolar contributions [49]. Depending on the sign of the surface magnetic charge in the domain given by $H = \boldsymbol{M}\cdot\boldsymbol{n}$, where $\boldsymbol{n}$ is the vector normal to the surface, the contrast of the MFM changes to light or dark. However, only a uniform stray magnetic field without spatial gradient $\frac{dH_z}{dz}$ is radiated from a perpendicularly magnetized infinite plane and only a coupled force that rotates the magnetic moment is produced. Therefore, a magnetic domain much larger than the MFM tip only has a small magnetic signal in the MFM image. In contrast, in DWs the stray magnetic field forms a flux loop that minimizes the magneto-static field energy with local large gradients in H_z, and provides a contrast between light and dark, which is called the dipolar contribution. The contrast of the MFM image is determined by a combination of these contributions. In any case, the magnetic domain structure is easier to estimate for perpendicularly magnetized films than for in-plane magnetized films because a perpendicularly magnetized film has only one easy axis, making it easier to distinguish the sign of the magnetic signal. Meanwhile, in-plane magnetized films have two-dimensional degrees of freedom in the

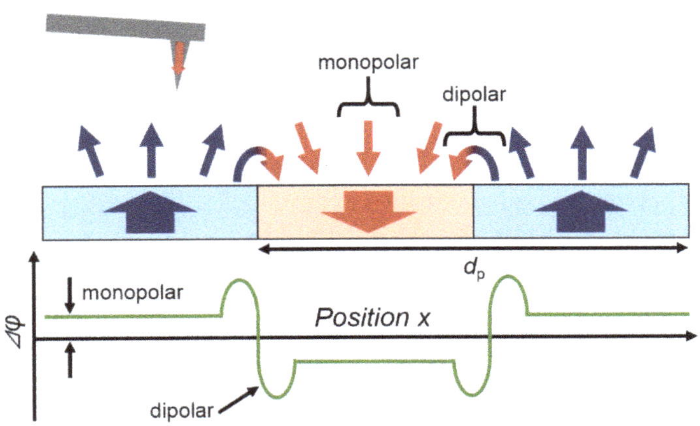

Figure 3.24. Stray field from PMA film and contribution to the MFM signal.

magnetization direction, but MFM can only provide information on the out-of-plane component of the stray magnetic field. Therefore, in-plane magnetized films require comparison of MFM images with micromagnetic simulation results to estimate detailed in-plane domain structure. MFM is independent of the wavelength of light and has much higher spatial resolution than MOKE imaging. However, it takes a longer time to scan a single region compared to MOKE microscopy, which responds instantly. Therefore, MOKE is more convenient for time-resolved and temporally-differential measurements, such as DW motion.

c. Equilibrium domain size

The equilibrium magnetic domain size of Mn_4N thin films is discussed here. It is obtained theoretically from the balance between the demagnetization energy and the DW energy $\sqrt{AK_u}$, where A is an exchange stiffness. According to reference [50], once a periodic stripe domain structure is assumed, the equilibrium domain size is given by,

$$\mathrm{Ln}\left(\frac{D}{t}\right) \approx \frac{\pi D_0}{2t} + \ln(t) - 1 + \mu\left(\frac{1}{2} - \ln(2)\right),\ D_0 = \frac{2\sqrt{AK_u}}{\mu_0 M_S^2},\ \mu = 1 + \frac{\mu_0 M_S^2}{2K_u} \tag{3.9}$$

Here, t is the magnetic layer thickness; D_0 is called 'dipolar length,' which is the ratio of the DW energy to the demagnetizing energy $\frac{\mu_0}{2}M_S^2$; and μ is the ratio of this demagnetizing energy to the anisotropy energy. From the actual parameters of Mn_4N thin films obtained by experiment (M_S =105 kA m^{-1}, K_u =0.11 MJ m^{-3}, t = 9.4 nm, A = 15 pJ m^{-1}, and T_N = 745 K), some basic parameters about the domain structure can be calculated. The DW energy is $\sqrt{AK_u}$ = 1.3 mJ m^{-2}, the DW width is $\sqrt{A/K_u}$=37 nm, D_0 = 0.19 μm, and μ = 1.06. As a result, the equilibrium domain size is calculated to be 2.5×10^5 m for this Mn_4N film. This value is much larger than the sample size. This result means that the demagnetizing energy is negligibly small relative to the DW energy due to the small M_S and large K_u, and that in fact the domain size and shape should rather be determined by the DW pinning on the extrinsic defects [51]. It is noted from equation (3.9) that the domain size is quite sensitive to the thickness and the magnetization. In any case, the Mn_4N thin film is expected to have an infinite-like large-sized domain corresponding to full remanence magnetization at zero magnetic field in nature.

d. Comparison of Mn_4N epitaxial films on MgO(001) and STO(001)

Domain observations are compared for (a) Mn_4N films on MgO with (b) those on STO prior to applying any magnetic field in figure 3.25. This state is often considered relatively close to equilibrium. The magnetic domain configuration is observed by MFM for Mn_4N films on MgO and MOKE microscopy for those on STO because the domain size is so different between them. The average domain period d_p, shown in figure 3.25, is estimated using a two-dimensional fast Fourier transformation method [52]. In the Mn_4N films on MgO, the magnetic domain size is as small as d_p = 0.28 μm, which is much smaller than the equilibrium size by theory. The irregular shape of the DWs is probably characteristic of strong disorder [53], and the

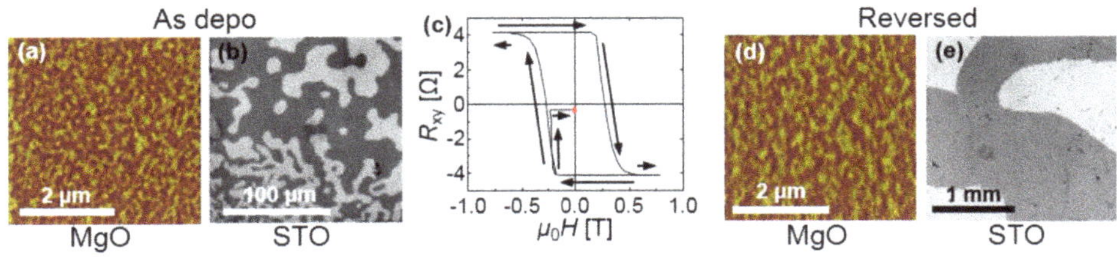

Figure 3.25. (a) MFM and (b) MOKE images of as-deposited Mn_4N films on MgO and STO substrates, respectively. (c) Hall resistance loop during partial reversal process for Mn_4N/STO used for (e). (d) MFM and (e) MOKE images after partial reversal process of Mn_4N films on MgO and STO substrates, respectively. Reproduced with permission from reference. Reproduced from [22]. © IOP Publishing Ltd. All rights reserved.

DW is considered to be pinned in the early stage of growth. Conversely, the magnetic domain size of the Mn_4N films on STO is larger than that on MgO by two orders of magnitude. The d_p is as large as 20 µm and a smooth magnetic domain is formed. This result indicates that the effect of extrinsic disorder is very small in the Mn_4N films on STO. The domain widths are comparable to those of the ultrathin CoFeB(1.1 nm)/MgO(1 nm) and Pt(2.4 nm)/Co(0.27 nm) systems, wherein the d_p values after thermal demagnetization are 14 and 6.5 µm, respectively.

Figure 3.25(c) shows the Hall voltage loop measured on the Mn_4N films on STO in a typical partial reversal process. Figures 3.25(d) and 3.25(e) show the domain structure of both samples in such a partially reversed state. In the case of Mn_4N films on MgO, the domain structure is almost identical to the as-deposited state, while the domain structure of Mn_4N films on STO evolves from micrometer size to millimeter size. The MOKE images show that few nucleation centers are present and that the magnetization reversal is caused by nucleation and subsequent easy propagation. This observation is in agreement with the squared hysteresis loops.

3.3.3 Correlation between lattice distortion and perpendicular magnetic anisotropy

PMA is essential for device applications using CIDWM [54]. As for various ferromagnets such as Mn–Ga alloys, FePt, and FePd, the origin of PMA has been studied and in-plane tensile distortion was confirmed [55–58]. In contrast, however, the origin of PMA for Mn_4N films is not fully understood. Mn_4N thin films have been grown on various kinds of substrates, such as glass, Si(001) [16], MgO(001) [6, 19–21], STO(001) [19, 22], $(LaAlO_3)_{0.3}(Sr_2TaAlO_6)_{0.7}$[LSAT](001) [59], $LaAlO_3$(001) [60], 6H-SiC(001) [18], and GaN(0001) [18] by a variety of growth methods. Assuming that the Mn_4N has a cubic structure, the lattice mismatch between Mn_4N and the (001)-oriented substrates (MgO, STO, and $LaAlO_3$) are approximately –7.6%, –0.4%, and +2%, respectively, as shown in figure 3.26. It should be noted that the value of the lattice mismatch changes slightly when using the lattice constant of Mn_4N, depending on whether the value assuming cubic crystal or the lattice constant in the in-plane direction under in-plane tensile strain is used.

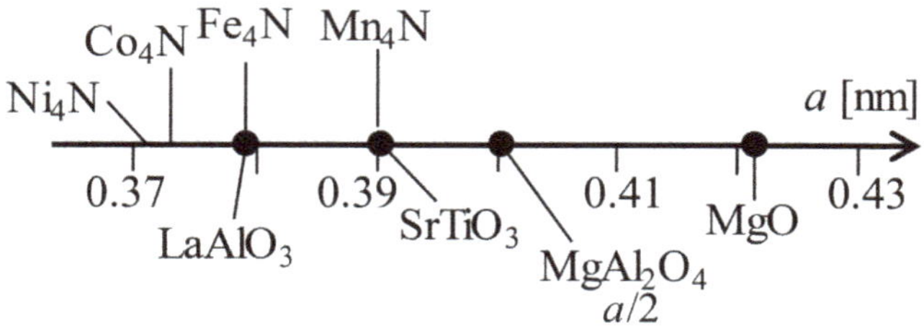

Figure 3.26. Lattice constant relationship between 3d nitrides and substrates.

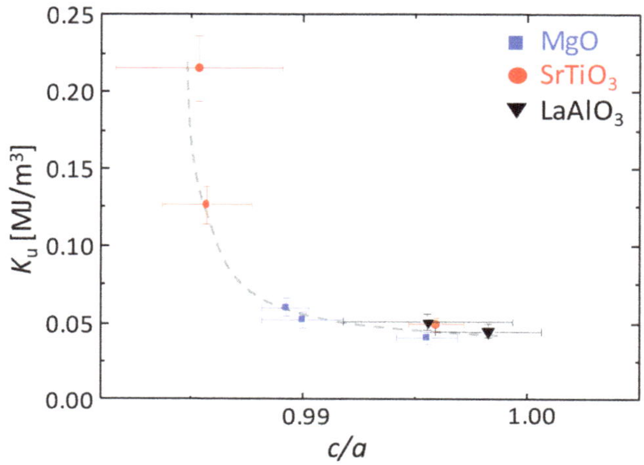

Figure 3.27. Relationship between K_u and c/a obtained from Mn_4N epitaxial films grown on MgO, $SrTiO_3$, and $LaAlO_3$ substrates. Blue-square points, orange-circle points, and black-triangular points show the data obtained from Mn_4N films on MgO, $SrTiO_3$, and $LaAlO_3$, respectively. Broken line is a guide to the eyes. Reproduced from [60]. CC BY 4.0.

Experiments using substrates having a larger lattice constant than Mn_4N have shown that the Mn_4N films formed on such substrates are under an in-plane tensile strain, meaning the ratio of the out-of-plane lattice constant c to the in-plane lattice constant a, c/a, is smaller than 1 ($c/a \sim 0.99$), regardless of the substrate [61–63]. Studies have been quite limited on Mn_4N films formed on substrates whose lattice constant is smaller than that of Mn_4N, and therefore might induce in-plane compressive strain ($c/a > 1$) in Mn_4N films. In reference [60], systematic studies were performed on various substrates including $LaAlO_3$ substrates and the following results were found:

(1) All the Mn_4N thin films show PMA and in-plane tensile distortion ($c/a < 1$) regardless of the layer thickness of Mn_4N and substrate used for the deposition.
(2) There is a significant correlation between K_u and c/a, although the magnitude of c/a depends on the Mn_4N layer thickness and substrate used; K_u increases significantly when c/a is deviated from 1, as shown in figure 3.27.

This result indicates that the tensile distortion in Mn_4N films is responsible for PMA; hence, it might be possible to control the K_u of Mn_4N films by tuning *c*/*a* through the selection of appropriate substrate.

3.3.4 Spin-resolved density of states of Mn_4N

Figure 3.28 shows the projected density of states (PDOS) of the $2 \times 2 \times 2$ Mn_4N by first-principles calculation using Vienna Ab initio Simulation Package (VASP) [62], wherein Mn atoms at face-centered sites (II sites) of Mn_4N are divided into Mn(II) A, Mn(II)B, and Mn(II)C. We first take a look at the PDOS of Mn_4N in figures 3.28(a–d). The structure is optimized while the cubic structure is kept. The PDOS in these figures are enlarged around the E_F. For Mn(I) sites; the PDOS of d_{xy}(red), d_{yz}(green), and d_{zx}(blue) overlap; and the PDOS of $d_{x^2-y^2}$ (black dotted line) also overlap with that of $d_{3z^2-r^2}$ (purple line) in figure 3.28(a). Thus the degeneracies of d_γ and $d_{\delta\varepsilon}$ orbitals are 2 and 3, respectively. For Mn(II) A sites, however, the PDOS of d_{zx} are different from those of d_{xy} and d_{yz} in figure 3.28(b). In the same way, the PDOS of d_{xy} are different from those of the other two orbitals for Mn(II)B in figure 3.28(c), and the PDOS of d_{yz} are different from those of the other two

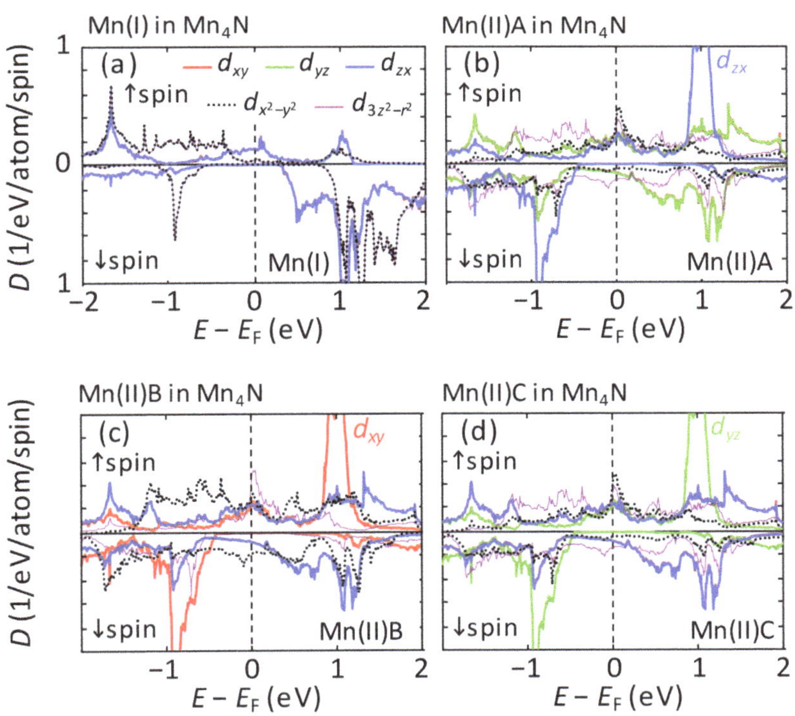

Figure 3.28. PDOS of cubic Mn_4N. Reproduced with permission from reference. Reprinted from [62], with the permission of AIP Publishing.

orbitals for Mn(II)C in figure 3.28(d). It is also noted that d_γ orbitals are not degenerated at Mn(II) sites, as shown in figures 3.28(b–d). Since the cubic Mn_4N has the same atomic arrangement in the *x*, *y*, and *z* directions, the PDOS of the three $d_{\delta\varepsilon}$ orbitals overlap, and those of the two d_γ orbitals also overlap when the PDOS are summed over the unit cell. However, when Mn(II) atoms are broken into three kinds of atoms, such as Mn(II)A, Mn(II)B, and Mn(II)C, neither the three $d_{\delta\varepsilon}$ orbitals nor the two d_γ orbitals are completely degenerated as described above,even in the cubic lattice of Mn_4N. We should note that the contribution of Mn(II) to the DOS at E_f is much higher than that of Mn(I), meaning that the electrical transport is mainly governed by Mn(II) atoms.

3.3.5 X-ray magnetic circular dichroism (XMCD) measurement on Mn_4N films

(1) **Principle of XMCD**

In chapter 4, we utilize x-ray magnetic circular dichroism (XMCD) to investigate the magnitude and the direction of magnetic moment of Mn and 3*d* impurity atoms in impurity-doped Mn_4N films. To reach a deeper understanding of XMCD spectra, we take some time for the basic principles of XMCD. One of the reasons for the utility of XMCD measurements for magnetic material research is that the integrated intensity of the XMCD spectrum is quantitatively related to the magnitude of the orbital and spin magnetic moments of a given atom by a simple relationship (sum rule) [63]. XMCD is the difference in the circular polarization response of the x-ray absorption spectrum (XAS) measured on a magnetic sample whose magnetization is aligned in one direct by applying an external magnetic field. Upon the x-ray irradiation, electrons such as photoelectrons and Auger electrons are excited and some of them come out from the sample's surface. To compensate for such electrons, electrons are supplied and thus the total amount of outgoing electrons is measured as current, which is proportional to the absorption coefficient of the material used at a given photon energy. This measurement is called the total electron yield method.

Circular polarization is a state in which the electromagnetic field associated with light is rotating. There are two types of circular polarization: clockwise rotation (μ^+) and counterclockwise rotation (μ^-). As shown in figure 3.29, x-rays are irradiated with clockwise or counterclockwise circular polarization parallel to the direction of the sample's magnetization, and the XAS spectrum changes according to the direction of polarization. The difference in these two XAS spectra is called MCD. Instead of inverting the direction of circular polarization, an equivalent MCD can be obtained by keeping the direction of circular polarization unchanged and inverting the direction of magnetization.

Here let us consider qualitatively why MCD is observed in the transition of electrons from 2*p* orbitals to 3*d* orbitals. The x-ray absorption coefficient is proportional to the product of the electron transition probability and the

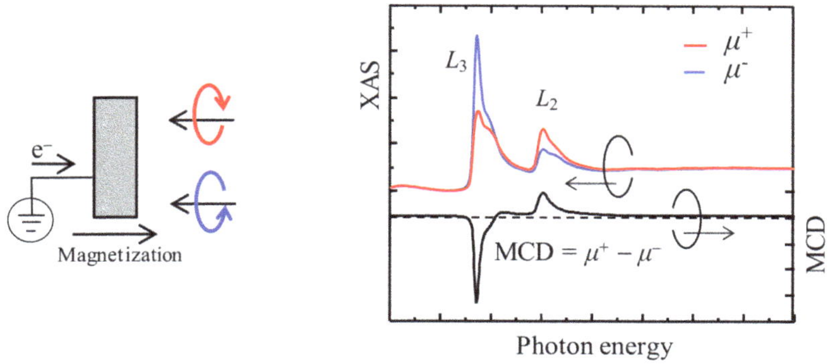

Figure 3.29. Schematic of MCD measurement using the total electron yield method.

number of empty level states in the 3d level. The dipole transition that we consider here is a transition from a state with azimuthal quantum number ℓ to a state with $\ell \pm 1$, and the electron spin is conserved. When the excitation is caused by circularly polarized light, the dipole transition is from the starting state with magnetic quantum number m to the $m + 1$ end state for right circularly polarized light and to the $m - 1$ end state for left circularly polarized light. For the 2p core level, the starting state of the transition can be divided into two states by the spin-orbit interaction: a state with total angular momentum $j = 3/2$ (spin angular momentum S is in parallel with orbital angular momentum L) and a state with $j = 1/2$ (S and L are antiparallel). The presence of two absorption structures denoted as L_3 and L_2 in figure 3.29 reflects the spin-orbit interaction of this inner-shell level. The dipole transition probability has a different magnitude depending on the magnetic quantum number. Thus, different MCDs appear between the $2p_{3/2}$ and $2p_{1/2}$ regions according to the vacant states of the 3d band as shown in figure 3.30.

Here we take two cases. When the empty state of the 3d orbital is the same for the up and down spins, i.e., zero spin polarization, the transition probabilities are the same between the up-spin electrons for right-hand circularly polarized light and down-spin electrons for left-hand circularly polarized light (and vice versa). Thereby, the x-ray absorption coefficients are the same for right-hand and left-hand circularly polarized light, leading to zero MCD signals. Next, we consider the state in which the spin magnetic moment is generated. The 3d level is energetically split into up-spin and down-spin states by exchange interactions, and the number of available empty states for electrons (number of holes) in each state is different. As a result, the x-ray absorption coefficient for right circularly polarized light is different from that for left circularly polarized light. This results in the absorption intensity in the $2p_{3/2}$ and $2p_{1/2}$ regions being weakened for one case, while being strengthened for the other, as shown in figure 3.29. If the

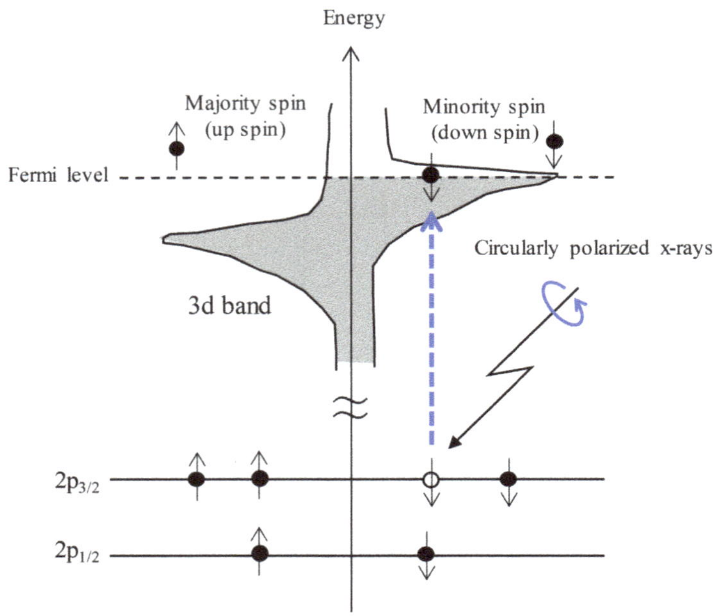

Figure 3.30. Schematic of electron transitions from $2p$ to $3d$ states.

hole number of the $3d$ orbital is the same for each magnetic quantum number, i.e., the orbital magnetic moment is zero, then the magnitude of the MCD is determined by the spin magnetic moment of the $2p_{3/2}$ and $2p_{1/2}$ regions. Thus, the magnitude of the MCD reflects the magnitude of the spin magnetic moment. Meanwhile, when the $3d$ level is split by the spin-orbit interaction and the hole number of $3d$ state becomes different for each magnetic quantum number (= generation of orbital magnetic moment), the magnitude of the MCD signal differs between the $2p_{3/2}$ and the $2p_{1/2}$ regions. The larger the $3d$ orbital magnetic moment, the larger the asymmetry of the MCD spectrum. (The sum of the integrated intensity of the MCD spectrum in the $2p_{3/2}$ region and the $2p_{1/2}$ region is proportional to the magnitude of the $3d$ orbital magnetic moment.)

(2) **XAS and XMCD spectra of Fe_4N ferromagnetic films**

We first take an example of XAS and XCD spectra of ferromagnets for reference. Figure 3.31 shows (a) XAS and (b) XMCD spectra at the Fe $L_{2,3}$ edges for 10 nm-thick Fe_4N epitaxial films observed at 300 K. The total electron yield mode was used at the BL23SU of SPring-8, Japan. Broken lines are a guide to the eyes [64].

The shoulders indicated by the arrow in figure 3.31(a) originate from the Fe atoms at the face-centered II sites, Fe(II). We see that the signs of the XMCD signals at the energies indicated by the dashed lines in

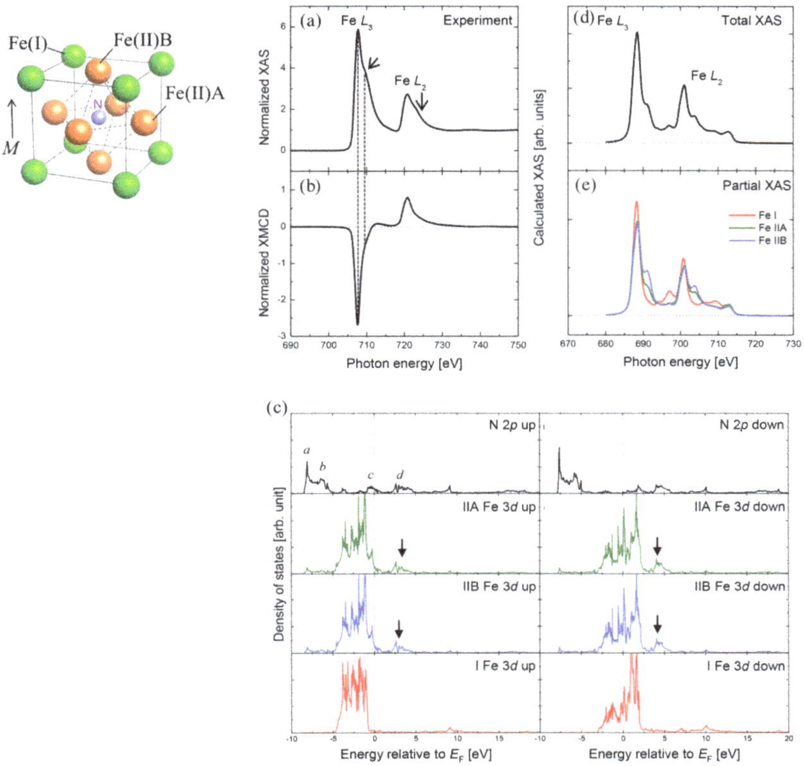

Figure 3.31. (a) XAS spectra at the Fe $L_{2,3}$ edges of a 10 nm thick Fe_4N epitaxial films and (b) XMCD spectra measured at 300 K. An external magnetic field ($\mu_0 H = \pm 3$ T) is applied normal to the sample plane. Dashed lines are eye guides. (c) Calculated projected DOS of the Fe 3*d* and N 2*p* states of Fe_4N majority and minority spins when the magnetization is oriented toward [001]. (d) Calculated total and (e) partial XAS spectra of Fe_4N at the Fe $L_{2,3}$ edges. Reproduced with permission from reference. Reprinted from [64], with the permission of AIP Publishing.

figures 3.31(a) and 3.31(b) are the same. This means that the magnetic moment of the Fe atoms at the corner I sites, Fe(I), is parallel to that of the Fe(II) atoms. They are therefore ferromagnetically coupled. Such a shoulder is an intrinsic property of antiperovskite nitrides and can be explained as follows. Figure 3.31(c) shows the calculation results of projected DOSs (PDOS) for the majority and minority spins of the N 2*p* and Fe 3*d* orbitals for the I, IIA, and IIB sites. Here, we discuss four peaks (a–d) that appear at −8, −6, −0.5, and 3 eV relative to the Fermi level (E_F), respectively, in the N 2*p* states. The 3*d*-state PDOS of Fe(II) sites shows peaks at the same energies as the N 2*p* state. In contrast, no such peaks are seen at these energies in the PDOS of the 3*d* state of the Fe(I) site. It is noted that the Fe (I) atoms are further away from the body-centered N atoms than the Fe(II) atoms. Therefore, the peaks a–d are interpreted to be due to states formed by band hybridization between the face-centered Fe(II) 3*d* and N 2*p*

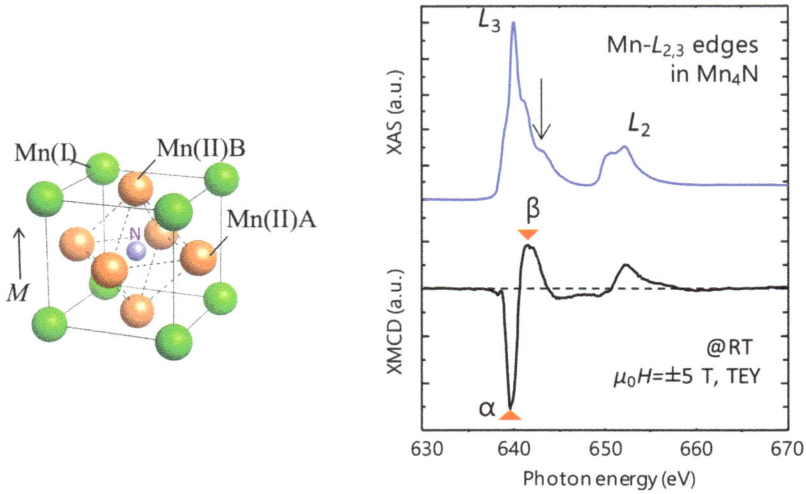

Figure 3.32. XAS and XMCD spectra of the Mn $L_{2,3}$ edges measured on 25 nm-thick Mn_4N films at 300 K. The external magnetic field ($\mu_0 H = \pm 5$ T) is applied perpendicular to the sample surface. Reproduced with permission from reference. Reproduced from [71]. © IOP Publishing Ltd. All rights reserved.

orbitals. They are produced by σ and π bonds and π* and σ* antibonds between these orbitals. The σ bonds (σ* antibonds) are formed by the N $2p_z$ and Fe $3d_{z^2-r^2}$ orbitals, and the π bonds (π* antibonds) are formed by N $2p_z$, Fe $3d_{yz}$, and Fe $3d_{zx}$ orbitals. The arrows indicate that there are electronic states in PDOS that are about 3 eV higher than E_F, and these are formed by σ* antibonds. However, at this energy, no electronic states exist in the PDOS for the Fe(I) 3*d* orbitals. A similar explanation is possible for PDOS with minority spins. The total XAS spectra and the site-specific XAS spectra are calculated based on the Fermi's golden rule, and they are shown in figures 3.31(d) and 3.31(e), respectively. These calculated spectra explain well the experimental results shown in figure 3.31(a). The shoulder structure above the main peak obtained experimentally is reproduced. In figure 3.31(e), the shoulder structure of the Fe(I) atom is not seen, which is different from the shoulder structure of the Fe(II) atom. These shoulders are explained by dipole transitions from the Fe(II) 2*p* inner-shell level to a hybrid state of Fe(II) 3*d* and N 2*p* orbitals (indicated by arrows in figure 3.31(c)). From the above discussion, we can say that the shoulder indicates a local electronic state of the Fe(II) site. Researchers have observed these shoulders in XAS spectra in Fe_4N films [65] and in other reports on various antiperovskite nitrides, including Co_4N [66], Co_3FeN [67], $Fe_{4-x}Ni_xN$ [68], and Mn_4N-based compounds [5, 69, 70]. Thus, these shoulders are an intrinsic property of 3*d* metal nitrides.

(3) **XAS and XMCD spectra of Mn_4N ferrimagnetic films**

Next, we move on to the XAS and XMCD spectra of Mn_4N films. The XMCD spectrum in figure 3.32 [71] is quite different from that for Fe_4N in figure 3.31(a) in particular around the L_3 region, while there is not much difference in the XAS spectra between them. We see a distinct shoulder at around 2 eV, which is higher than the main peaks of L_3 edge in the XAS spectra. Similar shoulders have been confirmed for other antiperovskite nitrides, as described previously. According to first-principles calculation for Mn_4N [5] and Fe_4N [64], the shoulder is interpreted as being due to the electric dipole transition from the Mn(Fe) $2p$ state to the hybrid state formed by the orbitals of the N $2p$ and Mn(Fe) $3d$ at II sites. This indicates that Mn(Fe) atoms at I sites are more localized than those at II sites. The XMCD spectra of Mn_4N can therefore be understood as the addition of a localized component featured by sharp peak α and an itinerant component characterized by broad peak β with a different sign from peak α.

These features are well reproduced by first-principles calculation using the all-electron full-potential linearized augmented-plane-wave method and Fermi's golden rule with $E1$ transitions [5]. The opposite sign in peaks α and β in figure 3.32 exhibits that Mn_4N is a ferrimagnet, i.e., the magnetic moment of the Mn(I) sublattice is aligned antiparallel to that of the Mn(II) sublattice. This difference comes from the difference in interatomic distance between N and Mn(I) atoms, and N and Mn(II) atoms. We should also note, however, that the calculation result does not fully explain the experimental result as follows [5].

In figure 3.33(a), narrow and broad bandwidths are obtained for Mn(I) and Mn(II), respectively.

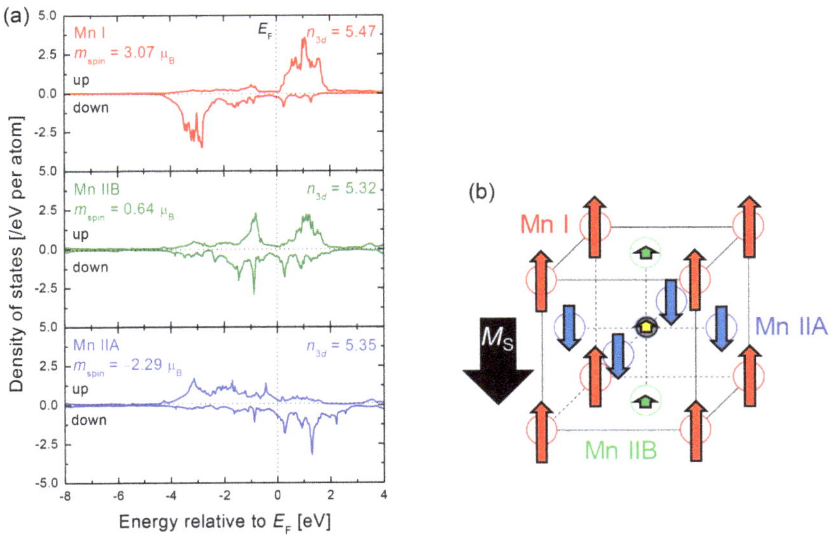

Figure 3.33. (a) Total DOS of Mn(I), Mn(IIA), and Mn(IIB) in Mn_4N calculated by the first-principles calculation, and (b) its ferrimagnetic spin structure with direction and magnitude of spin magnetic moment (red, blue, green, and yellow arrows) and M_S (black arrow). Reproduced with permission from reference. Reprinted figure with permission from [5], Copyright (2020) by the American Physical Society.

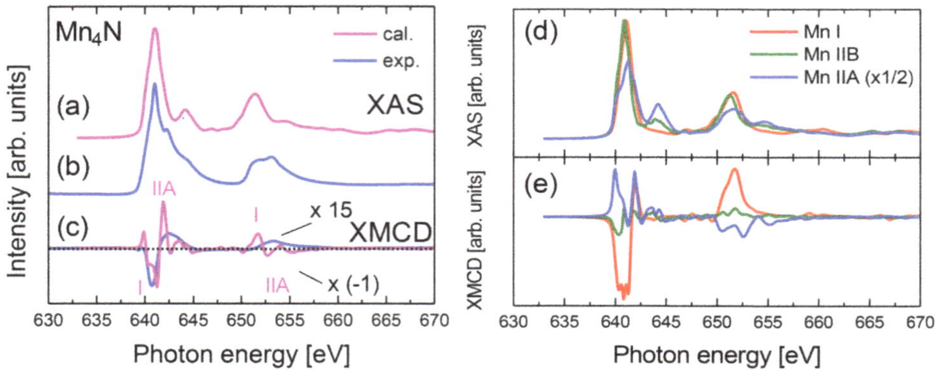

Figure 3.34. (a)–(c) Total XAS and XMCD spectra, and (d) and (e) partial XAS and XMCD spectra at the Mn $L_{2,3}$ edges of Mn_4N. Reprinted figure with permission from [5]. Copyright (2020) by the American Physical Society.

This suggests that the 3*d* electrons in the Mn(I) atoms are more localized and those in the Mn(II) atoms are more itinerant. As shown schematically in figure 3.33(b), the calculated spin magnetic moment (m_{spin}) for the Mn atoms at I, IIB, and IIA sites, and N atoms are 3.07, 0.64, −2.29, and 0.07 μ_B per atom, respectively. These calculated values are close to those reported in references [6, 61]. The m_{spin} is calculated to be −0.80 μ_B/f.u. in total. This value corresponds to an M_S of 131 kA m^{-1} and is consistent with the experimental value (=110 kA m^{-1}) for the Mn_4N films [19]. It is noted, however, that the Mn(I) moments in this theoretical magnetic structure are aligned antiparallel to the magnetic field. This is different from those obtained in the XMCD data by experiment. This reveals a quantitative discrepancy in the theoretical values. Figures 3.34(a)–(c) compare experimental and calculated XAS and XMCD spectra of Mn $L_{2,3}$ edges [5]. It is noted that the signs of the theoretical XMCD in figures 3.34(c) and (e) are inverted for comparison. The site-resolved XAS and XMCD spectra shown in figures 3.34(d) and (e) help us identify the sources of the theoretical spectral features. As shown in figure 3.34(c), the largest negative (positive) XMCD signal at the L_3 (L_2) edge is mainly ascribed to Mn(I), followed by the Mn(IIA) structure. Despite some discrepancies, the theoretical XMCD curves capture important qualitative features of the experimentally obtained L_3-edge XMCD spectra. This may support the aforementioned interpretation of the XMCD data that features α and β are mainly derived from Mn(I) and Mn(II), respectively. Furthermore, this comparison indicates that the previous theoretical treatment is still oversimplified [5] and that more sophisticated theoretical modeling is needed to account for core hole–3*d* and 3*d*–3*d* interactions, even in metallic Mn_4N.

3.4 Summary

In this chapter, we showed that the magnetic properties of Mn_4N thin films formed on lattice-matched STO substrates are found to be much better than those on MgO substrates. The Mn_4N films on STO exhibit several attractive features: a sharp magnetization switching, millimeter-scale giant magnetic domains with full remanence, and scarce nucleation. Therefore, when we discuss with the properties of Mn_4N films in the following chapters, we will only focus on the Mn_4N films on STO. These properties are associated with relatively small M_S and large PMA, indicating their potential for spintronics applications. CIDWM experiments confirming this potential will be presented in the next chapter.

Bibliography

[1] Guillaud C and Wyart J 1948 *Rev. Metall.* **45** 271
[2] Hägg G 1929 *Z. Phys. Chem.* B **4** 346
[3] Takei W J, Shirane G and Frazer B C 1960 *Phy. Rev.* **119** 122
[4] Lihl F, Ettmayer P and Kutzelnigg A 1962 *Z. Metallkd.* **53** 715
[5] Ito K *et al* 2020 *Phys. Rev.* B **101** 104401
[6] Isogami S, Masuda K and Miura Y 2020 *Phys. Rev. Mater.* **4** 014406
[7] Takei W J, Heikes R R and Shirane G 1962 *Phys. Rev.* **125** 1893
[8] Fruchart D, Givord D, Convert P, l'Heritier P and Senateur J P 1979 *J. Phys. F: Met. Phys.* **9** 2431
[9] Uhl M, Mater S F and Mohn P 1997 *Phys. Rev.* B **55** 2995
[10] Li C, Yang Y, Lv L, Huang H, Wang Z and Yang S 2008 *J. Alloys Compd.* **457** 57
[11] Gushi T 2019 Mn4N thin films for spintronics applications based on current-induced domain wall motion *PhD Thesis* (University of Tsukuba)
[12] Takenaka K and Takagi H 2005 *Appl. Phys. Lett.* **87** 261902
[13] Mekata M 1962 *J. Phys. Soc. Jpn.* **17** 796
[14] Zhang R, He Y, Fruchart D, Coey JMD and Gercsi Z 2022 *Acta Mater.* **234** 118021
[15] Nakagawa S and Naoe M 1994 *J. Appl. Phys.* **75** 6568
[16] Ching K M, Chang W D, Chin T S, Duh J G and Ku H C 1994 *J. Appl. Phys.* **76** 6582
[17] Ching K-M, Chang W-D and Chin T-S 1995 *J. Alloy Compd.* **222** 184
[18] Dhar S, Brandt O and Ploog K H 2005 *Appl. Phys. Lett.* **86** 112504
[19] Yasutomi Y, Ito K, Sanai T, Toko K and Suemasu T 2014 *J. Appl. Phys.* **115** 17A935
[20] Shen X, Chikamatsu A, Shigematsu K, Hirose Y, Fukumura T and Hasegawa T 2014 *Appl. Phys. Lett.* **105** 072410
[21] Kabara K and Tsunoda M 2015 *J. Appl. Phys.* **117** 17B512
[22] Gushi T *et al* 2018 *Jpn. J. Appl. Phys.* **57** 120310
[23] Kawasaki M, Takahashi K, Maeda T, Tsuchiya R, Shinohara M, Ishiyama O, Yonezawa T, Yoshimoto M and Koinuma H 1994 *Science* **266** 1540
[24] Kaufmann E N 2012 *Characterization of Materials* 2nd edn (New York: Wiely)
[25] Ito K, Higashikozono S, Takata F, Gushi T, Toko K and Suemasu T 2016 *J. Cryst. Growth* **455** 66
[26] Inaba K 2008 *Rigaku J.* **24** 1
[27] Mitsunaga T 2009 *Rigaku J.* **25** 1
[28] Fagaly R L 2006 *Rev. Sci. Instrum.* **77** 101101
[29] Josephson B D 1962 *Phys. Lett.* **1** 251
[30] van der Pauw L J 1958 *Philips Res. Repts* **13** 1
[31] van der Pauw L J 1958 *Philips Tech. Rev.* **20** 220

[32] Okamoto K and Magn J 1983 *Magn. Mater.* **35** 353
[33] Haan S, Lodder C and Popma T J A 1991 *J. Magn. Soc. Jpn* **15** 349
[34] Ejsing L, Hansen M F, Menon A K, Ferreira H A, Graham D L and Freitas P P 2004 *Appl. Phys. Lett.* **84** 4729
[35] Smit J 1955 *Physica* **21** 877
[36] Smit J 1958 *Physica* **24** 39
[37] Berger L 1970 *Phys. Rev.* B **2** 4559
[38] Chang M-C and Niu Q 1996 *Phys. Rev.* B **53** 7010
[39] Sundaram G and Niu Q 1999 *Phys. Rev.* B **59** 14915
[40] Saitoh E, Ueda M, Miyajima H and Tatara G 2006 *Appl. Phys. Lett.* **88** 182509
[41] Nagaosam N, Sinova J, Onota S, MacDonald A H and Ong N P 2010 *Rev. Modern Phys.* **82** 1539
[42] Kabara K, Tsunoda M and Kokado S 2017 *AIP Adv.* **7** 056416
[43] Chiba D, Kawaguchi M, Fukami S, Ishiwata N, Shimamura K, Kobayashi K and Ono T 2012 *Nat. Commun* **3** 888
[44] Yamada K, Kakizakai H, Shimamura K, Kawaguchi M, Fukami S, Ishiwata N, Chiba D and Ono T 2013 *Appl. Phys. Express* **6** 073004
[45] Hirai T, Koyama T, Obinata A, Hibino Y, Miwa K, Ono S, Kohda M and Chiba D 2016 *Appl. Phys. Express* **9** 063007
[46] Kerr J 1877 *Phyl. Mag* **3** 321
[47] Nederpel P Q J and Martens J W D 1985 *Rev. Sci. Instrum.* **56** 687
[48] Passeri D *et al* 2014 *Biomatter* **4** e29507
[49] Ito K, Rougemaille N, Pizzini S, Honda S, Ota N, Suemasu T and Fruchart O 2017 *J. Appl. Phys.* **121** 243904
[50] Gehanno V, Samson Y, Marty A, Gilles B and Chamberod A 1997 *J. Magn. Magn. Mater.* **172** 26
[51] Attané J P, Samson Y, Marty A, Toussaint J C, Dubois G, Mougin A and Jamet J P 2004 *Phys. Rev. Lett.* **93** 257203
[52] Yamanouchi M, Jander A, Dhagat P, Ikeda S, Matsukura F and Ohno H 2011 *IEEE Magn. Lett.* **2** 3000304
[53] Ando F, Kakizakai H, Koyama T, Yamada K, Kawaguchi M, Kim S, Kim K-J, Moriyama T, Chiba D and Ono T 2016 *Appl. Phys. Lett.* **109** 022401
[54] Emori S and Beach G S D 2011 *Appl. Phys. Lett.* **98** 132508
[55] Menshikov A, Tarnoczi T and Kren E 1975 *Phys. Status Solidi* A **28** K85
[56] Niida H, Hori T, Onodera H, Yamaguchi Y and Nakagawa Y 1996 *J. Appl. Phys.* **79** 5946
[57] Futamoto M, Nakamura M, Ohtake M, Inaba N and Shimotsu T 2016 *AIP Adv.* **6** 085302
[58] Yang B S, Jiang L N, Chen W Z, Tang P, Zhang J, Zhang X G, Yan Y and Han X F 2018 *Appl. Phys. Lett.* **112** 142403
[59] Hirose T, Komori T, Gushi T, Toko K and Suemasu T 2020 *J. Cryst. Growth* **535** 125566
[60] Hirose T, Komori T, Gushi T, Anzai A, Toko K and Suemasu T 2020 *AIP Adv.* **10** 025117
[61] Ito K, Yasutomi Y, Kabara K, Gushi T, Higashikozono S, Toko K, Tsunoda M and Suemasu T 2016 *AIP Adv.* **6** 056201
[62] Komori T, Mitarai H, Yasuda T, Ghosh S, Vila L, Attané J P, Honda S and Suemasu T 2022 *J. Appl. Phys.* **132** 143902
[63] Chen C T, Idzerda Y U, Lin H-J, Smith N V, Meigs G, Chanban E, Ho G H, ellegrin E P and Sette F 1995 *Phys. Rev. Lett.* **75** 152

[64] Ito K, Toko K, Takeda Y, Saitoh Y, Oguchi T, Suemasu T and Kimura A 2015 *J. Appl. Phys.* **117** 193906
[65] Takagi T, Isami K, Yamamoto I, Nakagawa T and Yokoyama T 2010 *Phys. Rev.* B **81** 035422
[66] Ito K, Harada K, Toko K, Ye M, Kimura A, Takeda Y, Saitoh Y, Akinaga H and Suemasu T 2011 *Appl. Phys. Lett.* **99** 252501
[67] Ito K *et al* 2013 *Appl. Phys. Lett.* **103** 232403
[68] Takata F, Ito K, Takeda Y, Saitoh Y, Takanashi K, Kimura A and Suemasu T 2018 *Phys. Rev. Mater.* **2** 024407
[69] Komori T, Hirose T, Gushi T, Toko K, Hanashima T, Vila L, Attané J P, Amemiya K and Suemasu T 2020 *J. Appl. Phys.* **127** 043903
[70] Mitarai H *et al* 2020 *Phys. Rev. Mater.* **4** 094401
[71] Yasuda T, Komori T, Mitarai H, Honda S, Ghosh S, Vila L, Attane S P, Amemiya K and Suemasu T 2022 *J. Phys.* D **55** 115003

Chapter 4

Ultrafast CIDWM in Mn_4N strips

Since Mn_4N, along with PMA, has a relatively small spontaneous magnetization ($\sim$100 kA m^{-1}), Mn_4N thin films are expected to have a large spin transfer torque (STT) efficiency. The purpose of this chapter is to demonstrate the current-induced domain wall motion (CIDWM) in Mn_4N by STT and evaluate the efficiency of STT-based phenomena. For this purpose, micro- or submicronic-devices must be fabricated to suppress the number of domains in the system. Here, we first see a typical nanofabrication process for patterning magnetic thin films into microstrip-like devices and we then go into the details of the CIDWM experiments.

4.1 Fabrication process of Hall bars

A flowchart of the nanofabrication process used in reference [1] is shown in figure 4.1. For more details, see reference [2]. Polymethyl methacrylate (PMMA) is coated on Mn_4N as a positive resist for electron beam (EB) and then an EB lithography is followed to make the pattern of microstrips. EB lithography is a very important nanofabrication technique for making micro- or nano-scale masks. First, PMMA is spin-coated at room temperature and is then post-baked at 100°C for 1 min in air. A positive resist such as PMMA becomes soluble after EB exposure. EB irradiation (JEOL 6300FS) is performed onto the PMMA resist coated on the Mn_4N films at an acceleration voltage of 100 kV. After the EB resist is developed and the resist pattern is formed, a vacuum deposited Al layer is deposited on the EB resist pattern and then lifted-off to form a mask pattern of Al. Ion milling etching with Ar ions is then performed with secondary ion mass spectrometry (SIMS) analysis and ion beam incidence angle of 70° while rotating the substrate. The endpoint of the etching process is detected using the presence of Sr atoms in the SIMS signal. Sr atoms are one of the elements that make up the $SrTiO_3$ substrate and are confirmed by *in situ* sheet resistance measurements. In contrast to semiconductors such as Si, transition metals such as Mn generally do not form volatile chemical products such as chloride and fluoride. It is therefore not appropriate to use reactive ion etching for

doi:10.1088/978-0-7503-5477-6ch4

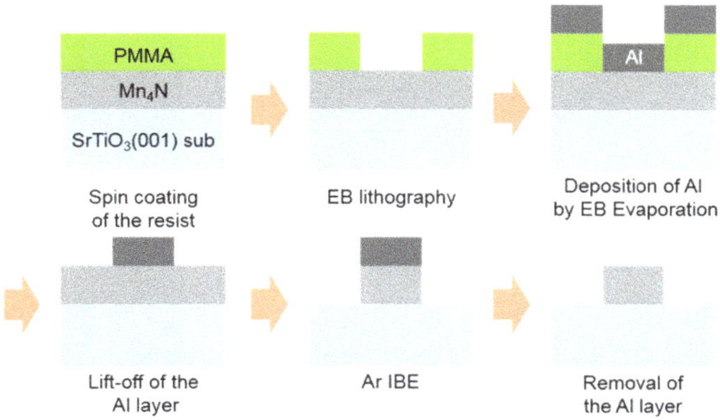

Figure 4.1. Fabrication process of Mn_4N films on STO(001) into microstrips. Reproduced from [2]. CC BY 4.0.

magnetic films. For this reason, a physical etching process called ion milling is often employed. Ar^+ plasma excited in a chamber is accelerated toward the sample by a bias voltage to remove atoms from the surface. Since it is a physical etching process, a wide variety of materials can be etched in principle, although the etching rate depends on the sample. One weakness of this ion etching process is that the sputtered spices redeposit on the sample surface since they are not volatile in contrast to chemical etching. The sputtered species are known to form high walls at the edges of the fabricated structures after the etch mask is removed, which hinders surface observations such as MFM measurements. Therefore, the ion beam etching system used for this purpose has a mechanism to tilt and rotate the sample during etching for enabling us to etch such re-deposited species. It also has an endpoint detector and SIMS analyzer to detect the etched elements, which reveals the atomic species on the topmost surface. This equipment helps us avoid over-etching of the sample, which may cause re-deposition and substrate damage. After the etching process, the Al mask is removed by wet chemical etching with a tetramethylammonium hydroxide (TMAH) solution that does not react with Mn_4N.

Figures 4.2 (a)–(c) show a Hall bar design, a scanning electron microscopy (SEM) image of the Al mask pattern after the lift-off process, and an optical microscope image after all fabrication processing, respectively.

The Mn_4N films are processed into microstrips of 0.5 or 1 μm width. There is a domain nucleation pad in the left-hand side of figure 4.2(a). This is to control the nucleation of an inverted domain. Three Hall crosses are formed for the detection of the magnetization state by AHE. The width of the strip at the notch is half of the wire width for pinning the DW.

4.2 Equivalency between current and magnetic field

The observation of STTs in Mn_4N films is first attempted by measuring the change in DW coercivity for various direct currents (DCs), I_{DC}. An alternating current (AC) i_{AC} is added to the DC. Therefore, the current flowing in the microstrip is given as

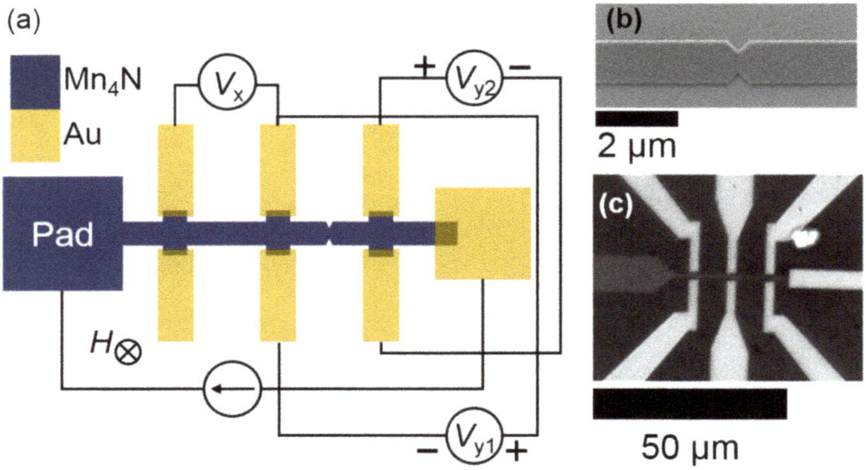

Figure 4.2. (a) Hall bar pattern with a nucleation pad, three Hall crosses with Au electrodes, and one notch to pin the DW between the center and right Hall crosses. (b) SEM image around the notch. (c) Hall bar pattern by an optical microscope. Reproduced from [2]. CC BY 4.0.

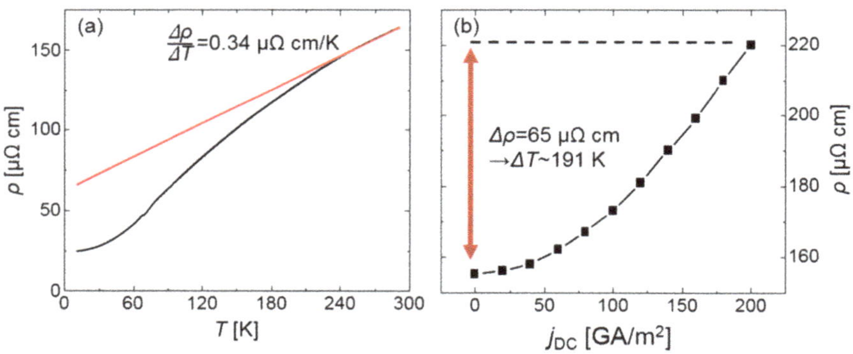

Figure 4.3. (a) Dependence of ρ on temperature. (b) Dependence of ρ on current density. The effect of Joule heating is estimated from the increase in ρ ($\Delta\rho$). Reproduced from [2]. CC BY 4.0.

$I = I_{DC} + i_{AC}\sin(\omega t)$. The i_{AC} can be used to detect the AHE using the lock-in technique. The magnitude of i_{AC} is about 10 μA, which is negligible for the much greater I_{DC} used to generate the STT. The center and left-hand Hall crosses are used for four-probe resistance measurements, while the center and right-hand crosses are for independent Hall measurements and surround a notch to trap DW.

Figures 4.3(a) and 4.3(b) show the dependence of longitudinal resistivity ρ on temperature and the dependence of ρ on injected current density j_{DC} at room temperature, respectively. At 10 K, ρ reaches 15% of the room temperature value, which is consistent with a previous report on the result obtained on MgO substrates [3]. In figure 4.3(a), ρ increases almost linearly with a slope of 0.34 μΩ cm K^{-1} as temperature increases above room temperature. The j_{DC} dependence of ρ in figure 4.3(b) shows an increase in resistivity up to 220 μΩ cm at the highest current density ($j_{DC} = 2 \times 10^{11}$ A m^{-2}). This resistivity increase accounts for the increase of temperature by 191 K. It is noted that the Néel temperature T_N is about 745 K for Mn_4N.

Hall measurements are performed on the 0.5 and 1 μm wide microstrips to measure the coercivity at the center and right-hand Hall crossings for various values of j_{DC}. Figure 4.4 shows the result of this measurement. 0.5 and 1 μm wide devices show similar behavior in DC variations. On the one hand, the negative current, which should assist the movement of the DW, does not affect the switching field. This can be understood as follows. The coercivity is governed by nucleation of magnetic domains rather than propagation of DW. Immediately after the nucleation of magnetic domains occurs, the DW propagates and the Hall cross magnetization switches. The magnetic field necessary for the nucleation (H_n) is greater than that necessary for the DW propagation (H_p), but it is also larger than the magnetic field for depinning the DW (H_d) from the notch located between the center and

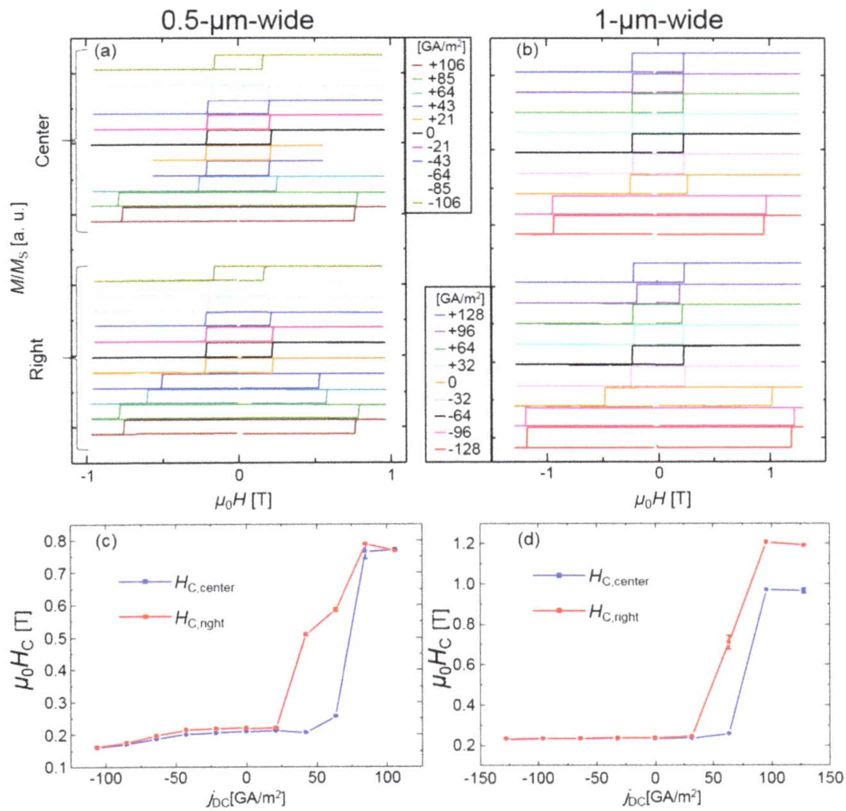

Figure 4.4. (a) and (b) M–H curves for different values of j_{DC}. (c) and (d) Dependence of switching field H_C on j_{DC} for (a), (c) 1 μm wide and (b), (d) 0.5 μm wide strips. Only positive currents affect the magnetic field for magnetization reversal, preventing the DW movement in the device. The switching magnetic field H_C of the right Hall cross increases for a smaller j_{DC} than that of the center Hall cross, due to the presence of the notch between them. Reproduced from [2]. CC BY 4.0.

Table 4.1. Reported STT efficiencies. Symbols * and † indicate the results obtained from CIDWM in the presence of a magnetic field and coercive force tuning measurement by bias current in the work, respectively [2]. The others are obtained from DW depinning experiments without magnetic field.

Material	CoFeB	$(Co/Ni)_N$	FePt	$SrRuO_3$	TbFeCo	Mn_4N
ε [10^{-14} Tm^2 A^{-1}]	0.8	60	30*	100	920	770†
Reference	[6]	[5]	[4]	[7]	[8]	This work

right-hand Hall crosses. On the other hand, the coercivity increases rapidly ($\Delta H_C >$ 0.5 T) when the DW motion is prevented by a positive bias. As seen in figure 4.4, it is possible to find a regime where the DW is trapped in the notch, where a large j_{DC} opposes the DW motion. At larger magnetic fields, the DW slips out of the notch and the magnetization of the right Hall cross switches. The positive current thus acts as an additional negative magnetic field [4] and prevents the motion of the DW. A rough estimate of the equivalency ε_{STT} between the field and current can be given as,

$$\varepsilon_{STT} = \frac{\Delta H_C}{\Delta j_{DC}} \tag{4.1}$$

They are calculated to be 6.5×10^{-12} and 7.7×10^{-12} Tm^2A^{-1} for 0.5 μm and 1.0 μm wide strips, respectively. These values are superior to well-known materials for DW motion, such as Co/Ni, CoFeB, Co/Pt, FePt, and $SrRuO_3$, and are comparable to the rare-earth ferrimagnet TbFeCo (see table 4.1). The relative efficiency of STT appears very large for Mn_4N due to its small spontaneous magnetization, which limits the effect of the magnetic field.

4.3 Spin-transfer torque using direct current (DC)

4.3.1 Pinning of the DW on the notch

As shown in figures 4.4(c) and (d), the coercive forces of the center and right crosses are different at around $j_{DC} = 5 \times 10^{10}$ A m^{-2}, and thus we are able to pin the DW at the notch. Figure 4.5 shows a schematic of the expected change in longitudinal voltage V_{xx}. In such a double Hall cross with perpendicular magnetization, it is known that there is an AHE contribution to the longitudinal voltage from the contact point [9]. In the saturated state, the Hall contribution to the transverse voltage is the same at each cross, so they cancel out. However, when DW is located in-between the Hall crosses, the magnetization direction is different between the two crosses, resulting in AHE contributions of the different sign. Therefore, the longitudinal voltage fluctuates abruptly when the magnetization of the crosses switches. Furthermore, the magnetization reversal of the strip induces resistance fluctuations due to magnon magnetoresistance (MMR) [10, 11] and possibly DW resistance [9, 12].

Figures 4.6(a) and (b) show the experimental results of the four-probe resistor ΔR_{xx} between the center and right crosses for 0.5- and 1-μm-wide devices,

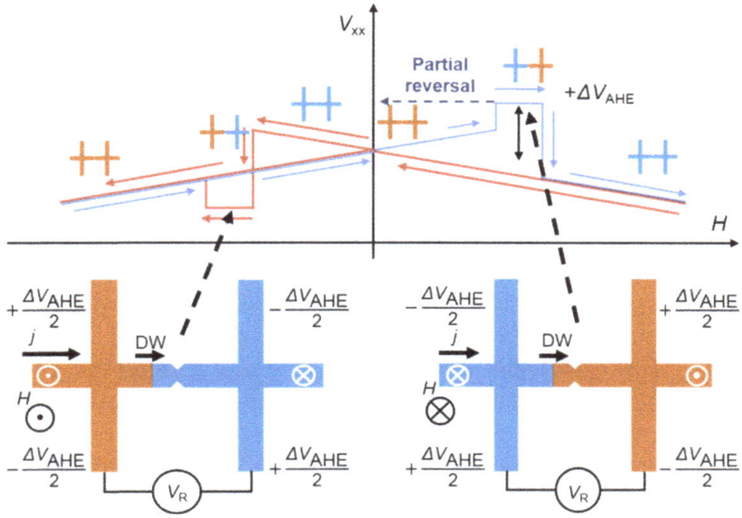

Figure 4.5. Diagram of the longitudinal voltage V_{xx} with respect to magnetic fields in the four-probe method and corresponding DW arrangements. The proportional slope of resistance with respect to the magnetic field is due to MMR. Sharp changes in V_{xx} corresponds to the difference of the Hall signal between the two Hall crosses. Reproduced from [2]. CC BY 4.0.

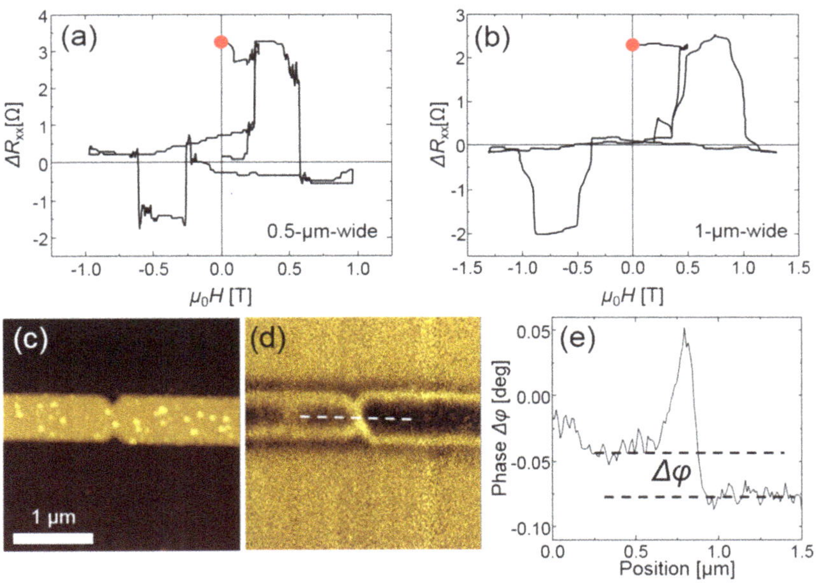

Figure 4.6. Four-probe resistances as a function of magnetic field for (a) 0.5 μm and (b) 1 μm wide strips. The resistances are measured between the right-hand and center Hall crosses. Red dots show the endpoint of the minor loop. (c) AFM and (d) MFM images of a 0.5 μm wide strip taken after partial reversal. The phase difference $\Delta\varphi$ on both sides of the notch shown in (e) suggests that the DW is indeed pinned at the notch. Reproduced from [2]. CC BY 4.0.

respectively. At the end of this partial reversal process, the voltage remains at a high value, and thus implies a DW being pinned at the notch. Figures 4.6(c) and (d) show AFM and MFM images of the notch after this field treatment for a 0.5 μm wide strip, respectively. The MFM signal is small due to the weak magnetization of Mn_4N but it is confirmed that there are two levels of contrast in the notch along the strip in figure 4.6(e), indicating the presence of DW in the notch.

4.3.2 CIDWM using DC without external field assistance

After the DW is pinned at the notch, DC (j_{DC}) is injected without an external magnetic field. We monitor the magnetic state at the central Hall cross by measuring the AHE signal. It is noted that a lock-in method with a small AC for detection is used again here and measurements are made for various values of j_{DC}. The direction of the positive current is defined as in the previous section, from the nucleation pad to the right-hand side. Since the direction of electron flow is opposite to the direction of current flow, the negative current is supposed to push the DW from the left-hand to the right-hand side. Figure 4.7 shows the anomalous Hall voltages at the central Hall crossing under several DC biases for (a) 0.5 μm and (b) 1 μm wide devices. After injecting 1.2×10^{11} A m^{-2}, the Hall voltage increases sharply to a higher value,

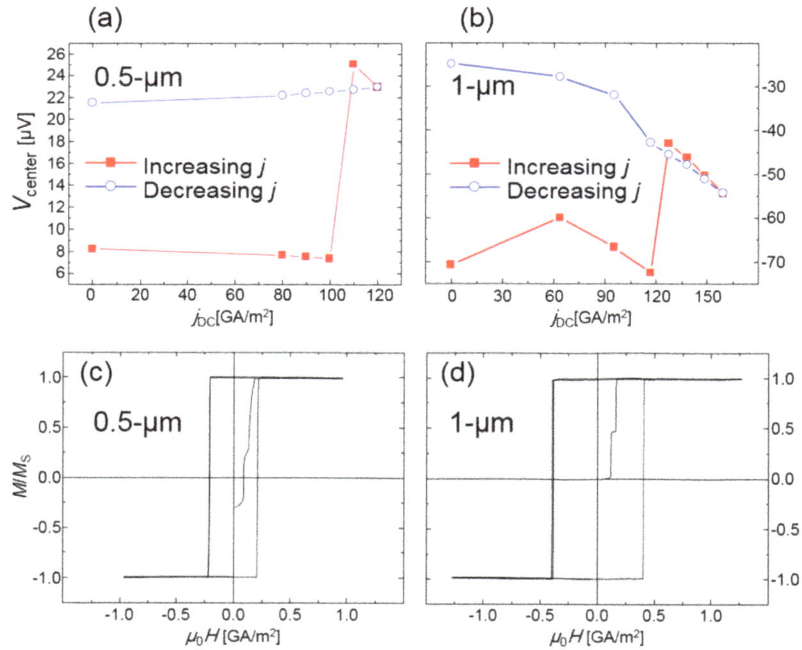

Figure 4.7. Hall voltages at the central Hall cross for positive j_{DC} sweeps in (a) 0.5 μm and (b) 1 μm wide strips. Red (blue) data points present increasing (decreasing) current sweeps. Initially, DW is fixed at a notch away from the cross. Major and minor hysteresis loops measured for in (c) 0.5 μm and (d) 1 μm wide strips by AHE after CIDWM measurements. From the initial magnetization ratio, the last position of DW at the hole cross can be estimated. Reproduced from [2]. CC BY 4.0.

indicating that the DW reaches the central Hall cross. The threshold current density necessary for the CIDWM can be compared to the propagation magnetic field. Figures 4.7 (c) and 4.7(d) show the hysteresis loops after the DC sweep experiment. These initial conditions are not saturated. In other words, during the CIDWM experiment, the DW stops in the middle of the Hall cross. The current density probably locally decreases due to the wide cross-sectional area of the strip. In contrast, the transverse voltage at the right-hand Hall cross does not change during the DC injection, indicating that the DW motion occurs along the electron flow direction, from the notch to the left-hand. This result indicates that the sign of the conduction electron's spin polarization is positive, which is consistent with the results of the previous section and with those in reference [13].

Thus far, transport measurements have been performed to observe the STT in Mn_4N. We show that DC can prevent the magnetic field-driven DW motion, and therefore increases the coercive force of the Hall bar. We also show that the DC current induces DW depinning from the notch in the absence of a magnetic field, followed by CIDWM by injected DC. The STT efficiency derived from the equivalence of the DC and external magnetic field is calculated to be 7×10^{-12} T m^2 A^{-1}, and the threshold current density for the DW motion is 1.2×10^{11} A m^{-2}. This high efficiency may also be a consequence of a small magnetization of Mn_4N. We next study the change in DW velocity as a function of injected j_{DC} in the absence of magnetic field.

4.4 Spin-transfer torque using pulsed current

In the previous section, we demonstrated CIDWM of Mn_4N microstrip using DC. In practical applications, however, threshold current density is not the only important parameter: DW velocity is a measure reflecting how fast information can be written or read in a DW-based device. Therefore, current pulses must be used to determine the DW speed. In this section, we demonstrate the potential of Mn_4N as a candidate for CIDWM applications by using current pulses coupled with MOKE microscopy observations to show that standard STT leads to very high DW rates reaching almost 1 km s^{-1} in this material. After describing the device fabrication, we present the results of DW velocity measurements. Next, the origin of the high DW mobility in Mn_4N is discussed. Finally, we estimate the spin polarization from these measurements.

4.4.1 Device fabrication

A 10 nm thick Mn_4N film is grown on STO by MBE using the process described in Chapter 3. Figure 4.8(a) shows the results of out-of-plane XRD, in-plane XRD, and RHEED patterns, indicating the epitaxial growth of Mn_4N films. The measured x-ray ω-scan rocking curve shown in figure 4.8(b) indicates that the Mn_4N crystal is well oriented in the [001] direction and has low mosaicism.

The magnetic and magnetotransport properties of the films are measured before nanofabrication. Figure 4.9 shows the hysteresis loops measured by (a) MPMS and (b) AHE using the magnetotransport measurement system. From the *M–H* loops,

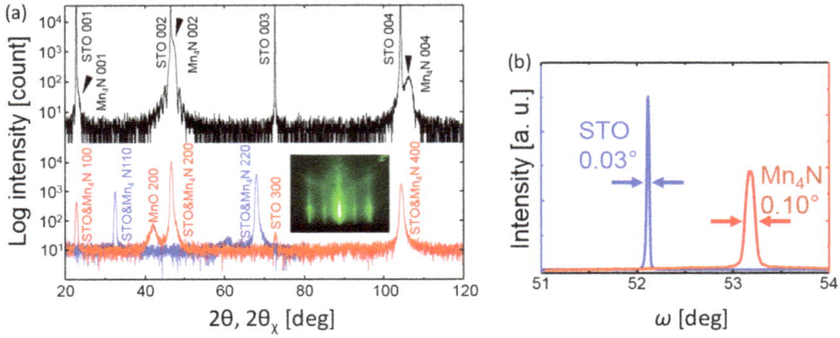

Figure 4.8. XRD patterns of scattering vectors along (a) (black) the [001] (out-of-plane), (red) [100], and (blue) [110] directions. The inset shows the RHEED pattern when the electron beam is injected in the STO [100] direction; the XRD pattern shows diffraction peaks only in the 001 orientation, and the RHEED streaks indicate successful epitaxial growth of Mn_4N. Reprinted with permission from [1]. Copyright (2019) American Chemical Society. (b) X-ray ω-locking curves for the (blue) STO 004 and (red) Mn_4N 004 diffraction peaks.

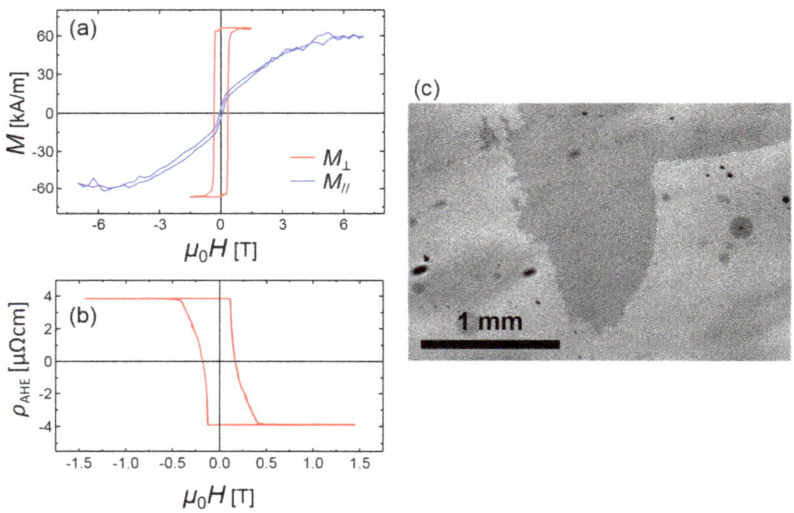

Figure 4.9. (a) *M–H* curves for in-plane (blue) and out-of-plane (red) magnetic fields, measured by MPMS. Reproduced with permission from reference. Reproduced from [2]. CC BY 4.0. (b) AHE loop under magnetic fields normal to the plane. Reproduced with permission from reference. (c) MOKE image of the thin film after partial reversal. Reprinted with permission from [1]. Copyright (2019) American Chemical Society.

the uniaxial anisotropy constant K_u is calculated to be 0.11 MJ m^{-3}, similar to the sample in chapter 3, with a smaller spontaneous magnetization (only 66.1 kA m^{-1}). It is noteworthy that the magnetization of Mn_4N is quite sensitive to the amount of nitrogen atoms [14]. The magnetotransport properties shown in figure 4.9(b) are typical for Mn_4N films on STO: ρ_{xx} = 178 μΩ cm, ρ_{AH} = 3.97 μΩ cm, and $\theta_{AH} = -2.23$ %. The domain structure observed in the MOKE shown in figure 4.9(c) also has millimeter-sized domains.

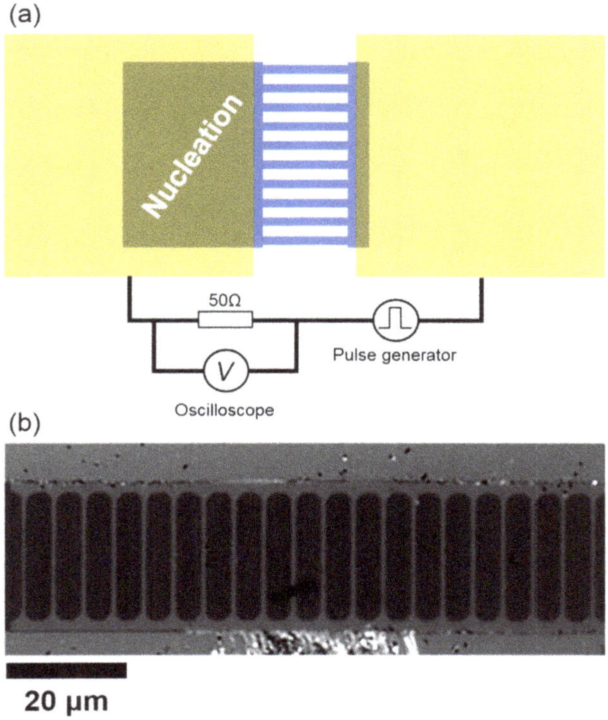

Figure 4.10. (a) Schematic and (b) optical microscope image of the multi-strips with electronic scheme. Reproduced from [2]. CC BY 4.0.

The Mn_4N film is processed to produce the multi-microstrips device shown in figure 4.10(a). This device consists of two electrical contacts connected with 20 strips. The current flows through the strips, where DW motion can be induced.

Under a perpendicularly applied magnetic field, the reversed domains nucleate first on the pads. This is likely to happen because nucleation is extrinsic and defects with low nucleation fields have a higher probability of being found in large areas [15]. Both electrical contacts covered with Ti/Au/Ti electrodes are formed using a lift-off process. This allows wire bonding to the sample and ensures that all strips are exposed to the same voltage difference, and thus the same current density. The topmost Ti layer suppresses the reflection of light on the Au surface and facilitates the detection of MOKE signals. Figure 4.10(b) is an optical microscope image of the multistrip pattern.

4.4.2 Speed of DWs driven by current pulses

The magnetic DW displacement during the pulse is measured using the differential Kerr imaging method. In other words, the MOKE image before pulse injection is subtracted from the post-injection image as a background to emphasize the difference between the two images corresponding to magnetization reversal, such as DW displacement. After magnetization saturation, magnetic field pulses are applied to nucleate domains in the pad and inject DW into the microstrip. DW is pushed by current pulses generated by a pulse generator capable of delivering up to 80 V and 200 ps pulse width. The velocity of the magnetic wall is obtained by dividing the displacement of the magnetic DW by the pulse length, which is measured by an oscilloscope connected in series with the device.

(1) DW speed measurement

Figure 4.11 shows the measurement setup. To monitor the pulse shape with an oscilloscope, the multistrip device is connected to a 50 Ω resistor in parallel with the pulse generator. The coil allows the magnetic field pulse to be applied under the sample stage and its power supply. The magnetic state of the device is monitored in differential mode with a MOKE microscope.

Figures 4.12(a)–(d) show typical results of the DW motion measurements. As shown in figure 4.12(a), the size and length of the field pulses are optimized to make the DW stop in the center of the strip. After the pulses are injected into the device, the displacement of the DW is also monitored (figure 4.12(b)). The direction of motion of the DW is opposite to the current, indicating that the conduction electrons in Mn_4N are positively spin polarized. [13]. The DW displacement Δx is obtained by measuring the full width at half maximum (FWHM) of the MOKE profile along the wire (see figure 4.12(c)).

As seen in figure 4.12(d), the pulse shape is not square. Thereby an analysis can be performed to derive the relationship between DW speed and current density on the assumption that the DW cannot move below a threshold current.

This analysis process consists of the following five steps:

(i) Calculate the current density by dividing the current measured through a 50 Ω resistor by 20 strips and cross-sectional area (1, 2 μm × 10 nm) assuming that the 20 strips are identical.

(ii) Integrate the j–t curve to find the area S over the threshold current ($j_{th} = 3.0 \times 10^{11}$ A m^{-2} for 1 ns pulse), as shown in figure 4.12(d).

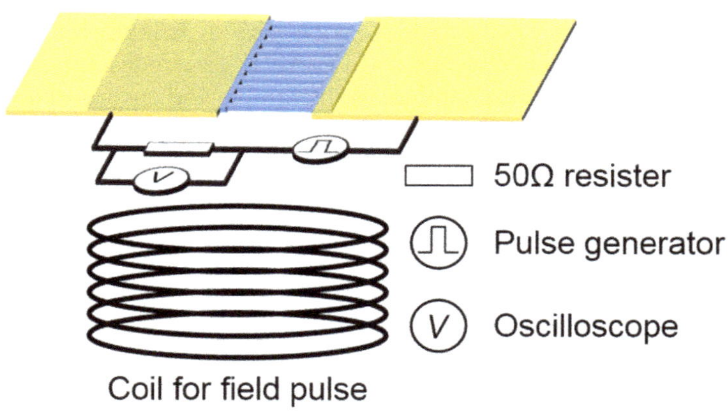

Figure 4.11. Measurement setup for current-induced DW velocities. The entire setup is placed under the objective lens of a MOKE microscope. Reproduced from [2]. CC BY 4.0.

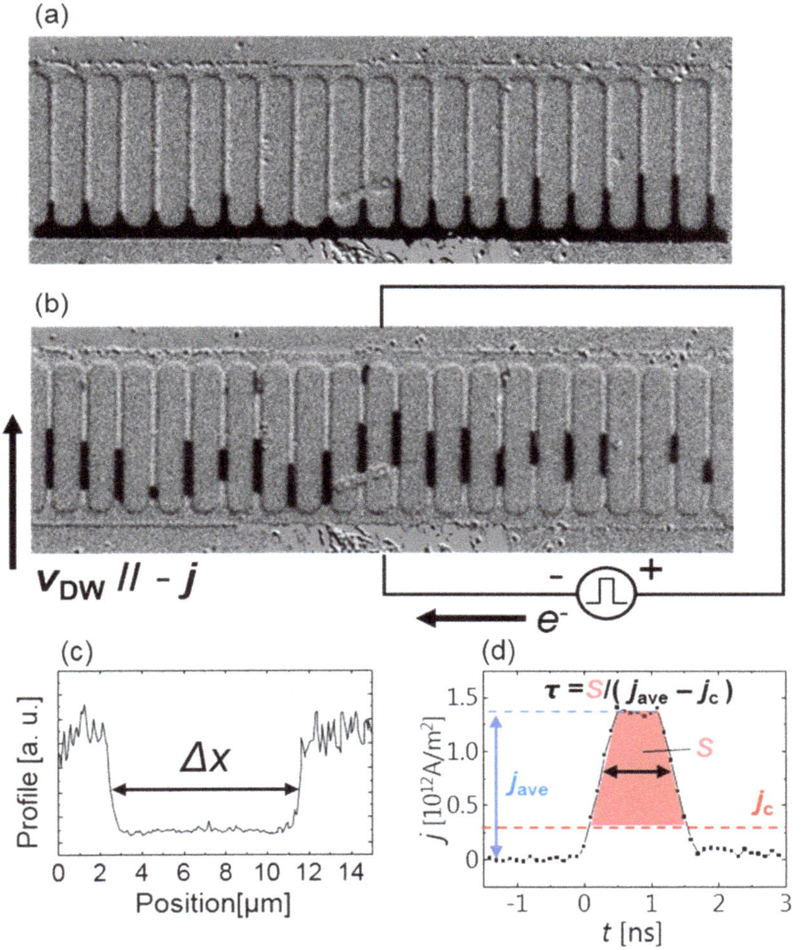

Figure 4.12. (a) Differential MOKE image after magnetic domain nucleation by applying a 0.6 T magnetic field pulse for 300 μs. (b) Differential MOKE image after eight injections of 1.2 A m^{-2} current pulses for 1 ns. The region where DW propagated is shown in black. The direction is the same as the direction of electron flow. (c) Profile of the MOKE image along the strip after DW motion. The DW displacement Δx is derived as the FWHM in the figure. (d) Shape of the incident pulse monitored by an oscilloscope. Reproduced from [2].

(iii) Calculate the pulse height j_{ave} by averaging the data points above 80 % of the maximum height.

(iv) Calculate the exact pulse length and DW velocity using the relation $\tau = S/(j_{\text{ave}} - j_{\text{th}})$ and $v_{\text{DW}} = \Delta x/N\tau$, respectively. Here, N is the number of pulses injected between the two images.

(v) Average the obtained v_{DW} values for all the DWs. v_{DW} is smaller than one-third of the maximum v_{DW} at the current densities studied.

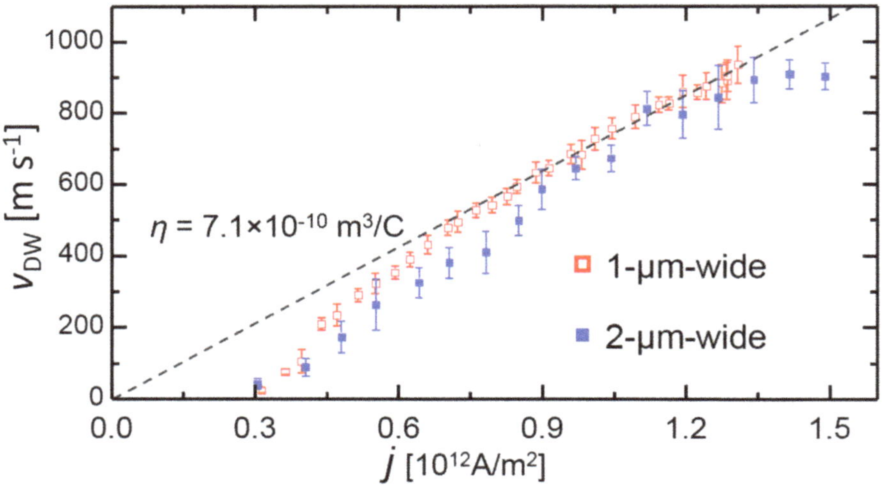

Figure 4.13. Current density dependence of DW velocity. The DW mobility has been derived from the linear fitting in the high current regime. Reprinted with permission from [1]. Copyright (2019) American Chemical Society.

(2) Current dependence of DW speed

Figure 4.13 shows the current density versus v_{DW}. As noted above, the threshold current density for a 1 ns pulse is 3.0×10^{11} A m^{-2}. When it exceeds 8×10^{11} A m^{-2}, the data points follow a straight line for a 1 μm wide strip. We will show later that this behavior is similar to the theory of CIDWM in the one-dimensional model. In the range of current densities studied, the v_{DW} reached a maximum of 900 m s^{-1} at $j_{DC} = 1.3 \times 10^{12}$ A m^{-2}. This value is one of the highest for all materials and structures where the DW is driven purely by STT.

The wide DW widths (>100 nm) in these in-plane systems limit their interest for logic and storage device applications. Considerably higher spin transfer efficiencies have been measured in systems with PMA, such as Co/Ni and FePt [4], but in these systems the DW motion is usually hindered by pinning, leading to large coercivity and stochastic behavior. In Pt/Co/Pt, the limited spin polarization in the Co layer precludes moving DW by STT [16]. Meanwhile, in magnetic semiconductor films such as (Ga,Mn)(As,P), very high DW mobility is observed at very low-current densities [17, 18]. However, these measurements are performed at low temperatures, current densities are limited to 10^{10} A m^{-2}, and the maximum velocity at 10^{10} A m^{-2} was 30 m s^{-1}.

The velocities observed in Mn_4N can be considered to be a milestone in the history of STT. In recent years, there has been almost total focus on DW dynamics driven by spin-orbit torque (SOT), but the giant DW velocities obtained in Mn_4N indicate that STT can compete with SOT; SOT is an efficient DW-driving mechanism [19, 20]. A typical example is Pt/Co(0.6 nm)/AlO_x [21], where the chiral Néel structure acquired by the DW in the presence of Dzyaloshinskii-Moriya interaction (DMI) [22] at the interface resulted in high mobility (v_{DW} = 400 m s^{-1}

for $j = 3 \times 10^{12}$ A m^{-2}) with satisfactory reproduction of the DW motion. Recently, higher SOT-driven DW mobility has been obtained in multilayer structures with two Co/Ni layers antiferromagnetically coupled via a Ru layer [23] and in GdCo ferrimagnetic layers deposited on top of Pt [24]. In these systems, it is clearly observed that the DW velocity varies with the total magnetization. In the former experiment, a maximum velocity of v = 750 m s^{-1} was obtained for $j = 3 \times 10^{12}$ A m^{-2} at RT with a Ru spacer layer, and the stack magnetization is reduced to 8% of the spontaneous magnetization of the Co/Ni single layer. In the latter, a peak velocity of 1300 m s^{-1} is reported at the compensation temperature for a current density of 2×10^{12} A m^{-2} and is still as high as 850 m s^{-1} at 300 K.

From this perspective, the room temperature velocity of DW in Mn_4N is comparable to the best results obtained using SOT. Furthermore, very efficient CIDWM by pure STT is demonstrated for the first time in a PMA-based system. Another advantage of STT efficiency is the DW mobility $\eta = \frac{dv_{\mathrm{DW}}}{dj}$, which can be derived from the v_{DW}–j curve and is equal to 7.1×10^{-10} m^3 C^{-1}. This mobility is the highest for all STT-driven systems and is superior to most SOT-driven records. When $j > 1.3 \times 10^{12}$ A m^{-2}, many domains are observed to nucleate everywhere probably caused by the sample reaching 745 K, the Curie temperature of Mn_4N, and thus Mn_4N strips temporarily become paramagnetic.

4.4.3 Field driven DW motion

DWs with chiral Néel structures stabilized by the DMI at the interface can be driven by SOT, which arises from the spin Hall effect and Rashba effect [25]. To rule out a contribution of SOT to the DW dynamics in our system, measurements aimed at establishing the nature of the DW structure are performed. Magnetic DW dynamics are driven by an out-of-plane magnetic field in the presence of a continuous longitudinal in-plane magnetic field $\mu_0 H_x$. The DW velocity in the in-plane field direction depends on the nature of the internal DW structure. For Bloch walls, the DW velocity increases as a function of $\mu_0 H_x$ and is the same for magnetic fields of opposite sign (velocity versus $\mu_0 H_x$ curve is symmetric). However, in the presence of DMI, the velocity versus $\mu_0 H_x$ curve should be asymmetric and the velocity is minimized when the $\mu_0 H_x$ field compensates the DMI field [26, 27]. Figures 4.14(a)(b) show examples of DW displacements in the absence and presence of a 400 mT in-plane magnetic field. The white contrast in the Kerr image represents the displacement of the DW by the magnetic field pulse. This displacement is isotropic in the presence of an in-plane magnetic field, indicating the absence of DMI. Figure 4.14(c) shows the magnetic DW velocity versus $\mu_0 H_x$ curve.

4.4.4 Origin of the high DW mobility in Mn_4N

To understand the origin of such a high DW mobility, we consider the analytical expression for STT in the adiabatic limit. As described in chapter 1, when j is much higher than j_c, the DW velocity v_{DW} in a precessional mode is given by,

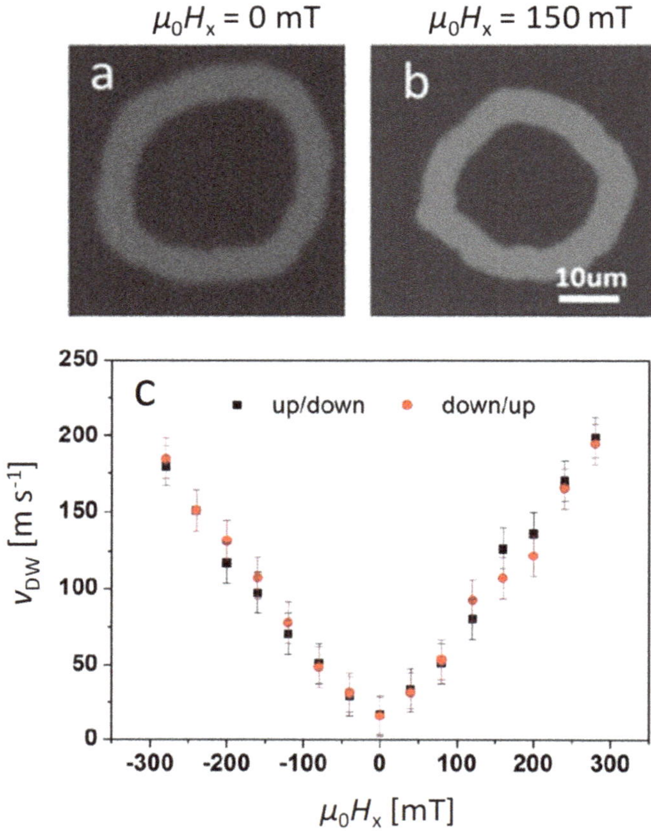

Figure 4.14. Differential Kerr images of the displacement of domain walls driven by an out-of-plane magnetic field pulse with a magnitude of 400 mT and a pulse width of 30 ns (white contrast) in the absence (a) and in the presence (b) of an in-plane continuous magnetic field $\mu_0 H_x$. (c) DW velocity as a function of the in-plane field amplitude $\mu_0 H_x$ for an up/down and down/up domain wall. A perpendicular magnetic field pulse $\mu_0 H_z$ is 320 mT. The symmetric curve confirms the presence of achiral domain walls. Reproduced from [2]. CC BY 4.0.

$$| v_{\mathrm{DW}} | \approx \frac{1}{1+\alpha^2} | \boldsymbol{u} | \tag{4.2}$$

Here, $\boldsymbol{u} = \frac{g\mu_B}{2eM_S} P\boldsymbol{j}$ is the spin-drift velocity, parallel to the flow of electrons. Experimentally, the data points in the $v_{\mathrm{DW}}-j$ curve follow a straight line above the stationary region. Since the mobility, expressed as dv_{DW}/dj, is expected to be proportional to P/M_{S}, both small M_{S} and high P lead to such large v_{DW} being achieved.

Micromagnetic simulations are performed using the finite difference software MuMax3 [28] to confirm the above interpretation. Simulations at 0 K are performed on a 6000 × 120 × 10 nm^3 defect-free strip with a cell size of 2.5 × 2.5 × 10 nm^3. Figure 4.15 shows the results of the one-dimensional model along with the parameters used, showing the overall agreement of the one-dimensional model

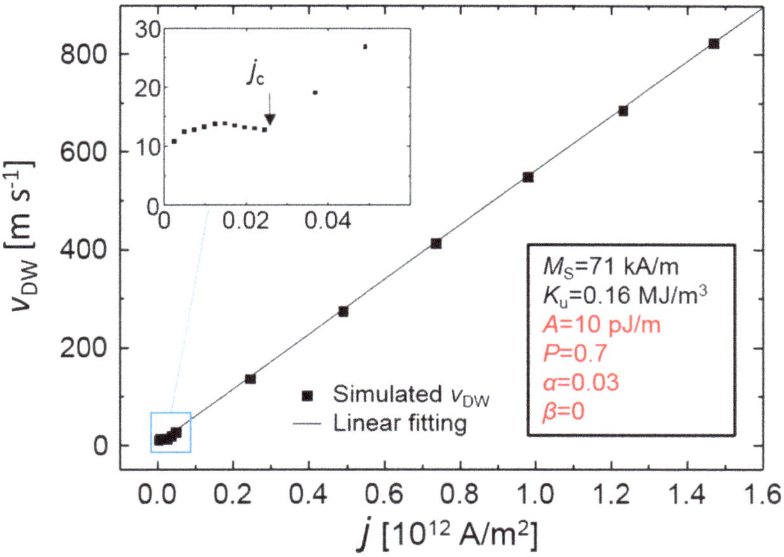

Figure 4.15. DW velocity versus current density in an adiabatic one-dimensional model. The DW velocity near the theoretical threshold current j_c is shown in the inset. Black and red letters in the lower right represent measured values and assumed values, respectively. Reproduced from [2]. CC BY 4.0.

with the experimentally obtained results in the high current density region. Meanwhile, the trend of the v_{DW} in the low-current regime is quite different from the experiment, showing a relatively large critical current density of 3.0×10^{11} A m^{-2}. This is considered to be due to the pinning of the DWs by defects in the actually fabricated devices. In the simulation, the system is assumed to be perfect without any defects.

4.4.5 Estimation of the spin polarization

In the previous section, the DW velocity v_{DW} varies linearly with current density j, corresponding to the spin-drift velocity u. We can derive the spin polarization P of Mn_4N from the slope of the high current region, which corresponds to the DW mobility η. In fact, the P is calculated in the one-dimensional model by,

$$P = \eta \frac{2eM_S}{g\mu_B} = 0.81. \tag{4.3}$$

Here, $e = 1.60 \times 10^{-19}$ C, $g = 2.0$, $M_S = 66.1$ kA m^{-1}, $\mu_B = 9.27 \times 10^{-24}$ Am^2, and $\eta = 7.1 \times 10^{-10}$ C m^{-3}. The obtained P value of 0.81 is higher than expected from first-principles calculations ($P = 0.75$) from the total DOS at the E_F. We should note that the experimentally obtained $P = 0.81$ is the spin polarization of electrical conductivity P_σ [29], whereas the calculated $P = 0.75$ is the one at the E_F in the DOS (P_{DOS}) [30]. As an example of the difference between P_σ and P_{DOS}, the ferromagnetic Fe_4N has a fully spin-polarized conduction electron $P_\sigma = -1.0$, whereas P_{DOS}

is only −0.6 [31]. Such a high spin polarization of electrical conductivity in Mn_4N could be useful not only for the development of CIDWM devices but also for magnetoresistive applications, such as GMR and TMR devices.

4.4.6 Influence of the non-adiabatic torque and the damping parameter on CIDWM

The non-adiabatic torque has to be taken into account to explain the experimentally observed DW motion in in-plane magnetized systems, such as permalloy strips below the intrinsic j_c. However, this term has been neglected thus far in PMA systems because the large DW velocities observed in Mn_4N stripes at a low-current density can be explained by the adiabatic term alone. It is shown analytically that the non-adiabatic torque leads to steady-state regime motion with velocity $v = (\beta/\alpha)u$ below the critical current density j_c. Meanwhile, in the precessional regime, a term $\alpha\beta/(1 + \alpha^2)u$ is added to the velocity [27]. However, for Mn_4N, given that α is small and β is expected to have the same order of magnitude as α, this term can be neglected. To justify our assumption to ignore the non-adiabatic torque, we perform micromagnetic simulations and consider two extreme cases, i.e., $\beta = 5\alpha$ and $\beta = 0.5\alpha$ for $\alpha = 0.03$. As shown in figure 4.16(a), when non-adiabatic torque is considered, DW moves in the steady-state region until it reaches the Walker spin-drift velocity u_W, which is expressed as,

$$u_W = u_C \cdot \frac{\alpha}{\mid \beta - \alpha \mid}. \tag{4.4}$$

Neglecting the non-adiabatic torque, the Walker spin-drift velocity coincides with the critical spin-drift velocity. When $j > j_c$, DW begins to move linearly with respect to j, entering the precessional motion regime, and the mobility of DW is no longer dependent on the value of β as shown in figure 4.16(b). From these results, we

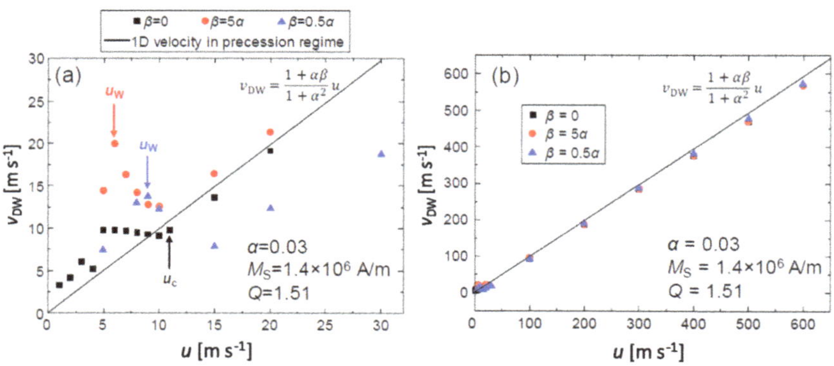

Figure 4.16. Micromagnetic simulation results showing DW velocities under spin-polarized current for different values of $\beta = 0$ (black squares), $\beta = 5\alpha$ (red circles), and $\beta = 0.5\alpha$ (blue triangles). The solid black lines represent 1D model velocities. (a) Enlarged results for the low velocity region around the Walker flow. Each arrow indicates the point at which the magnetization in the DW begins precession, corresponding to the Walker current. (b) Results for a wide range of current regimes, mainly showing the precession regime. Reproduced from [2]. CC BY 4.0.

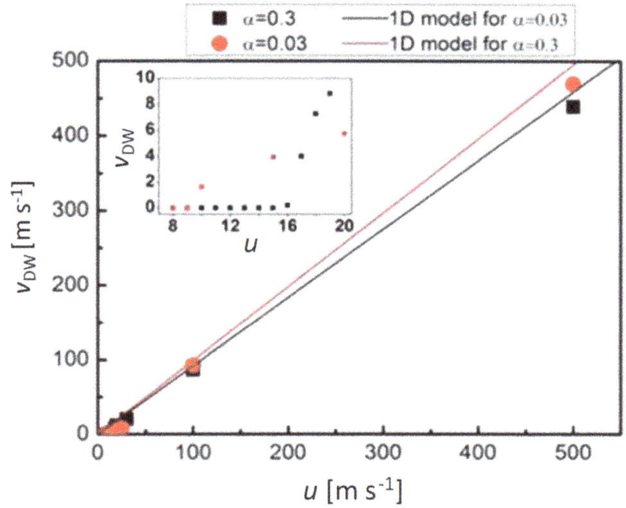

Figure 4.17. Micromagnetic simulation results that show the DW velocity v_{DW} versus spin-drift velocity u for two different α values: $\alpha = 0.03$ (red circles), $\alpha = 0.3$ (black squares). The solid line represents the one-dimensional model velocity. The inset presents an enlarged view near the critical current j_c. Reproduced from [2]. CC BY 4.0.

conclude that non-adiabatic torque is not necessary to explain the experimentally observed velocities. Note that this does not mean to exclude the presence of the non-adiabatic term. This happens because the regime below j_c, where the non-adiabatic torque really matters, is hidden by the thermally activated regime.

Gilbert damping α may play an important role in the DW motion. To evaluate the effect of the α value on DW mobility, we perform micromagnetic simulations for the two cases: low $\alpha = 0.03$ and high $\alpha = 0.3$ and compare the results with the one-dimensional model. Although a low value of $\alpha = 0.03$ is plausible, it has not yet been verified by experiment. The value $\alpha = 0.3$ is a typical high Gilbert damping value obtained for Co layers formed on a large spin-orbit Pt layer. Figure 4.17 shows the result. The difference in velocity is only about 6% at the maximum current density. The larger the damping, the larger the u_c, but the difference is negligibly small (see inset in figure 4.17). Therefore, we can state that the uncertainty in α does not affect the interpretation.

4.4.7 Damping and non-adiabatic torque: the particular case at $\alpha = \beta$

So far, we have classified the DW motion behavior into two regimes: the steady-state regime below the Walker breakdown and the precessional regime for currents above the Walker breakdown. By simulating the DW motion in the precessional regime and modeling our system as a low-damping material implying the relation of $v_{\text{precession}} \approx u$, we are able to reproduce the experimentally obtained DW velocity.

However, when non-adiabatic torque and Gilbert damping compensate each other, i.e., the $\alpha = \beta$, the Walker spin-drift velocity is predicted to go to infinity in the

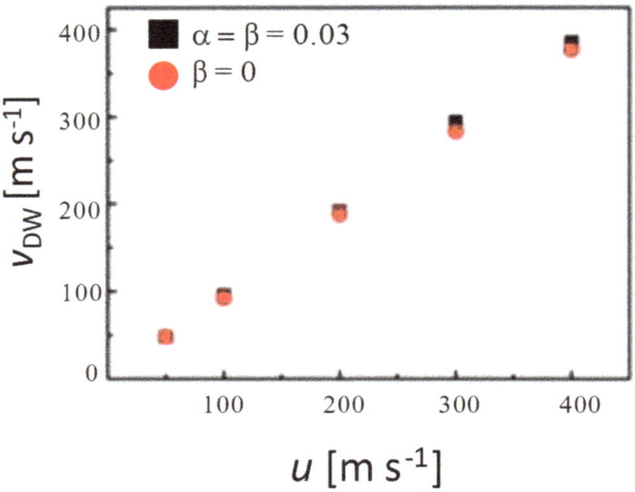

Figure 4.18. Micromagnetic simulation results that show the DW velocity v_{DW} versus spin-drift-velocity u for different conditions: $\alpha = \beta = 0.03$ (black squares) and $\beta = 0$ (red circles). Reproduced from [2]. CC BY 4.0.

one-dimensional model. In this special case, DW moves at a speed equal to the spin-drift velocity $v_{\text{steady}} = u$ in the steady-state regime, regardless of the injected current density. To confirm this result, we simulate a PMA strip with $Q = K_u/K_d = 1.51$, and $\alpha = \beta = 0.03$. We show the results in figure 4.18 and compare them with the case of $\beta = 0$. It is noted that there is not much difference between the above two cases. Meanwhile, the DW moves in the precessional regime for $\beta = 0$ and in the steady-state regime for $\alpha = \beta$ [2].

It is noted that experimental evidence for the case $\beta = \alpha$ has been reported [17, 18]. Here, the difference in DW velocity between the cases where the non-adiabatic and damping torques are balanced and the cases where they are not balanced is so small that the two cases cannot be distinguished experimentally.

4.5 Summary

In this chapter, the STT efficiency in Mn_4N is evaluated by measuring the DW velocity driven by injecting current pulses. The DW velocity reaches a record value of 900 m s^{-1} at a current density of 1.3×10^{12} A m^{-2}. The DW mobility η derived from the slope of the v_{DW}–j curve is as high as 7.1×10^{-10} m^3 C^{-1}. This value is the highest for an STT-based system. Data fitting using a one-dimensional analytical model yields a spin polarization of 0.81 for the conduction electrons, indicating that Mn_4N is suitable for obtaining a large MR. In the future, we plan to study CIDWM in the flow regime. We also plan to study CIDWM in the flow regime, i.e., CIDWM with longer current pulses and below the threshold current density. SOT-driven DW motion measurements are also planned to evaluate SOT properties after harmonic Hall measurements in W/Mn_4N/STO stack structures [32].

Bibliography

[1] Gushi T *et al* 2019 *Nano Lett.* **19** 8716
[2] Gushi T 2019 Mn_4N thin films for spintronics applications based on current-induced domain wall motion *PhD Thesis* University of Tsukuba
[3] Shen X, Chikamatsu A, Shigematsu K, Hirose Y, Fukumura T and Hasegawa T 2014 *Appl. Phys. Lett.* **105** 072410
[4] Burrowes C *et al* 2010 *Nat. Phys.* **6** 17
[5] Koyama T *et al* 2011 *Nat. Mater.* **10** 194
[6] Fukami S, Suzuki T, Nakatani Y, Ishiwata N, Yamanouchi M, Ikeda S, Kasai N and Ohno H 2011 *Appl. Phys. Lett.* **98** 082504
[7] Feigenson M, Reiner J W and Klein L 2007 *Phys. Rev. Lett.* **98** 247204
[8] Li S, Nakamura H, Kanazawa T, Liu X and Morisako A 2010 *IEEE Trans. Mag.* **46** 1695
[9] Nguyen V D, Vila L, Marty A, Warin P, Vergnaud C, Jamet M, Notin L, Beigné C and Attané J P 2013 *J. Appl. Phys.* **113** 183906
[10] Mihai A P, Attané J P, Marty A, Warin P and Samson Y 2008 *Phys. Rev.* B **77** 060401(R)
[11] Nguyen V D, Vila L, Laczkowski P, Marty A, Faivre T and Attané J P 2011 *Phys. Rev. Lett.* **107** 136605
[12] Danneau R *et al* 2002 *Phys. Rev. Lett.* **88** 157201
[13] Kabara K, Tsunoda M and Kokado S 2017 *AIP Adv.* **7** 056416
[14] Kabara K and Tsunoda M 2015 *J. Appl. Phys.* **117** 17B512
[15] Nguyen V D, Vila L, Marty A, Pillet J C, Notin L, Beigné C, Pizzini S and Attané J P 2012 *Appl. Phys. Lett.* **100** 252403
[16] Cormier M, Mougin A, Ferré J, Thiaville A, Charpentier N, Piéchon F, Weil R, Baltz V and Rodmacq B 2010 *Phys. Rev.* B **81** 024407
[17] Curiale J, Lemaître A, Ulysse C, Faini G and Jeudy V 2012 *Phys. Rev. Lett.* **108** 076604
[18] Adam J-P, Vernier N, Ferré J, Thiaville A, Jeudy V, Lemaître A, Thevenard L and Faini G 2009 *Phys. Rev.* B **80** 193204
[19] Emori S, Bauer U, Ahn S-M, Martinez E and Beach G S D 2013 *Nat. Mater.* **12** 611
[20] Ryu K-S, Thomas L, Yang S-H and Parkin S 2013 *Nat. Nanotech.* **8** 527
[21] Miron I M *et al* 2011 *Nat. Mater.* **10** 419
[22] Thiaville A, Rohart S, Jué É, Cros V and Fert A 2012 *Europhys. Lett.* **100** 57002
[23] Yang S-H, Ryu K-S and Parkin S 2015 *Nat. Nanotechnol.* **10** 221
[24] Caretta L *et al* 2018 *Nat. Nanotech.* **13** 1154
[25] Khvalkovskiy A V, Zvezdin K A, Gorbunov Y V, Cros V, Grollier J, Fert A and Zvezdin A K 2009 *Phys. Rev. Lett.* **102** 067206
[26] Je S-G, Kim D-H, Yoo S-C, Min B-C, Lee K-J and Choe S-B 2012 *Phys. Rev.* B **88** 1440
[27] Hrabec A, Porter N A, Wells A, Benitez M J, Burnell G, McVitie S, McGrouther D, Moore T A and Marrows C H 2014 *Phys. Rev.* B **90** 020402
[28] Vansteenkiste A, Leliaert J, Dvornik M, Helsen M, Garcia-Sanchez F and Waeyenberge B V 2014 *AIP Adv.* **4** 107133
[29] Thiaville A, Nakatani Y, Miltat J and Suzuki Y 2005 *Europhys. Lett.* **69** 990
[30] Miao M S, Herwadkar A and Lambrecht W R L 2005 *Phys. Rev.* B **72** 033204
[31] Kokado S, Fujima N, Harigaya K, Shimizu H and Sakuma A 2006 *Phys. Rev.* B **73** 172410
[32] Lau Y-C and Hayashi M 2017 *Jpn. J. Appl. Phys.* **56** 0802B5

Chapter 5

Growth and characterization of compensated Mn_4N epitaxial films

Mn_4N-based mixed crystals, in which Mn is partially replaced by other elements, can have zero magnetization when the net magnetization of the two antiparallel sublattices reaches compensation, leading to the development of new compensated ferrimagnets for spintronics. In bulk crystals and epitaxial thin films, magnetization compensation has been reported to occur at room temperature.

5.1 $Mn_{4-x}Z_xN$ (Z=Ge, Ga, Zn, Cu, Ni, Sn, In, Cd, Ag, Pd) powders

The possibility of magnetic compensation in Cr-, Fe-, or Ni-doped Mn_4N was demonstrated in 1959 from the temperature dependence of magnetization using powder samples [1]. In reference [2], a comprehensive study of modification of the spin structure of Mn_4N by various metal substitution ions Z to achieve compensation at room temperature was performed on powder samples [2]. It has been reported that the anomalous temperature dependence of magnetization obtained in impurity-doped Mn_4N powders strongly indicates the noncollinear nature of magnetic order.

Figures 5.1(a) and (b) show the ferrimagnetic prototypes [2]. In figure 5.1(a), the collinear ferrimagnetic spin structure is a combination of ferromagnetic and antiferromagnetic spin structures, while in figure 5.1(b) the noncollinear ferrimagnetic structure is a combination of noncollinear antiferromagnetic and ferromagnetic structures. Figure 5.1(c) shows the XRD pattern of the Mn_4N powder. The inset shows the crystal and magnetic structures with Γ_{4g} triangular ferrimagnetism in the (111) plane. Gray, blue, and red atoms represent N, Mn 1a (corner sites), and Mn 3c (face-center sites), respectively. The Kagome lattice of Mn 3c in the (111) plane in figure 5.1(c) shows the Γ_{4g} magnetic structure. Figures 5.1(e)–(h) show schematic temperature-dependent magnetization curves for four ferrimagnetic materials with two sublattices (1a and 3c in Mn_4N). Figure 5.1(e) shows the case

doi:10.1088/978-0-7503-5477-6ch5

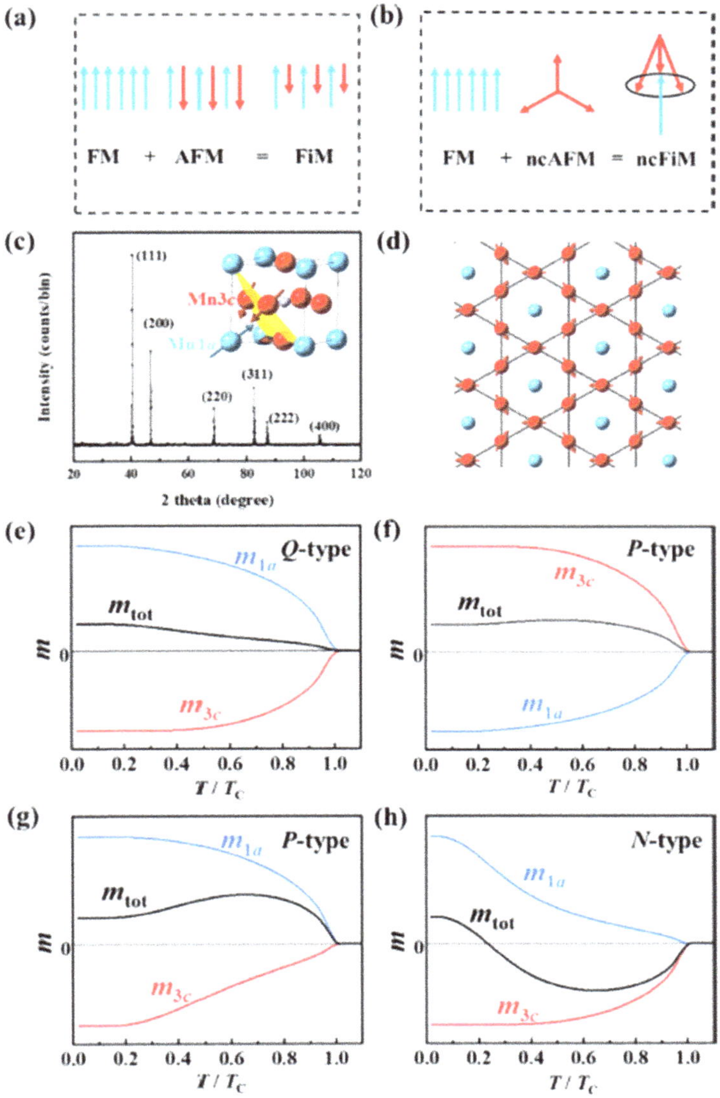

Figure 5.1. Ferrimagnetic prototypes. (a) Collinear ferrimagnetic (FiM) spin structure combined with ferromagnetic (FM) and antiferromagnetic (AFM) spin structures. (b) Noncollinear ferrimagnetic (ncFiM) structure composed with noncollinear antiferromagnetic (ncAFM) and FM spin structures. (c) XRD pattern of Mn_4N. The inset shows crystal and magnetic structure with Γ_{4g} triangular ferrimagnetism in the (111) plane. (d) Kagome lattice of Mn 3*c* in a (111) plane showing the Γ_{4g} magnetic structure. Four types of temperature-dependent magnetization curves are shown schematically for a ferrimagnet with two sublattices (1*a* and 3*c* for Mn_4N) in panels (e)–(h). (e) is for $m_{1a} > 3m_{3c}$ with n_{1a3c} as the main molecular field coefficient; (f) is for $m_{1a} < 3m_{3c}$ with n_{1a3c} as the main molecular field coefficient; (g) is for $m_{1a} > 3m_{3c}$ with n_{1a1a} as the main molecular field coefficient; and (h) $m_{1a} > 3m_{3c}$ and n_{3c3c} as the main molecular field coefficient. A collinear model is used here. Reproduced from [2]. CC BY 4.0.

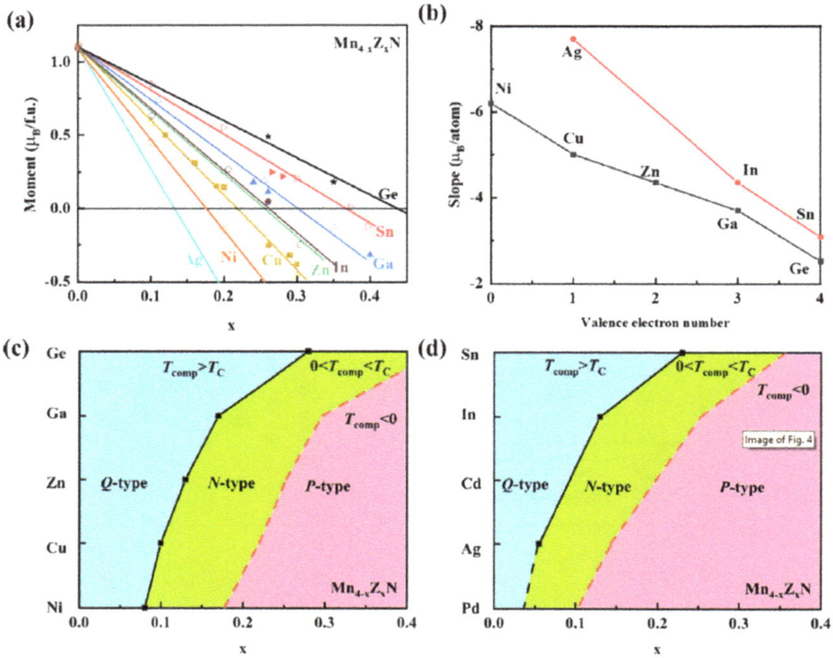

Figure 5.2. (a) Dependences of net magnetic moment on composition x in $Mn_{4-x}Z_xN$. Solid points are data obtained by the authors in reference [1] and open points are those reported previously. (b) The slope in (a) showing different efficiencies. Magnetic diagram for *Q, N* and *P* type *M-T* curves, depending both on x and Z for elements from the fourth (c) and fifth (d) periods. Reproduced from [2]. CC BY 4.0.

where the magnetic moment of Mn at the corner site (m_{1a}) is greater than three times the magnetic moment of Mn at the face-center site (m_{3c}) (Q-type), where $m_{1a} > 3m_{3c}$. Figure 5.1(f) shows the case where $m_{1a} < 3m_{3c}$ and n_{1a3c} is the main molecular field coefficient (P-type), (g) where $m_{1a} > 3m_{3c}$ and n_{1a1a} is the main molecular field coefficient (P-type), and (h) where $m_{1a} > 3m_{3c}$ and n_{3c3c} is the main molecular field coefficient (N-type).

Figure 5.2(a) shows an overview of the net magnetization of $Mn_{4-x}Z_xN$ as a function of impurity composition x at 4 K. The solid lines are data newly obtained by the authors in reference [2], and the open dots are previously reported data. As shown in figure 5.2(b), it is interesting to note that the slope changes significantly with Ni, Cu, Zn, Ga, and Ge as the valence electron of the dopant impurity increases. Figures 5.2(c) and (d) are magnetic diagrams based on these trends; magnetic compensation is expected to occur in the 'N-type' region.

5.2 $Mn_{4-x}Ni_xN$ epitaxial films

5.2.1 Growth of $Mn_{4-x}Ni_xN$ films on $SrTiO_3(001)$ by MBE

$Mn_{4-x}Ni_xN$ ($x = 0.1, 0.25$) epitaxial films with a thickness of 10 nm are grown on STO(001) substrates using the MBE method [3]. Solid sources of Mn and Ni in

K-cells, and a radio-frequency nitrogen plasma source are used. The substrate temperature is set at 450°C during deposition. The composition ratios of Mn and Ni are calculated using the deposition rate ratios from the crucible of each K-cell. After growth, the samples are capped with Ti layers to prevent oxidation of sample surface. The crystallinity of the grown films is characterized by RHEED and XRD. The polar Kerr effect is measured by MOKE microscopy and the sign of the measured hysteresis loop is used to determine whether each sample composition is below or above the MCP. The magnetic field $\mu_0 H$ is applied perpendicular to the plane in the range $\mu_0 H = -1.25$ to 1.25 T. XAS and XMCD measurements are performed at the twin APPLE-II undulator beamline BL-16A at the KEK Photon Factory in Japan [4]. A magnetic field of 3 T and circularly polarized x-rays are irradiated perpendicular to the plane. The polarization of x-rays is switched at each energy point using five kicker magnets at a frequency of 10 Hz. Spectra are acquired at room temperature in total fluorescence yield mode and total electron yield mode at the absorption edges of Ni $L_{2,3}$ and Mn $L_{2,3}$, respectively, so that the Ni $L_{2,3}$ spectrum does not overlap with the background signal.

Figure 5.3 shows the XRD patterns of the $Mn_{4-x}Ni_xN$ films and RHEED streaks taken along the STO[100] azimuth [3]. Since $Mn_{4-x}Ni_xN$ and STO(001) have excellent lattice matching, the diffraction peaks of the $Mn_{4-x}Ni_xN$ films overlap with those of the STO substrate. In the θ–2θ XRD patterns in figures 5.3(a)–5.3(d), the $Mn_{4-x}Ni_xN$ peak appears at a slightly larger 2θ than the cubic STO peak, but it disappears in the φ–$2\theta_\chi$ XRD in figures 5.3(e) and (f). In figures 5.3(c) and (d), the presence of the 002 peak and satellite peaks can be seen. Supposed that STO is a perfect cubic structure, the lattice constants determined for $Mn_{4-x}Ni_xN$ have the relation $c/a < 1$, indicating the presence of in-plane tensile strain. The sharp $Mn_{4-x}Ni_xN$ 004 diffraction peaks and RHEED streaks, as well as the Kikuchi lines, suggest that the grown epitaxial thin films are highly c-axis oriented. The lines indicated by the white arrows in the RHEED image originate from the superlattice diffraction, meaning that the N atoms are correctly positioned. Furthermore, in each φ–$2\theta_\chi$ XRD pattern, a diffraction peak is observed around $2\theta \approx 41°$. This peak is considered to be a Mn-related oxide [5].

5.2.2 Characterization of $Mn_{4-x}Ni_xN$ films

(1) Magnetic properties

The polar MOKE hysteresis loops for $Mn_{4-x}Ni_xN$ films at (a) $x = 0.1$ and (b) $x = 0.25$ are shown in figure 5.4 [3]. The distortion of the loops is seen when the magnetic field is high. This comes from non-linear Faraday rotations of the polarized light that occur in the objective lens [5]. As shown by arrows, when the magnetic field exceeds a coercivity, the magnetization reverses, and thus the MOKE image reverses accordingly. This behavior is different in samples between $x = 0.1$ and 0.25, as indicated by arrows. This result shows the presence of MCP between $x = 0.1$ and 0.25. Similar results have been reported in ferrimagnets such as CoGd and Co/Ti/(Gd-Co) multilayers [6, 7]. The sharp corners of the loop, which are

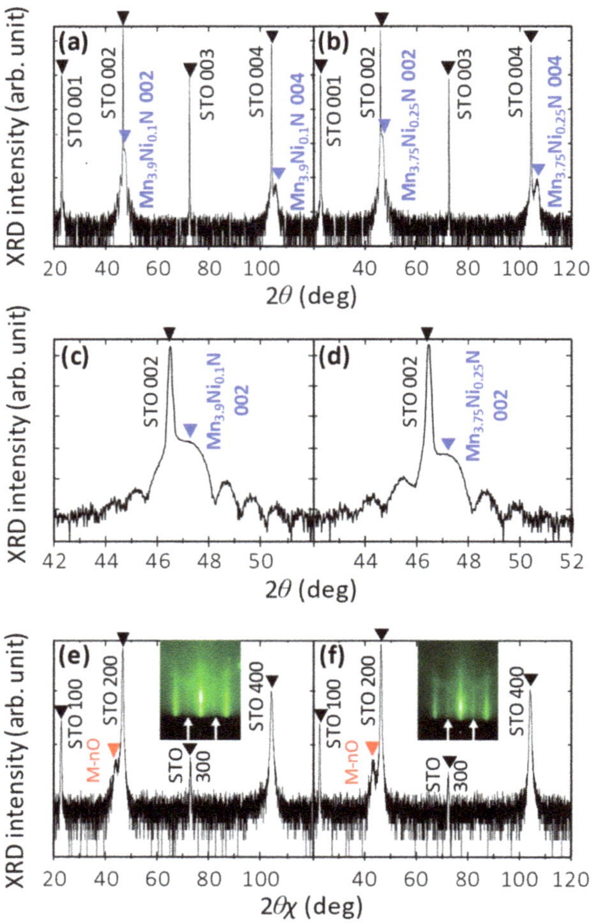

Figure 5.3. XRD patterns of $Mn_{4-x}Ni_xN$ films on STO for (a,c,e) $x = 0.1$ and (b,d,f) $x = 0.25$. (a and b) Out-of-plane XRD patterns and (c and d) enlarged ones at $2\theta \approx 46°$. In-plane XRD patterns with RHEED images observed along the STO[100] azimuth for (e) $x = 0.1$ and (f) $x = 0.25$. White arrows indicate superlattice diffraction. Reprinted from [3], with the permission of AIP Publishing.

typical for high-quality Mn_4N films on STO(001) [8], suggest good magnetic properties. This result is attributable to the excellent lattice matching between grown films and STO substrate. It is noted that the coercivity increases from $x = 0.1$ to $x = 0.25$. This probably happens because the composition is close to the MCP.

Figure 5.5(a) shows the magnetization loops of $Mn_{4-x}Ni_xN$ films with various values of Ni compositions x, measured by VSM at room temperature. The magnetic fields are applied normal to the sample plane [9]. The M_S is 24.2 kA m^{-1} for $Mn_{3.9}Ni_{0.1}N$/MgO and is 47.4 kA m^{-1} for $Mn_{3.9}Ni_{0.1}N$/STO. The introduction of Ni into Mn_4N reduces the M_S by 45% at $x = 0.1$ and by 75% at $x = 0.25$. Such a sharp reduction in M_S can be explained by a simple calculation [9]. PMA appears in the $Mn_{3.5}Ni_{0.5}N$ film on STO but not in the $Mn_{3.5}Ni_{0.5}N$ film on MgO, a kink

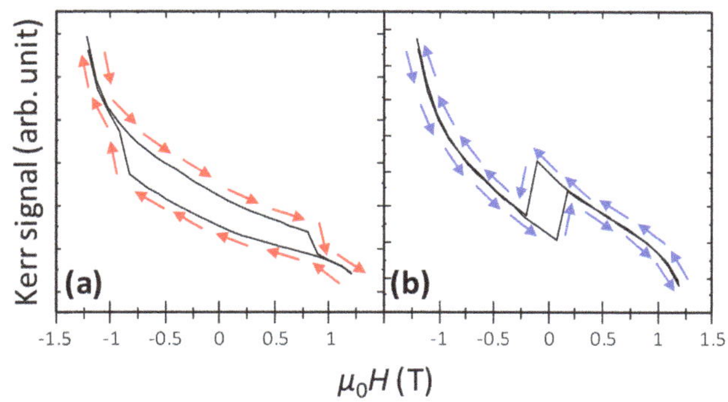

Figure 5.4. Polar MOKE hysteresis loops of $Mn_{4-x}Ni_xN$ films on STO for (a) $x = 0.1$ and (b) $x = 0.25$. The magnetic field is scanned in the direction of arrows. The reversal of the hysteresis loops occurs. Reprinted from [3], with the permission of AIP Publishing.

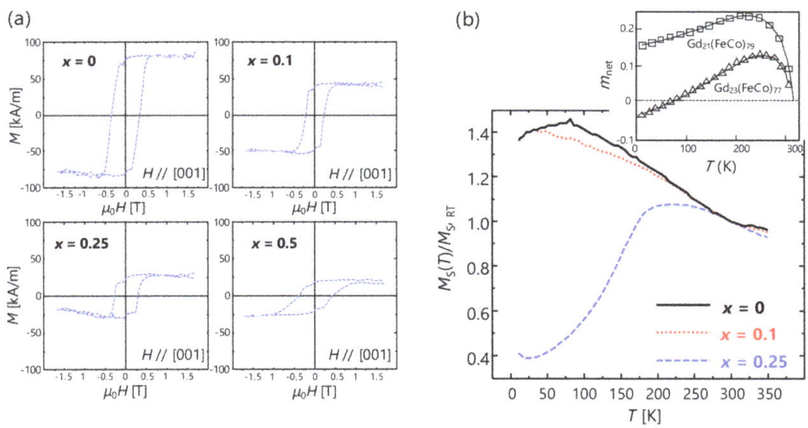

Figure 5.5. (a) *M–H* curves of $Mn_{4-x}Ni_xN$ films on STO by VSM under magnetic fields normal to the plane. (b) Temperature dependences of the normalized magnetization of $Mn_{4-x}Ni_xN$ films for different Ni compositions. Samples are measured under field cooling, and the magnetic field, $\mu_0 H$, is applied normal to the plane. Temperature dependences of magnetization measured on $Gd_{23}(FeCo)_{77}$ and $Gd_{21}(FeCo)_{79}$ are inserted. Reprinted from [9], with the permission of AIP Publishing

appears in the magnetization loop of the $Mn_{4-x}Ni_xN$ film on STO, and the magnetization loop of the $Mn_{3.5}Ni_{0.5}N$ film on MgO is not affected by the introduction of Ni. This result may be due to the mixed hard ($Mn_{4-x}Ni_xN$) and soft magnetic phases.

Figure 5.5(b) shows the dependence of magnetization for $Mn_{4-x}Ni_xN$ films with $x = 0$, 0.1, and 0.25, normalized by the room temperature value of M_S. For comparison, the dependences of M_S on temperature for $Gd_{21}(FeCo)_{79}$ and $Gd_{23}(FeCo)_{77}$ are inserted [10]. The data points of $Gd_{21}(FeCo)_{79}$ exhibit that the FeCo behavior is dominant over Gd at all temperatures. Meanwhile, a transition from FeCo-dominant to Gd-dominant is observed for $Gd_{23}(FeCo)_{77}$. In $Gd_{23}(FeCo)_{77}$, the net magnetization m_{net} decreases with increasing temperature

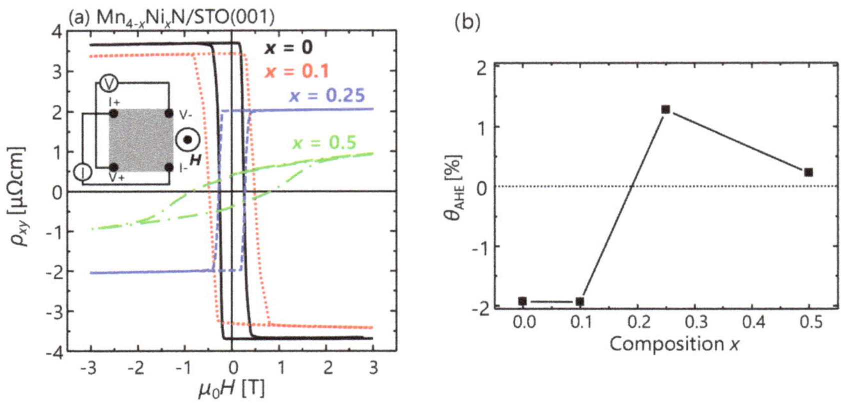

Figure 5.6. (a) AHE loops of $Mn_{4-x}Ni_xN$ thin films on STO under magnetic fields normal to the plane. (b) Dependence of θ_{AHE} on Ni content x. Reprinted from [9], with the permission of AIP Publishing.

and finally disappears at 75 K. In $Mn_{4-x}Ni_xN$ films, the M_S increases with decreasing temperature for samples at x = 0 and 0.1, but decreases at x = 0.25. According to reference [11], Mekata found similar temperature dependences of magnetization in In- and Sn-doped Mn_4N [11]. Therefore, the temperature dependences of magnetization for the two Mn sublattice in $Mn_{4-x}Ni_xN$ differ from those in Mn_4N. This happens because the spin structure of Mn atoms might change depending on the Ni composition x in $Mn_{4-x}Ni_xN$.

(2) Magneto-transport properties

Figure 5.6(a) shows the AHE loops of the $Mn_{4-x}Ni_xN$ thin films on STO(001) when a magnetic field is applied perpendicular to the plane. The transverse voltage (V_y) is expressed as,

$$V_y = \left(R_H\frac{B_Z}{t} + \frac{\rho_{AHE}}{t}\right)I_x = \frac{\rho_{xy}}{t}I_x. \tag{5.1}$$

In contrast from equation (3.4), the terms corresponding to the offset voltage and the planar Hall effect are excluded for simplicity. The slope in the high magnetic field region, where the magnetization saturates, corresponds to the first term in equation (5.1). Thus, we can subtract the contribution of this first term from the measured slope. It is remarkable to see the ratio of remnant magnetization (M_r) to M_S, M_r/M_S, being almost 1 for x = 0, 0.1, and 0.25 for $Mn_{4-x}Ni_xN$ films on STO(001). In contrast, $Mn_{3.5}Ni_{0.5}N$ films show a slow magnetization reversal and require a greater magnetic field for magnetization reversal. This behavior is attributed to the degraded crystallinity of the $Mn_{3.5}Ni_{0.5}N$ films. Furthermore, the growth of $Mn_{4-x}Ni_xN$ films becomes more difficult with increasing the Ni content x because the films tend to significantly decompose into Mn-Ni alloys or Ni_8N [12].

For the sign of ρ_{xy} in figure 5.6(a), ρ_{xy} becomes negative in the positive magnetic field region for $Mn_{4-x}Ni_xN$ films at x = 0 and 0.1. However, the sign of ρ_{xy} reverses between x = 0.1 and 0.25. Figure 5.6(b) shows the dependence of anomalous Hall angle given by $\theta_{AHE} = \rho_{AHE}/\rho_{xx}$ on Ni composition x. Here, ρ_{xx} is the longitudinal resistivity of the $Mn_{4-x}Ni_xN$ films. The abrupt sign reversal of θ_{AHE} from negative at $x = 0.1$ to positive at $x = 0.25$ is associated with magnetic compensation. At the MCP, the direction of the magnetic moment of each sublattice flips with respect to the external magnetic field. Below the MCP ($x = 0$ and 0.1), the magnetic moment of Mn atoms at I sites determines the direction of magnetization. However, above the MCP ($x = 0.25$), the magnetic moment of Mn atoms at II sites determines the direction of magnetization. Thus, such a sign change is commonly used to identify the compensation point for rare-earth ferrimagnets [13, 14]. We should note that the sign reversal of ρ_{AHE} is commonly found in alloys of various compositions. Isogami *et al* investigated the magneto-transport properties of $Mn_{4-x}Fe_xN$ thin films [15] and found that the sign of ρ_{AHE} flips around $x = 1$ and ρ_{AHE} gradually changes with x at 300 K. However, this is not due to the compensation but due to a shift in the E_F and a resultant change in the exchange coupling splitting energy due to Fe doping [16]. The sign reversal of ρ_{AHE} is also reported in $Mn_{4-x}Dy_xN$ films at $x = 0.08$ and 0.2 [17]. This result is explained by the changes of the band structure and the position of E_F. In the $Mn_{4-x}Ni_xN$ system, the sign reversal of ρ_{AHE} with varying Ni content is quite sharp and is different from the smooth change observed in the $Mn_{4-x}Fe_xN$ thin films [15]. Therefore, this reversal is expected to be caused by the crossing of MCP.

(3) XAS and XMCD properties

Figures 5.7(a) and (b) are XAS and XMCD spectra at the Ni $L_{2,3}$ absorption edge for the $Mn_{3.9}Ni_{0.1}N$ and $Mn_{3.75}Ni_{0.25}N$ films, respectively. We compare these spectra with those reported for Ni_3FeN in figure 5.7(c) and $NiFe_3N$ in figure 5.7(d) [18]. For the XAS spectra of the $Mn_{3.9}Ni_{0.1}N$ and $Mn_{3.75}Ni_{0.25}N$ films, the L_3 edge (~852 eV), and L_2 edge (~870 eV), only sharp main peaks are observed. However, shoulders are observed in the XAS spectrum of the Ni_3FeN film shown by arrows in figure 5.7(c). These shoulders are observed at about 2 eV above the Ni $L_{2,3}$ main peak. Such shoulders have been reported for ferromagnetic nitrides such as Fe_4N and Co_4N films above the Fe(Co) $L_{2,3}$ main peak [19, 20]. As discussed in chapter 3, these shoulders are interpreted as originating from an electric dipole transition from the metal $2p$ inner-shell level to a hybrid state between the N $2p$ and metal $3d$ orbitals at the II site, indicating that the target element is at the II site [19]. However, we notice that these shoulders are missing in the XAS spectrum of the $NiFe_3N$ film in figure 5.7(d). In $Ni_xFe_{4-x}N$, the preferential sites of Ni atoms determined by XAS measurements are in good agreement with those obtained by Mössbauer measurements and ab-initio calculations [18, 21, 22]. Therefore, there is reasonable confidence in the results obtained for $Mn_{4-x}Ni_xN$ as well. On the basis of these results, we reach a conclusion that Ni atoms dominantly occupy the I site of Mn_4N when $x \leqslant 0.25$. Figures 5.7(a)–5.7(d) show other apparent satellites, labeled A and A', in addition to the shoulders mentioned above. Let us now discuss their origin. Satellites A and A' are seen at about 7 eV above the L_3 and L_2 edges of Ni. We initially thought that these originated from a hybrid state of Ni $3d$ and O $2p$ orbitals. However, they are reported to be only 1.5 eV above the Ni $L_{2,3}$ edge [23].

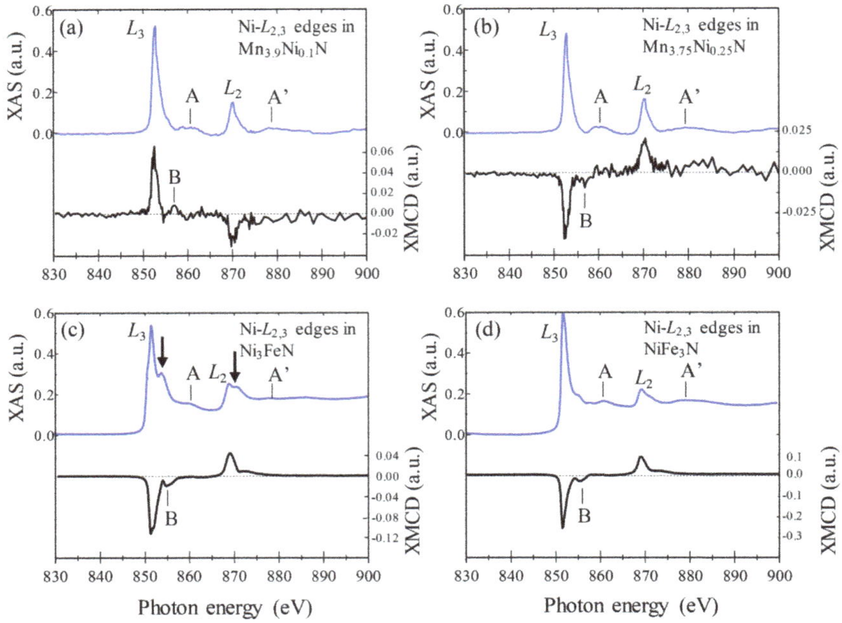

Figure 5.7. XAS (blue) and XMCD (black) spectra at Ni-$L_{2,3}$ edges of $Mn_{4-x}Ni_xN$ films on STO for (a) $x = 0.1$ and (b) $x = 0.25$. XAS spectra at Ni-$L_{2,3}$ edges for (c) Ni_3FeN films and (d) $NiFe_3N$ films. Reprinted from [3], with the permission of AIP Publishing

Therefore, these values are not consistent with the results of this study. They are quite close to the positions of the satellites reported for the Ni $L_{2,3}$ edge of $Ni_xFe_{4-x}N$. Such satellites have also been reported for pure fcc-Ni both experimentally and in one-electron relativistic tight-binding calculations [24, 25].

The XMCD spectrum of Ni is simple: the signs of the peaks are opposite at the L_2 and L_3 edges for $Mn_{3.9}Ni_{0.1}N$ and $Mn_{3.75}Ni_{0.25}N$. Thus, the orientation of the magnetic moment of Ni(I) in the $Mn_{3.9}Ni_{0.1}N$ films is different from that in the $Mn_{3.75}Ni_{0.25}N$ films. For satellite B in figures 5.7(a)–5.7(d), this has only been found experimentally and has not been reproduced by calculation [18]. This can be understood by considering that satellite B arises from the interaction between the multiplet of final states described by $2p^53d^9$ and $2p^53d^{10}$. No clear satellite is observed in the XMCD spectrum above the L_2 edge of Ni, but it has been confirmed experimentally elsewhere [19]. We conclude that the satellites are buried in the background signal due to the small atomic ratio of Ni.

Next, XAS and XMCD spectra of the Mn $L_{2,3}$ absorption edges of $Mn_{3.9}Ni_{0.1}N$ (figure 5.8(a)) and $Mn_{3.75}Ni_{0.25}N$ (figure 5.8(b)) films are shown. It is noted that similar features are observed in the XAS spectra of Mn-related oxides such as $MnFe_2O_4$ [26], which are characterized by a multiplicity of final states from a highly localized Mn^{2+} ground state to a $2p^53d^6$ configuration.

The XMCD spectra of these thin films also show complexity: in $Mn_{3.9}Ni_{0.1}N$, the signs of the XMCD signals near the Mn L_3 edge are positive, negative, and positive

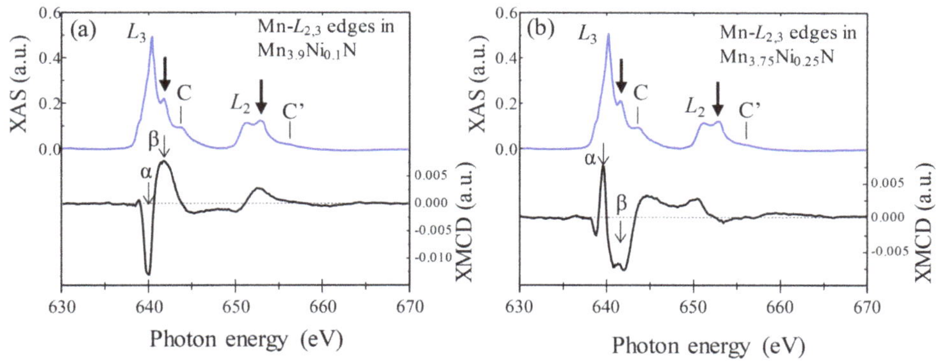

Figure 5.8. XAS (blue) and XMCD (black) spectra at the Mn-$L_{2,3}$ absorption edges for $Mn_{4-x}Ni_xN$ films on STO for (a) $x = 0.1$ and (b) $x = 0.25$. Reprinted from [3], with the permission of AIP Publishing

from low to high photon energy. Such a complicated spectrum implies a superposition of signals of opposite signs. This happens because the magnetic moments of Mn(I) and Mn(II) are antiparallel. In our previous work on Fe_4N thin films, the transition metal at the I(II) site has been studied using the all-electron full-potential linearized enhanced plane wave (FLAPW) method and the E1 transition [19], which is a localized (itinerant) state supported by ab-initio calculations using Fermi's golden rule. Similar results have been obtained for Mn_4N by the FLAPW method [27]. Therefore, the XMCD spectra of Mn $L_{2,3}$ edges for the $Mn_{3.9}Ni_{0.1}N$ and $Mn_{3.75}Ni_{0.25}N$ films can be recognized as the superposition of localized and itinerant components characterized by the opposite sign of features α and β, respectively. In $Mn_{3.9}Ni_{0.1}N$, the broad positive peak (β) near 642 eV is likely to originate from the Mn(II) atoms because of the hybridization between the 3*d* orbital of Mn(II) and the 2*p* orbital of N. In contrast, the sharp negative peak (α) around 640 eV is more likely to originate from Mn(I), away from N, and therefore less hybridization. The sign of the XMCD signal near the Mn L_3 absorption edge of $Mn_{3.75}Ni_{0.25}N$ in figure 5.8(b) is negative, positive, and negative from low to high photon energy. It is noted that this is opposite to the results obtained for the $Mn_{3.9}Ni_{0.1}N$ film in figure 5.8(a). We ascribe this result to a magnetization reversal of Mn atoms at both I and II sites. Such spectral behavior has been reported for ferrimagnets near the compensation temperature [28, 29]. The XMCD signal at the Ni L_3 absorption edge has the opposite sign to the sharp peak α at the Mn L_3 absorption edge regardless of x and the same sign as the broad spectrum β. The former is mainly attributed to Mn(I) atoms, while the latter is attributed to Mn(II) atoms. Thus, the magnetic moment of Ni(I) is always parallel to the magnetic moment of Mn(II) or antiparallel to the magnetic moment of Mn(I) in the range of x studied.

From these considerations, we conclude that the site preferentially occupied by Ni in the $Mn_{4-x}Ni_xN$ thin film is the I site and that the magnetic moment of the Ni atom is parallel to the magnetic moment of Mn(II). Based on these conclusions, figure 5.9 shows a simplified schematic diagram of the magnetic structure in $Mn_{4-x}Ni_xN$ ((a) below and (b) above the magnetic compensation composition).

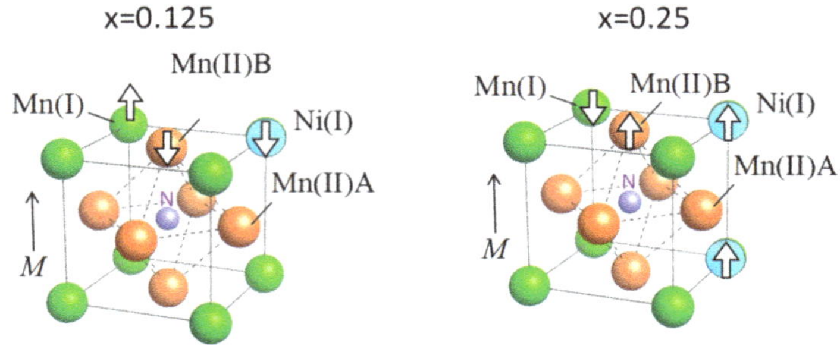

Figure 5.9. Magnetic structures in $Mn_{4-x}Ni_xN$ for $x = 0.125$ (below MCP) and $x = 0.25$ (above MCP) with respect to magnetization. The magnetic moment of Mn(I) is aligned antiparallel to those of Ni(I) and Mn(II).

Here, all Ni atoms are assumed to occupy the I site, which seems reasonable from figure 5.7 and the discussion above. Samples with $x = 0.1$ and 0.25 are used in the experiment, but the composition ratios in figures 5.9(a) and (b) are set to 0.125 and 0.25, respectively, to simplify the expressions. Mn_4N is ferrimagnetic, and the magnetic moment of Mn(I) atoms are aligned parallel to the magnetic field and antiparallel to the magnetic moment of Mn(II) atoms. Therefore, the magnetic moment of the Mn(I) atoms determines the direction of magnetization. At $x = 0.1$, some of the Mn(I) atoms are occupied by Ni atoms, and the magnetic moment of the Ni(I) atoms aligns parallel to the magnetic moment of Mn(II), resulting in a decrease of magnetization [12]. The Mn(I) atom still determines the magnetization direction. As x is further increased to 0.25, more Mn(I) atoms are replaced by Ni atoms, and the sum of magnetic moments antiparallel to the magnetic field becomes larger than the sum of magnetic moments parallel to the magnetic field. In this state, the net magnetization and the magnetic field become antiparallel, resulting in an unstable Zeeman energy state. Therefore, to reach the stable state shown in figure 5.9(b), all magnetic moments must be inverted. Consequently, the magnetic moment of the Mn(II) site determines the direction of magnetization at $x = 0.25$. Therefore, we can state that magnetic compensation occurs at values of x between 0.1 and 0.25.

5.3 $Mn_{4-x}Co_xN$ epitaxial films

5.3.1 MBE growth of $Mn_{4-x}Co_xN$ films on $SrTiO_3(001)$ by MBE

We next move on the properties of Co-doped Mn_4N epitaxial films on STO. Co-doped Mn_4N thin films of about 20 nm thickness are epitaxially grown on STO(001) substrates by MBE at 450°C. The composition ratios of Mn and Co are calculated using the deposition rate ratios from the crucible of each K-cell. The composition ratio is controlled by the crucible temperature of Co and Mn K-cells, and are controlled to be in the range 0–4. The nitrogen plasma is provided simultaneously using an RF plasma gun. RHEED images are observed along the STO[100] azimuth to evaluate the crystallinity of the grown films. Finally, a capping layer (Pt or SiO_2) is formed by sputtering to prevent oxidation of sample surface. Out-of-plane (θ–2θ)

XRD patterns are obtained and the crystallinity of the grown films is evaluated using a Cu Kα radiation source. Ge(220) single crystals are used to make x-rays monochromatic. The thickness of the grown film is measured by x-ray reflectivity. Magnetization curves are measured by VSM.

Figure 5.10(a) shows the out-of-plane XRD patterns of $Mn_{4-x}Co_xN$ ($x = 0$, 0.4, 0.8, 1.0, 1.2, 2.0, 3.0, 3.6, 3.9, 4) films on STO(001) and RHEED patterns taken before SiO_2 capping layers are deposited. $Mn_{4-x}Co_xN$ films with x up to 1.0, clear reflections of 001, 002, and 004 are observed. For $Mn_{4-x}Co_xN$ films with $x \leqslant 0.8$, superlattice diffraction (indicated by white arrows) due to body-centered N atoms is evident. These results indicate that c-axis oriented $Mn_{4-x}Co_xN$ films are epitaxially grown. It is noted, however, the RHEED streaks become blurred at $x = 1.2$–3.6. As x is further increased to 3.9, the RHEED streaks are again observed. At $x = 4.0$, superlattice diffraction is also observed along with clear 002 reflections. From these results, we can state that the crystallinity of $Mn_{4-x}Co_xN$ epitaxial films degrades as

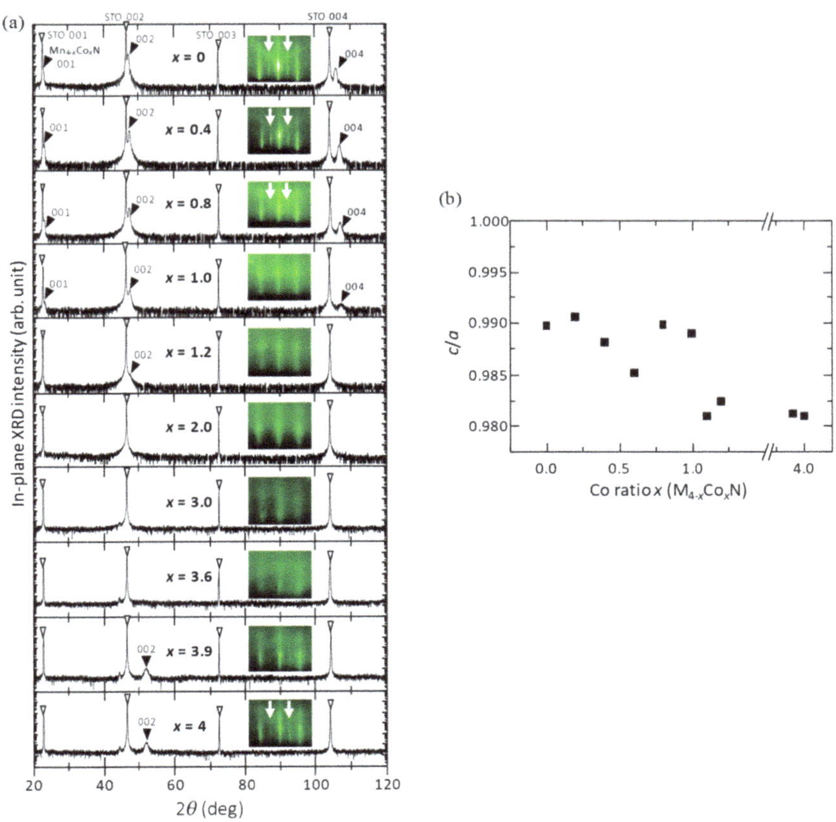

Figure 5.10. (a) Out-of-plane XRD patterns of $Mn_{4-x}Co_xN$ films, and RHEED patterns taken along the $SrTiO_3$[100] azimuth. The white arrows show the positions of the superlattice diffraction. (b) Dependence of the c/a of $Mn_{4-x}Co_xN$ films on Co composition x. Reprinted from [30], Copyright (2022), with permission from Elsevier.

x moves away from 0 or 4. Similar results have been obtained for the epitaxial growth of Ni-doped Mn_4N films [12]. The c/a ratio of the grown films shown in figure 5.10(b) indicates that the grown film is subject to tensile strain regardless of x [30].

5.3.2 Characterization of $Mn_{4-x}Co_xN$ films

(1) Magnetic properties

Figures 5.11(a)–(h) show the magnetization curves of $Mn_{4-x}Co_xN$ films ($x = 0$, 0.4, 0.6, 0.8, 0.9, 1.0, 1.1, 1.2) measured by applying a magnetic field perpendicular to the plane. The introduction of Co in Mn_4N causes the saturation magnetization (M_S) to decrease. Figure 5.11(i) summarizes the Co composition x dependence of saturation magnetization. The M_S decreases with increasing x, reaching a minimum when $x = 1.0$ and increasing rapidly thereafter. These results suggest the presence of MCP at $x \approx 1.0$. It should be noted, however, that several factors may influence the M_S summarized in figure 5.11(i). One is the occurrence of an inverse phase boundary (APB) in $Mn_{4-x}Co_xN$ epitaxial films. According to the literature [31, 32], APBs in Fe_3O_4 epitaxial films, confirmed by transmission electron microscopy, induce very

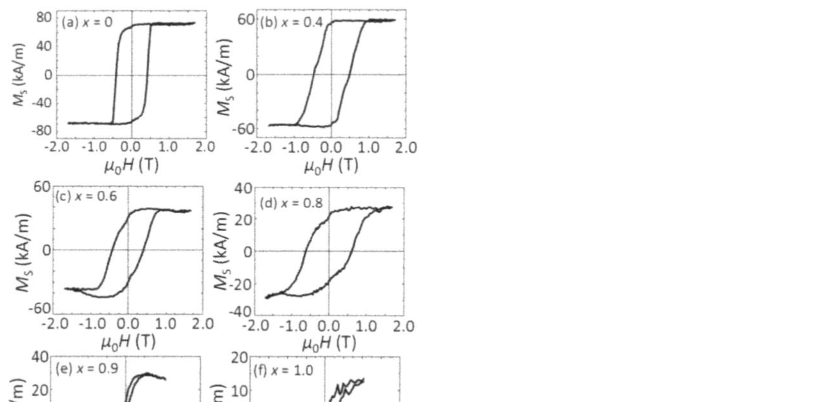

Figure 5.11. (a) M–H curves of $Mn_{4-x}Co_xN$ films under magnetic fields normal to the sample plane. Reproduced with permission from reference [33]. Reprinted figure with permission from [33], Copyright (2020) by the American Physical Society. (b) Dependence of the saturation magnetization on the Co composition ratio x in $Mn_{4-x}Co_xN$. Reprinted from [30], Copyright (2023), with permission from Elsevier.

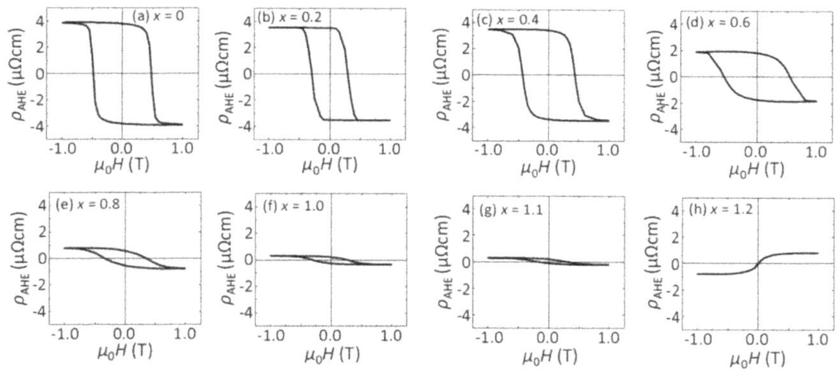

Figure 5.12. AHE loops of $M_{n4-x}Co_xN$ films on STO under magnetic fields normal to the plane. Reprinted figure with permission from [33], Copyright (2020) by the American Physical Society.

large saturation fields, which is significantly different from the saturation fields expected for Fe_3O_4 single crystals with bulk behavior. Another is the presence of other phases, such as Mn oxides. To discuss these issues in more detail, anomalous Hall resistivity (ρ_{AHE}) curves are measured for $Mn_{4-x}Co_xN$ films as shown in figures 5.12(a)–(f) [33].

(2) Magneto-transport properties

AHE measurements are performed by PPMS (Quantum Design) under magnetic fields applied normal to the sample surface.

Figure 5.12 shows the out-of-plane AHE loops of $Mn_{4-x}Co_xN$ thin films with different Co concentrations ($x = 0$, 0.2, 0.4, 0.6, 0.8, 1.0, 1.1, 1.2) [33]. The shape of these hysteresis curves is more distinct than the magnetization curves. This effect can be attributed to the fact that this is a magneto-transport measurement and the effect of STO substate and Mn oxides are not included in such magneto-transport properties of grown films. It is also important to note that the AHE hysteresis loops saturate below 1T. From this observation, as in the case of large APBs interactions, we can infer that the role of APBs is minimal. A negative to positive sign reversal of AHE between $x = 1.1$ and 1.2 is observed, which implies the magnetic compensation point crossing observed for $Mn_{4-x}Ni_xN$. Thus, with the help of the AHE and the magnetization hysteresis loops, it is reasonable to consider that the MCP is somewhere between $x = 1.1$ and 1.2.

(3) XAS and XMCD properties

XAS and XMCD measurements are performed at the twin APPLE-II undulator beamline BL-16A at the KEK Photon Factory in Japan [4]. For more details, please visit the description about the same measurement for $Mn_{4-x}Ni_xN$ films in this chapter. The energy of the soft x-rays is adjusted so that the L_3 and L_2 absorption edges of the Mn and Co elements are included. The total electron yield (TEY) method is employed. All measurements are performed at room temperature.

Figures 5.13(a)–5.13(f) show the XAS and XMCD spectra at the Mn $L_{2,3}$ absorption edge of the $Mn_{4-x}Co_xN$ film, and figures 5.13(g)–5.13(k) show those at the Co $L_{2,3}$ absorption edge [12]. Since the shape of the XAS spectra does not change substantially regardless of x, we focus on the XMCD spectra here. In figure 5.13(a), the XMCD spectrum of Mn_4N has a sharp downward peak α at 640 eV and an upward broad peak β at the L_3 absorption edge at 642 eV. As described earlier, the interatomic distance between Mn(I) and N atoms is longer than that between Mn(II) and N atoms. Therefore, the sharp peak α is mainly derived from more isolated Mn(I) atoms at the corner sites, while the more itinerant electronic states of the Mn(II) atoms at the face-centered sites contribute to the broad peak β, which is due to the hybrid state formed by the N 2*p* state and the Mn 3*d* orbital at the II site [4]. The different signs of the peaks α and β indicate that the magnetic moment of Mn(I), $\mu_{Mn(I)}$, is antiparallel to the magnetic moment of Mn(II), $\mu_{Mn(II)}$. The signs of these peaks reverse twice, between $x = 0.2$ and 0.8, and between $x = 0.8$ and 1.3 [33], indicating the presence of two MCPs. For the XMCD spectra at the Co $L_{2,3}$ absorption edge, the Co atoms preferentially occupy the I site, Co(I) atoms [26]. The different sign of peak α in figure 5.13(b) and peak A in figure 5.13(g) means that the Co(I) magnetic moment $\mu_{Co(I)}$ is antiparallel to $\mu_{Mn(I)}$ at $x = 0.2$. As x increases to 0.8, figure 5.13(h) shows that the direction of $\mu_{Co(I)}$ reverses, and the sum of $\mu_{Mn(II)}$ and $\mu_{Co(I)}$ becomes larger than $\mu_{Mn(I)}$, thereby

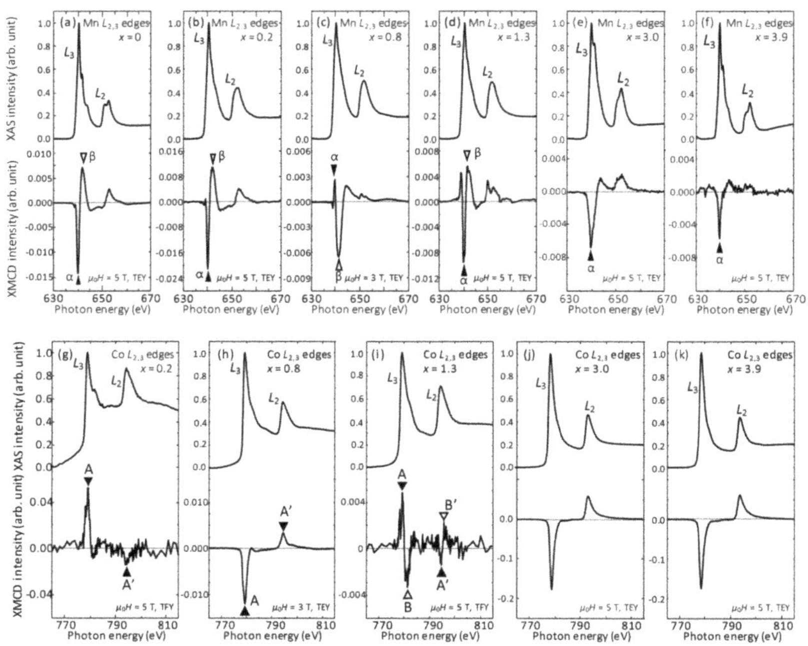

Figure 5.13. XAS and XMCD spectra at the (a)-(f) Mn $L_{2,3}$ absorption edges and (g)-(k) Co $L_{2,3}$ absorption edges for $Mn_{4-x}Co_xN$ films. Reproduced with permission from reference. Reprinted from [30], Copyright (2022), with permission from Elsevier.

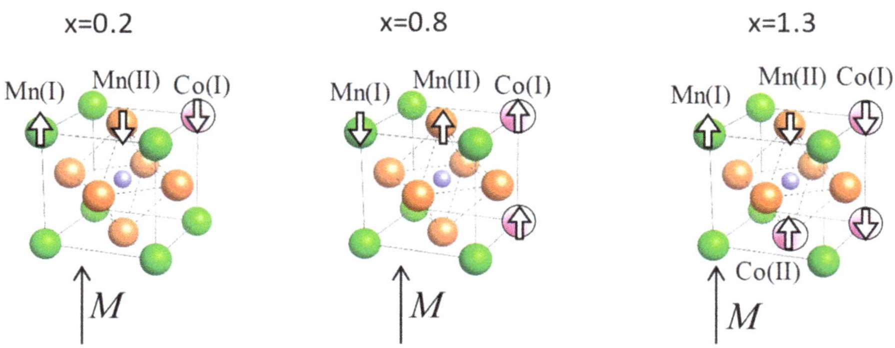

Figure 5.14. Magnetic structures of Mn(I), Mn(II), Co(I), and Co(II) with respect to magnetization.

making $\mu_{\mathrm{Mn(II)}}$ parallel to the magnetization. As x further increases to 1.3, some of the Co atoms begin to occupy the II site, Co(II), which means that the magnetic moment $\mu_{\mathrm{Co(II)}}$ parallel to $\mu_{\mathrm{Mn(I)}}$. Therefore, the spectrum in figure 5.11(i) can be interpreted as the superposition of two components with different signs (A-A' and B-B'). The peak position of B(B') is slightly shifted from that of A(A'), similar to peaks α and β in figure 5.11(a). Therefore, peaks B and B' in figure 5.11(i) are interpreted as originating from Co(II) atoms with $\mu_{\mathrm{Co(II)}}$ aligned antiparallel to $\mu_{\mathrm{Co(I)}}$ [33]. At $x = 1.3$, the sum of $\mu_{\mathrm{Mn(I)}}$ and $\mu_{\mathrm{Co(II)}}$ exceeds the sum of $\mu_{\mathrm{Mn(II)}}$ and $\mu_{\mathrm{Co(I)}}$. Thus, the second reversal of the magnetic moment occurs between $x = 0.8$ and 1.3. As x increases further to 3.0 and 3.9, the direction of $\mu_{\mathrm{Co(I)}}$ reverses at some x between $x = 1.3$ and 3.0, and $\mu_{\mathrm{Co(I)}}$ becomes parallel to $\mu_{\mathrm{Co(II)}}$. As for the Mn $L_{2,3}$ absorption edges, only one sharp downward peak α at 640 eV is observed in the $Mn_{4-x}Co_xN$ films at $x = 3.0$ and 3.9 in figures 5.13(e) and (f), indicating that the Mn atoms in these films are located mainly at the I site rather than the II site. These results show that $\mu_{\mathrm{Mn(I)}}$ is parallel to both $\mu_{\mathrm{Co(I)}}$ and $\mu_{\mathrm{Co(II)}}$ for $x \geqslant 3.0$ in the $Mn_{4-x}Co_xN$ films. Based on these results, the magnetic structure of $Mn_{4-x}Co_xN$ can be viewed as follows: as x increases (>1.3), the Co atoms occupy more of the II sites, the ferromagnetic interaction between the Co(I) and Co(II) atoms causes the direction of $\mu_{\mathrm{Co(I)}}$ to reverse at some x and $\mu_{\mathrm{Co(I)}}$ becomes parallel to $\mu_{\mathrm{Co(II)}}$ and $\mu_{\mathrm{Mn(I)}}$, and magnetization increases as x increases. Therefore, we can say that there is no magnetic compensation composition point in $Mn_{4-x}Co_xN$ at $x > 1.3$. Figure 5.14 summarizes the magnetic structure of $Mn_{4-x}Co_xN$.

5.4 Summary

In this chapter, 10 nm thick $Mn_{4-x}Ni_xN$ thin films and 30 nm thick $Mn_{4-x}Co_xN$ films were epitaxially grown on STO(001) substrates at 450°C by MBE. The out-of-plane and in-plane XRD profiles and superlattice RHEED streaks confirm very good epitaxial growth of both films. For $Mn_{4-x}Ni_xN$, the saturation magnetization decreases with increasing Ni concentration. The sign of the AHE changes between $x = 0.1$ and 0.25, confirming that the magnetic compensation point lies between these points. The XMCD spectrum of Mn in $Mn_{3.9}Ni_{0.1}N$ shows that the opposite sign between peaks α (sharp, from site I) and β (broad, from site II) suggests the antiparallel alignment of magnetic moment between Mn(I) and Mn(II) atoms. The sign of the peaks in the XMCD spectrum of Ni confirms that the direction of the magnetic moment of Ni is oriented toward Mn(II). In $Mn_{3.75}Ni_{0.25}N$, the signs of the α and β peaks are reversed along with the sign of the Ni peak, confirming that the presence of MCP between $x = 0.1$ and 0.25 and the net magnetization is oriented toward Mn(II) at $x = 0.25$.

For $Mn_{4-x}Co_xN$, the saturation magnetization essentially decreases with increasing Co concentration. This is followed by a sharp increase near $x = 1.1$. The sign of the AHE reverses from negative to positive after $x = 1.1$, confirming the presence of MCP around this concentration. XAS and XMCD measurements are performed on $Mn_{4-x}Co_xN$ thin films at Co concentrations $x = 0.2$, 0.8, and 1.3. Between $x = 0.2$ and 0.8, the sign reversal is observed for the α and β peaks in the Mn XMCD spectra, as well as the A and A' peaks in the Co XMCD spectra. These sign reversals confirm the presence of MCP between $x = 0.2$ and 0.8. In the Mn XMCD spectrum, a sign reversal of the α and β peaks is also observed between $x = 0.8$ and 1.3. In the case of the Co XMCD spectrum, a combination of A-A' and B-B' peaks is observed. The B-B' peak indicates that Co is also present at Mn site II. Thus, at this concentration, Co atoms also occupy site II. At $x = 1.3$, the reversal of the Co A-A' peak, along with the reversal of the Mn α and β peaks, supports the presence of another MCP between $x = 0.8$ and 1.3. This is also seen in the magnetization loops and the sign change from negative to positive in the AHE loop.

Bibliography

[1] Juza R, Deneke K and Puff H 1959 *Z. Elektrochem.* **63** 551

[2] Zhang R, He Y, Fruchart D, Coey J M D and Cercsi Z 2022 *Act. Mater.* **234** 118021

[3] Komori T, Hirose T, Gushi T, Toko K, Hanashima T, Vila L, Attané J P, Amemiya K and Suemasu T 2020 *J. Appl. Phys.* **127** 043903

[4] Amemiya K *et al* 2013 *J. Phys.: Conf. Ser.* **425** 152015

[5] Soldatov I V and Schäfer R 2017 *J. Appl. Phys.* **122** 153906

[6] Svalov A V, Fernandez A, Vas'kovskiy V O, Kurlyandskaya G V, Banrandiarán J M, Anton R L and Tejedor M 2006 *J. Alloys Compd.* **419** 25

[7] Li F, Yang J, Xue D and Zhou R 1995 *Appl. Phys. Lett.* **66** 2343

[8] Gushi T *et al* 2019 *Nano Lett.* **19** 8716

[9] Komori T, Gushi T, Anzai A, Vila L, Attané J-P, Pizzini S, Vogel J, Isogami S, Toko K and Suemasu T 2019 *J. Appl. Phys.* **125** 213902

[10] Zhu Z, Fong X and Liang G 2018 *J. Appl. Phys.* **124** 193901

[11] Mekata M 1962 *J. Phys. Soc. Jpn.* **17** 5

[12] Komori T, Anzai A, Gushi T, Toko K and Suemasu T 2019 *J. Cryst. Growth* **507** 163

[13] Kim K-J *et al* 2017 *Nat. Mater.* **16** 1187

[14] Hirata Y *et al* 2018 *Phys. Rev.* B **97** 220403(R)

[15] Isogami S, Anzai A, Gushi T, Komori T and Suemasu T 2018 *Jpn. J. Appl. Phys.* **57** 120305

[16] Chen L 2006 *J. Appl. Phys.* **100** 113717

[17] Meng M, Wu S X, Zhou W Q, Ren L Z, Wang Y J, Wang G L and Li S W 2015 *J. Appl. Phys.* **118** 053911
[18] Takata F, Ito K, Takeda Y, Saitoh Y, Takanashi K, Kimura A and Suemasu T 2018 *Phys. Rev. Mater.* **2** 024407
[19] Ito K, Toko K, Takeda Y, Saitoh Y, Oguchi T, Suemasu T and Kimura A 2015 *J. Appl. Phys.* **117** 193906
[20] Ito K, Sanai T, Yasutomi Y, Zhu S, Toko K, Takeda Y, Saitoh Y, Kimura A and Suemasu T 2014 *J. Appl. Phys.* **115** 17C712
[21] Monachesi T, Björkman T, Gasche and Eriksson O 2013 *Phys. Rev.* B **88** 054420
[22] Alders D, Voogt F C, Hibma T and Sawatzky G A 1996 *Phys. Rev.* B **54** 7716
[23] Chen C T, Sette F, Ma Y and Modesti S 1990 *Phys. Rev.* B **42** 7262
[24] Amemiya K, Sakai E, Matsumura D, Abe H and Ohta T 2005 *Phys. Rev.* B **72** 201404(R)
[25] Smith N V, Chen C T, Sette F and Mattheiss L F 1992 *Phys. Rev.* B **46** 1023
[26] Ito K *et al* 2020 *Phys. Rev.* B **101** 104401
[27] Goering E, Gold S and Schütz G 2001 *J. Synchrotron Rad.* **8** 422
[28] Maruyama H and Nakamura N 2004 *J. Electron. Spectrosc. Relat. Phenom.* **136** 135
[29] Thole B T, Carra P, Sette F and van der Laan G 1992 *Phys. Rev. Lett.* **68** 1943
[30] Komori T, Horiuchi T, Mitarai H, Yasuda T, Amemiya K and Suemasu T 2022 *J. Mag. Mag. Mater.* **564** 170050
[31] Margulies D T, Parker F T, Rudee M K, Spada F E, Chapman J N, Aichison P R and Berkowitz A E 1997 *Phys. Rev. Lett.* **79** 5162
[32] Hibma T, Voogt F C, Niesen L, van der Heijden P A A, De Jonge W J M, Donkers J J T M and van der Zaag P J 1999 *J. Appl. Phys.* **85** 5291
[33] Mitarai H *et al* 2020 *Phys. Rev. Mater.* **4** 094401

Chapter 6

Growth and characterization of nonmagnetic element doped Mn_4N epitaxial films

Here we look at four examples of nonmagnetic element doped Mn_4N epitaxial films: Cr-, In-, Au-, and Sn-doped Mn_4N epitaxial films grown by MBE.

6.1 $Mn_{4-x}Cr_xN$ epitaxial films

6.1.1 MBE growth of $Mn_{4-x}Cr_xN$ films on $SrTiO_3(001)$ by MBE

Approximately 20 nm thick Cr-doped Mn_4N thin films are formed on STO(001) by MBE at 450°C. The Cr composition x is set to 0–1.5 by controlling the crucible temperature of each K-cell. After the growth, a capping layer is formed by sputtering to prevent oxidation of sample surface.

Figure 6.1(a) shows the out-of-plane XRD patterns [1]. RHEED images are also shown. Clear 001, 002, and 004 reflections are observed at $x = 0.4$ and 0.8, indicating that c-axis oriented $Mn_{4-x}Cr_xN$ thin films are grown epitaxially. However, as x is further increased to 1.2, the RHEED streaks become blurred, suggesting that the crystallinity has degraded. The c/a ratio is also smaller than 1, as shown in figure 6.1(b), which is the same as that of the $Mn_{4-x}Co_xN$ epitaxial film.

6.1.2 Characterization of $Mn_{4-x}Cr_xN$ films

(1) Magnetic properties

The Cr composition x dependence of the M_S measured by VSM at room temperature is shown in figure 6.2. The M_S of $Mn_{4-x}Cr_xN$ films decreases monotonously with increasing x [1].

(2) Magneto-transport properties

Figure 6.3 shows the AHE curves measured on $Mn_{4-x}Cr_xN$ films [1]. No sign reversal is observed in the ρ_{AHE}. Considering the results on Ni-, Co-, and In-doped Mn_4N

doi:10.1088/978-0-7503-5477-6ch6

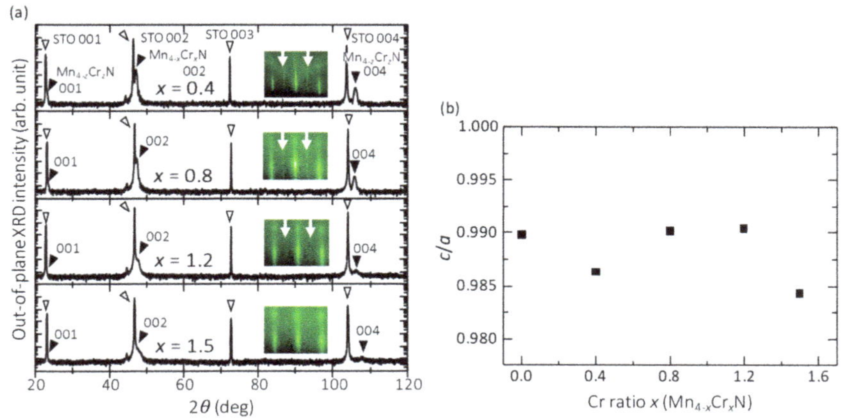

Figure 6.1. (a) Out-of-plane XRD patterns of $Mn_{4-x}Cr_xN$ films with RHEED streaks observed along the STO [100] azimuth. White arrows show the superlattice diffraction lines. (b) Dependence of c/a of $Mn_{4-x}Cr_xN$ films on x. Reprinted from [1], Copyright (2022), with permission from Elsevier.

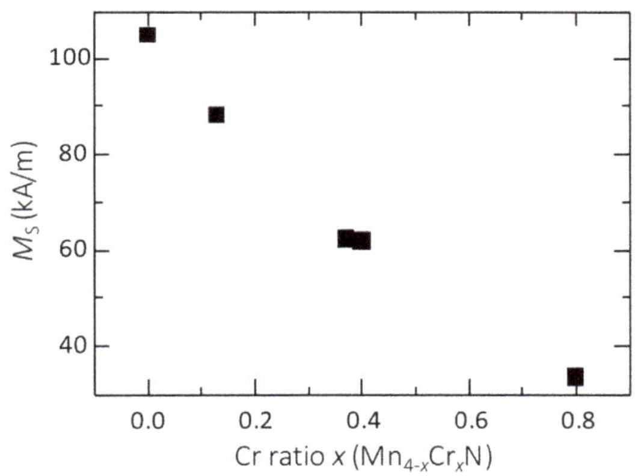

Figure 6.2. Dependence of saturation magnetization (M_S) on x in $Mn_{4-x}Cr_xN$ by VSM. Reprinted from [1], Copyright (2022), with permission from Elsevier.

described in the former chapters, the lack of sign change in the AHE curves implies that the direction of the magnetic moment of Mn(II) atoms does not flip in the range $x = 0$–1.4.

(3) XAS and XMCD properties

XAS and XMCD measurements are performed at the twin APPLE-II undulator beamline BL-16A at the KEK Photon Factory in Tsukuba [2]. Both a magnetic field of ±5 T and circularly polarized x-rays with left-hand or right-hand polarization are applied normal to the plane. The x-ray polarization is switched at every energy point using five kicker magnets with a frequency of 10 Hz. Spectra are acquired at room temperature in total electron yield mode.

Figures 6.4(a)–(c) and 6.4(d)–(f) show the XAS and XMCD spectra at the Mn $L_{2,3}$ and Cr $L_{2,3}$ absorption edges of $Mn_{4-x}Cr_xN$ films (x = 0.13, 0.37, 0.66),

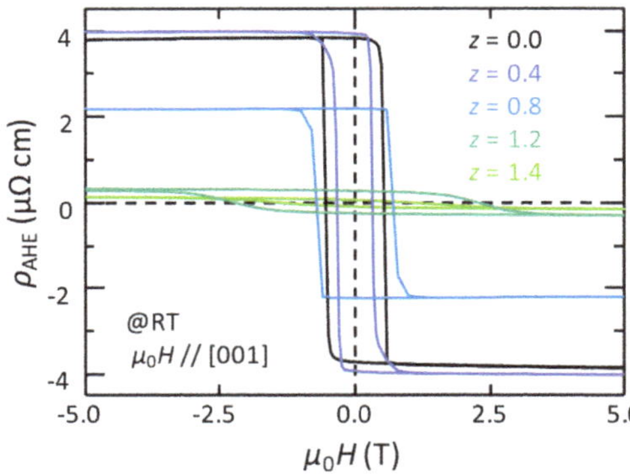

Figure 6.3. AHE loops measured on $Mn_{4-z}Cr_zN$ films. Reprinted from [1], Copyright (2022), with permission from Elsevier.

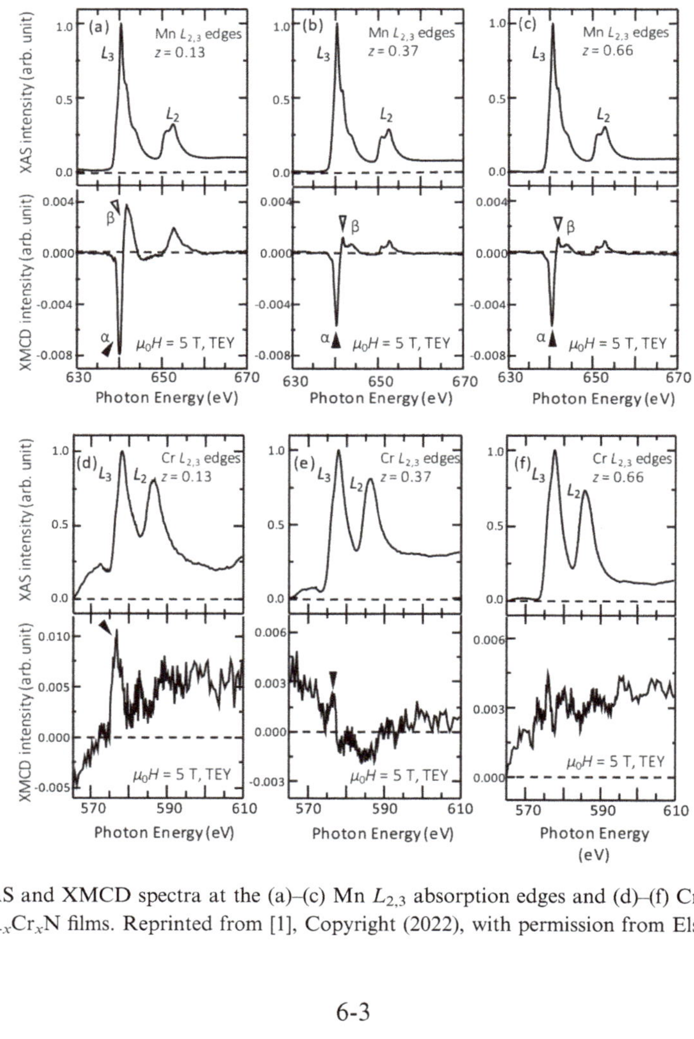

Figure 6.4. XAS and XMCD spectra at the (a)–(c) Mn $L_{2,3}$ absorption edges and (d)–(f) Cr $L_{2,3}$ absorption edges for $Mn_{4-x}Cr_xN$ films. Reprinted from [1], Copyright (2022), with permission from Elsevier.

respectively. At $x = 0.13$, the obtained XAS and XMCD spectra are similar to those of Mn_4N that are shown in Fig. 3.32. However, as x is increased, the intensity of both peaks α and β decreases [1], as shown in figures 6.4(b) and 6.4(c), indicating that the Cr atoms are considered to occupy both the I and II sites. However, the sign reversal is not observed at both α and β peaks in the x range measured. Regarding the direction of the Cr magnetic moment at $x = 0.13$, the sign of the peak at around 576 eV in the XMCD spectrum (black arrow) in figure 6.4(d) is different from that of Mn(I)-related peak α in figure 6.4(a). This result shows that the magnetic moment of Cr atoms, μ_{Cr}, is antiparallel to $\mu_{Mn(I)}$. However, when x is increased much further, such a distinct peak of Cr disappears, as shown in figure 6.4(f). This result means that the net magnetic moment of Cr has become negligible, which is probably caused by the presence of Cr atoms whose magnetic moment are both parallel and antiparallel to $\mu_{Mn(I)}$ regardless of their sites. From these results, magnetic compensation is not likely to occur in $Mn_{4-x}Cr_xN$.

6.2 $Mn_{4-x}In_xN$ epitaxial films

6.2.1 MBE growth of $Mn_{4-x}In_xN$ films on $SrTiO_3(001)$ by MBE

Approximately 25 nm thick $Mn_{4-x}In_xN$ (x = 0.15, 0.27, 0.41) films are formed on MgO(001) at 450°C by MBE, followed by a sputter-deposited 2 nm thick Pt capping layer. Since the addition of In is expected to expand the lattice constant of grown films, MgO is used as a substrate instead of STO. The atomic composition of the grown films is evaluated by Rutherford backscattering spectroscopy (RBS) using He^+ ions accelerated at 2.0 MeV.

The XRD patterns for the grown films are shown in figures 6.5(a) and (b) [3]. In the out-of-plane XRD patterns in figure 6.5(a), 002 and 004 reflections appear with RHEED streaks. This result shows that single-crystalline $Mn_{4-x}In_xN$ epitaxial films are formed. Superlattice diffraction lines in the RHEED images, indicated by arrows, indicates the presence of N atoms in the body center of each lattice. In

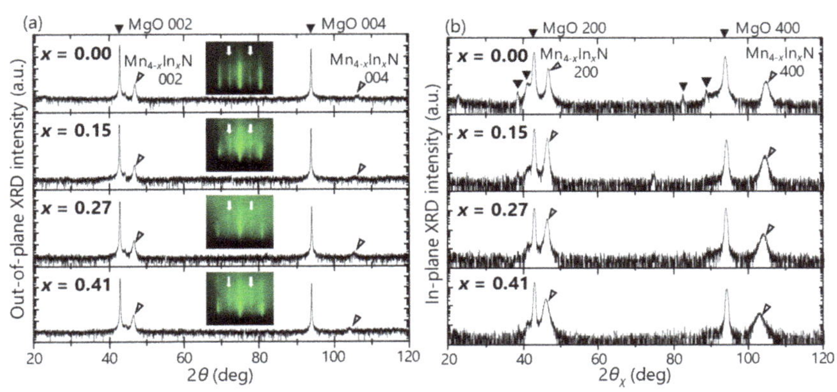

Figure 6.5. (a) Out-of-plane and (b) in-plane XRD patterns of $Mn_{4-x}In_xN$ films. RHEED streaks are observed along the MgO[100] azimuth. White arrows indicate the superlattice diffraction lines. Reproduced from [3].

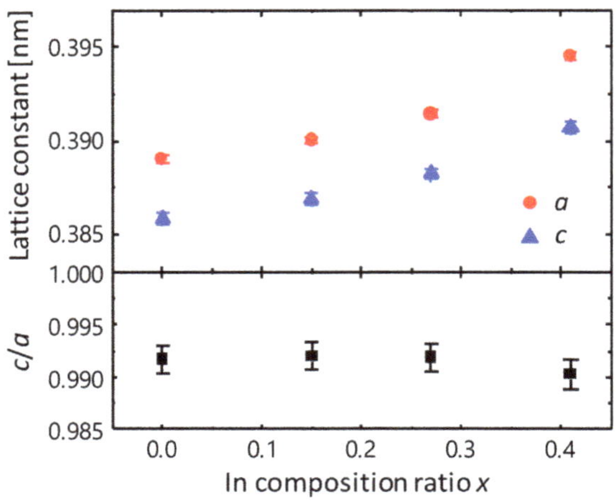

Figure 6.6. Dependence of c/a of $Mn_{4-x}In_xN$ films on x. Reproduced from [3].

figure 6.5(b), 200 and 400 reflections are observed. In both the XRD patterns, the peak position related to $Mn_{4-x}In_xN$ shifts to a lower angle as the In composition x increases. Figure 6.6 shows the In composition x dependent in-plane lattice constant (a), out-of-plane lattice constant (c), and c/a ratios. When x increases from 0 to 0.41, c increases from 0.3858 to 0.3907 nm, and a increases from 0.3890 nm to 0.3945 nm, while keeping the $c/a \sim 0.99$.

6.2.2 Characterization of $Mn_{4-x}In_xN$ films

(1) Magneto-transport properties

AHE measurements are performed at room temperature by PPMS (Quantum Design).

The AHE loops of $Mn_{4-x}In_xN$ films measured at room temperature are shown in figure 6.7. A magnetic field, $\mu_0 H$, is applied normal to the sample plane. The coercive force of $Mn_{4-x}In_xN$ layers becomes much higher than that of Mn_4N layers. Similar results are also reported in $Mn_{4-x}Ni_xN$ epitaxial films [4]. We should note that the sign of ρ_{AHE} reverses between $x = 0.15$ and 0.27. This kind of sign reversal of ρ_{AHE} in ferrimagnets is usually interpreted as the presence of the MCP [5, 6]. Similar sign reversal in ρ_{AHE} has been observed in $(Mn,Ni)_4N$ and $(Mn,Co)_4N$ [7, 8]. Therefore, the result shown in figure 6.7 implies that the direction of magnetic moment of Mn(II) atoms changes with respect to the magnetization between $x = 0.15$ and 0.27.

(2) XAS and XMCD properties

To confirms the reversal of the magnetic moment of Mn(II) atoms, XAS and XMCD spectra are measured at BL-16A of KEK Photon Factory in Tsukuba. For comparison, XAS and XMCD measurements are also done on 20 nm thick Mn_4N

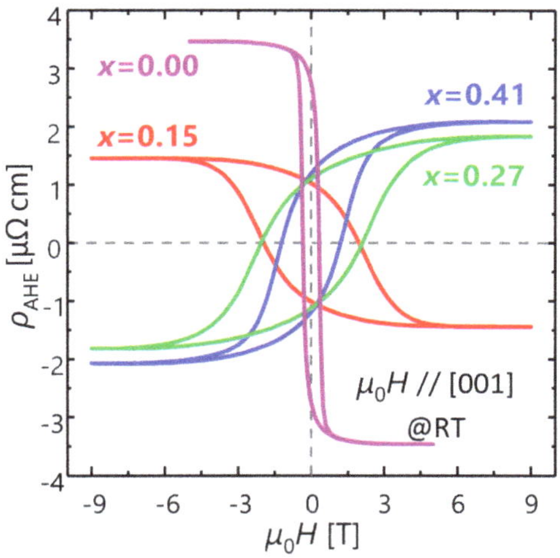

Figure 6.7. AHE loops measured on $Mn_{4-x}In_xN$ films. Reproduced from [3]. © IOP Publishing Ltd. All rights reserved.

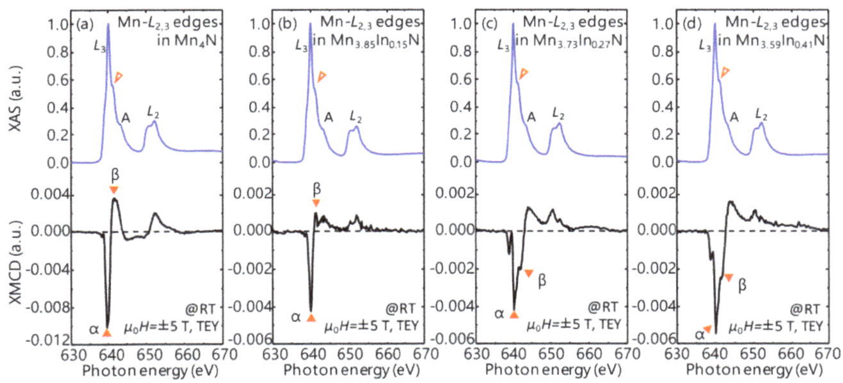

Figure 6.8. XAS (blue) and XMCD (black) spectra at the Mn $L_{2,3}$ absorption edges for $Mn_{4-x}In_xN$ films at x = (a) 0, (b) 0.15, (c) 0.27, and (d) 0.41. Reproduced from [3]. © IOP Publishing Ltd. All rights reserved.

epitaxial films. Both circularly polarized x-rays and ±5 T magnetic fields are applied at the magic angle (54.7°) from the sample normal [9]. The XAS spectra are acquired in total electron yield (TEY) mode.

The XAS and XMCD spectra at the Mn $L_{2,3}$ absorption edges of $Mn_{4-x}In_xN$ epitaxial films are shown in figures 6.8(a)–(d). Regarding the XAS spectra, there is not much difference among samples. The obtained XAS spectra at the Mn $L_{2,3}$ absorption edges are quite similar to those of $(Mn,Ni)_4N$ [7] and $(Mn,Co)_4N$ [8] epitaxial films. In the XAS spectra, a distinct shoulder is observed in all the films at 2 eV higher than the main peak at the L_3 edge, as indicated by the arrow. Such shoulders have been reported for other antiperovskite nitrides, such as Fe_4N [10] and Co_4N [11]. First-principles calculations of Mn_4N [12] and Fe_4N [10] suggest that this shoulder originates from a hybrid state formed by the N 2*p* and Mn(Fe) 3*d* orbitals at the II site. The Mn(Fe) atoms at site I are considered to exhibit a more localized state than those at site II. This arises from the difference in the interatomic distance between N and Mn(I) atoms and between N and Mn(II) atoms. Therefore, the XMCD spectrum can be understood as the addition of an itinerant component characterized by broad peak β in addition to the localized component characterized by sharp peak α, each with a different sign in Mn_4N, as shown in figure 6.8(a). These features are well reproduced by first-principles calculation [12]. Unlike the XAS spectra, the XMCD spectra are quite complex and highly dependent on x. For $x = 0.15$, the peak intensities of α and β are reduced in figure 6.8(b) compared to those of Mn_4N in figure 6.8(a), suggesting that In atoms partially replace Mn atoms at both I and II sites. When x is further increased to 0.27, the sign of peak β reverses in figure 6.8(c), but the sign of peak α remains unchanged. This means that the magnetic moment of the Mn(II) atom has flipped and is now parallel to the magnetic moment of Mn(I). This relationship between peaks α and β is also observed at $x = 0.41$ in figure 6.8(d). These results indicate that a ferrimagnetic to ferromagnetic phase transition appears when the In composition is increased to $x = 0.27$ and 0.41.

Next, we used first-principles calculations to study the effect of In substitutions into some of the Mn(II) sites on the behavior of $Mn_{4-x}In_xN$. The Mn(II) site is chosen for the In substitution because the peak intensity and its sign changes more clearly for peak β than for peak α in the XMCD spectrum in figure 6.8. The calculated structure is a supercell in which the Mn(II) site of $Mn_{32}N_8$ is partially replaced by an In atom. The Vienna *ab initio* simulation package code (VASP) is used to calculate the cell lattice energy and the magnetic moment of each atomic site. Figure 6.9(a) shows the energy per $Mn(I)Mn(II)_{3-x}In_xN$ as a function of the lattice constant, which is multiplied by σ. The value of σ giving the minimum energy increases with increasing In composition x. This result is consistent with the lattice constants obtained from the XRD pattern in figure 6.6. Figure 6.9(b) shows the results of the magnetic moment calculations. The net magnetic moment of the Mn(I) site and the net magnetic moment of the Mn(II) site are plotted. For $x = 0.125$ $\{Mn(I)_8Mn(II)_{23}In_1N_8\}$, as shown in figure 6.9(c), the magnetic moments of the eight Mn(II) sites are aligned parallel to the magnetic moment of Mn(I), but the net magnetic moment of the Mn(II) sites is still antiparallel to the magnetic moment of Mn(I). Of these eight Mn(II) sites, six are nearest-neighbor atoms of the In site. When the In content is further increased to 0.25 $\{Mn(I)_8Mn(II)_{22}In_2N_8\}$ and 0.5 $\{Mn(I)_8Mn(II)_{20}In_4N_8\}$, the magnetic moments of the 16 Mn(II) sites closest to the In site align parallel to the magnetic moment of the Mn(I) site. For $x = 0.000$ and 0.125, the number of Mn(II) sites with magnetic moments aligned parallel to the Mn(I) sites is smaller than the number of other Mn(II) sites. However, as x further increases to 0.250 and 0.500, the number of such Mn(II) sites exceeds the number of other Mn(II) sites. It is for these reasons that the σ dependence of the net magnetic moment of Mn(II) is different from that of $Mn(I)Mn(II)_{3-x}In_xN$ at $x = 0.000$, 0.125, 0.250, and 0.500. The absolute value of the magnetic moment of each Mn(II) site increases with increasing σ. As a result, the net magnetic moment of the Mn(II) site is also aligned parallel to that of Mn(I). The magnitude of ρ_{AHE} reflects the

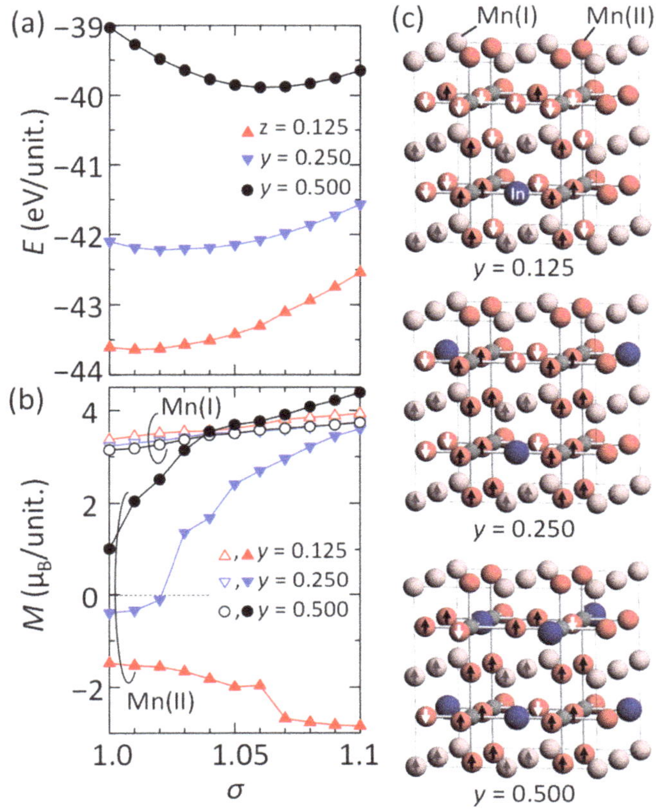

Figure 6.9. First-principles calculations of (a) energy E and (b) net magnetic moment M of Mn(I) and Mn(II) for $Mn(I)Mn(II)_{3-x}In_xN$ as a function of σ, a factor of unit cell size. (c) Schematics of magnetic moment at each Mn site. At $\sigma = 1$, the unit cell size equals to that of Mn_4N. Reproduced from [3].

spontaneous magnetization of the $Mn_{4-x}In_xN$ layer. The increase in ρ_{AHE} with In composition x is well explained by the parallel arrangement of the magnetic moments of the Mn atoms at the I and II sites, as shown in figure 6.9.

6.3 $Mn_{4-x}Au_xN$ epitaxial films

6.3.1 MBE growth of $Mn_{4-x}Au_xN$ films on $SrTiO_3(001)$ by MBE

Approximately 23 nm thick $Mn_{4-x}Au_xN$ thin films (x = 0–0.3) are made at 450°C on STO(001) substrates by MBE [13, 14], followed by a 3 nm SiO_2 capping layer. The Au/Mn atomic ratio of the grown thin films is calculated from the deposition rates of Mn and Au.

Figure 6.10(a) shows the out-of-plane XRD patterns with RHEED images. In addition to reflections of the STO substrate, 002 and 004 reflections of $Mn_{4-x}Au_xN$ are visible. Therefore, the $Mn_{4-x}Au_xN$ film can be said to have grown epitaxially. As Au composition x is increased, these peaks shift to lower angles and overlaps

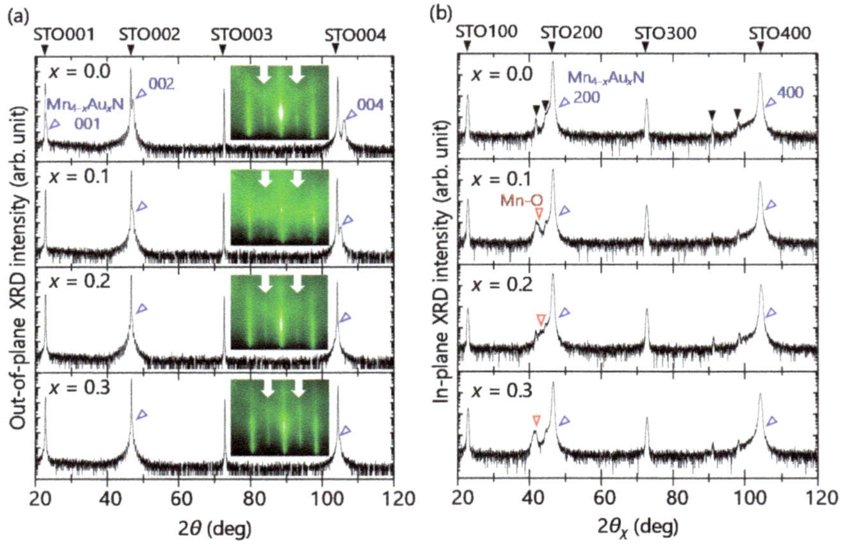

Figure 6.10. (a) Out-of-plane and (b) in-plane XRD patterns of $Mn_{4-x}In_xN$ films. RHEED streaks are observed along the MgO[100] azimuth. White arrows indicate the superlattice diffraction lines. Reproduced from [13]. CC BY 4.0.

with the STO peaks. Thus, the vertical lattice constant c of $Mn_{4-x}Au_xN$ increases with composition x. However, in figure 6.10(b), the reflection corresponding to $Mn_{4-x}Au_xN$ is too close to the STO reflection to be observed. The RHEED image shows a streaky pattern and superlattice diffraction (white arrows) originating from body-centered N atoms. In the low composition sample, the Kikuchi line [15], which is found in high quality crystals, is also observed. However, as x is increased, the diffraction lines of the superlattice become blurred. This result suggests that the crystallinity of the $Mn_{4-x}Au_xN$ film degrades with increasing x.

6.3.2 Characterization of $Mn_{4-x}Au_xN$ films

(1) Magneto-transport properties

The anomalous Hall effect of $Mn_{4-x}Au_xN$ films is measured by PPMS.

Figure 6.11 shows the AHE loop measured for the $Mn_{4-x}Au_xN$ films at room temperature. Loops show excellent squareness ratios close to 1 in the composition range of $x = 0$–0.3. Thus, it is assumed that PMA is preserved in these films. The sign of the AHE loop is reversed between $x = 0.1$ and 0.2. A similar sign reversal occurs in the impurity-doped Mn_4N film. In the case of $Mn_{4-x}Ni_xN$ [16], a sign reversal of the AHE loop is observed between $y = 0.1$ and 0.25, and XMCD [17] confirms the magnetic moment reversal of Mn(I), Mn(II), and Ni(I) atoms. For $Mn_{4-x}In_xN$ [3], a similar sign reversal occurs in the AHE loop between $x = 0.15$ and 0.27, which is accompanied by a reversal of the magnetic moment of the Mn(II) atom. The flip of the magnetic moment of the Mn(II) atom is common to both cases.

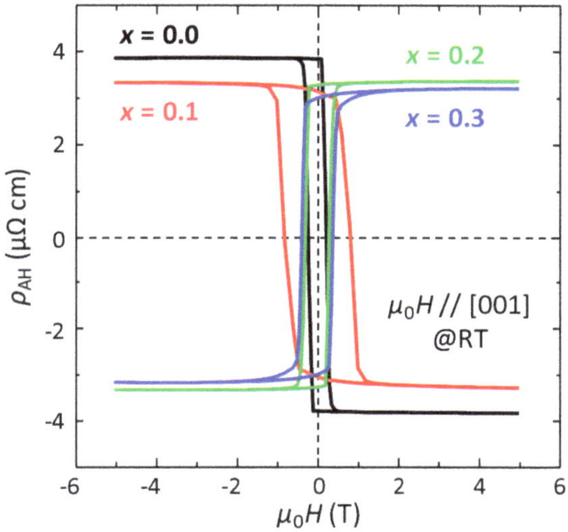

Figure 6.11. AHE loops of $Mn_{4-x}Au_xN$ films. Reproduced from [13]. CC BY 4.0.

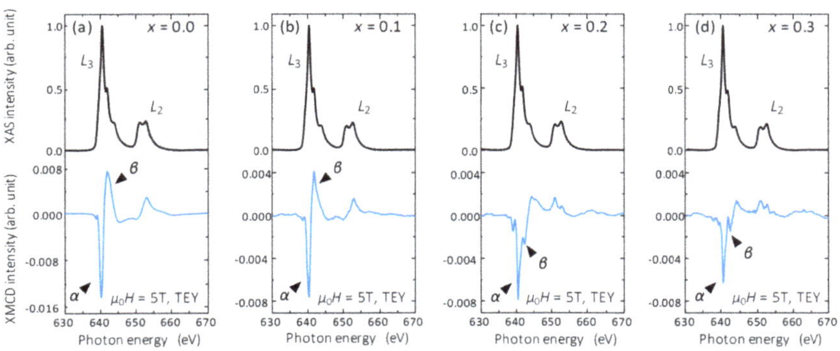

Figure 6.12. XAS (black) and XMCD (blue) spectra at the Mn $L_{2,3}$ absorption edges of (a) Mn_4N, (b) $Mn_{3.9}Au_{0.1}N$, (c) $Mn_{3.8}Au_{0.2}N$, and (d) $Mn_{3.7}Au_{0.3}N$ films. The XAS spectra are normalized. Reproduced with permission from reference [13]. Reproduced from [13]. CC BY 4.0.

(2) XAS and XMCD properties

XAS and XMCD measurements are performed at the Mn-$L_{2,3}$ absorption edge of $Mn_{4-x}Au_xN$ to investigate the origin of the sign reversal of the AHE loop and to see if a magnetic composition exists. They are measured in total electron yield (TEY) mode using a 5 T magnetic field and circularly polarized x-rays incident at an angle of 54.7° from the sample normal at BL-16A at the KEK Photon Factory in Japan.

The results are shown in figure 6.12. The intensities of the XAS spectra are normalized. Since there is no substantial difference in the XAS spectrum despite the value of x, we focus here on the XMCD spectrum. In figure 6.12(a), the XMCD spectrum of Mn_4N features a sharp downward peak α at 640 eV and an upward broad peak β at 642 eV. The difference in the sign of peaks α and β means that the magnetic moment of Mn(I) is aligned antiparallel to that of Mn(II), as described earlier. For $x = 0.1$, the magnetic structure in figure 6.12(b) is the same as for Mn_4N. However, when x increases to 0.2 in figure 6.12(c), a reversal of the magnetic moment of the Mn(II) atom, indicated by peak β, is observed, while the direction of the magnetic moment of the Mn(I) atom remains unchanged with respect to the magnetization. This magnetic structure is retained at $x = 0.3$, as shown in figure 6.12(d). These results mean that the magnetic moments of Mn(II) atoms are parallel to those of Mn(I) atoms at $x = 0.2$ and 0.3, suggesting that a ferrimagnetic-ferromagnetic phase transition occurs in $Mn_{4-x}Au_xN$ films, which is the same as in $Mn_{4-z}In_zN$ films [3]. Magnetic compensation does not occur in the x region of this study because the sign of the magnetic moment of Mn(I) does not change.

6.4 $Mn_{4-x}Sn_xN$ epitaxial films

6.4.1 MBE growth of $Mn_{4-x}Sn_xN$ films on $SrTiO_3(001)$ by MBE

Since the lattice constant of Sn-doped Mn_4N is expected to expand, we use MgO (001) instead of STO(001). The composition of the films is controlled by the crucible temperature of Sn and Mn K-cells. Approximately 25 nm thick $Mn_{4-x}Sn_xN$ (x = 0.0–1.4) films are formed on MgO(001) substrates at 450°C by MBE. After growth, a 3 nm thick SiO_2 capping layer is formed by sputtering to prevent oxidation of the grown film.

Figure 6.13 shows the XRD profiles of $Mn_{4-x}Sn_xN$ on MgO(001) substrate [18]. The black triangles indicate diffraction peaks of the MgO(001) substrate, including Cu-Kβ contributions from the x-ray source. The inset in figure 6.13(a) shows the

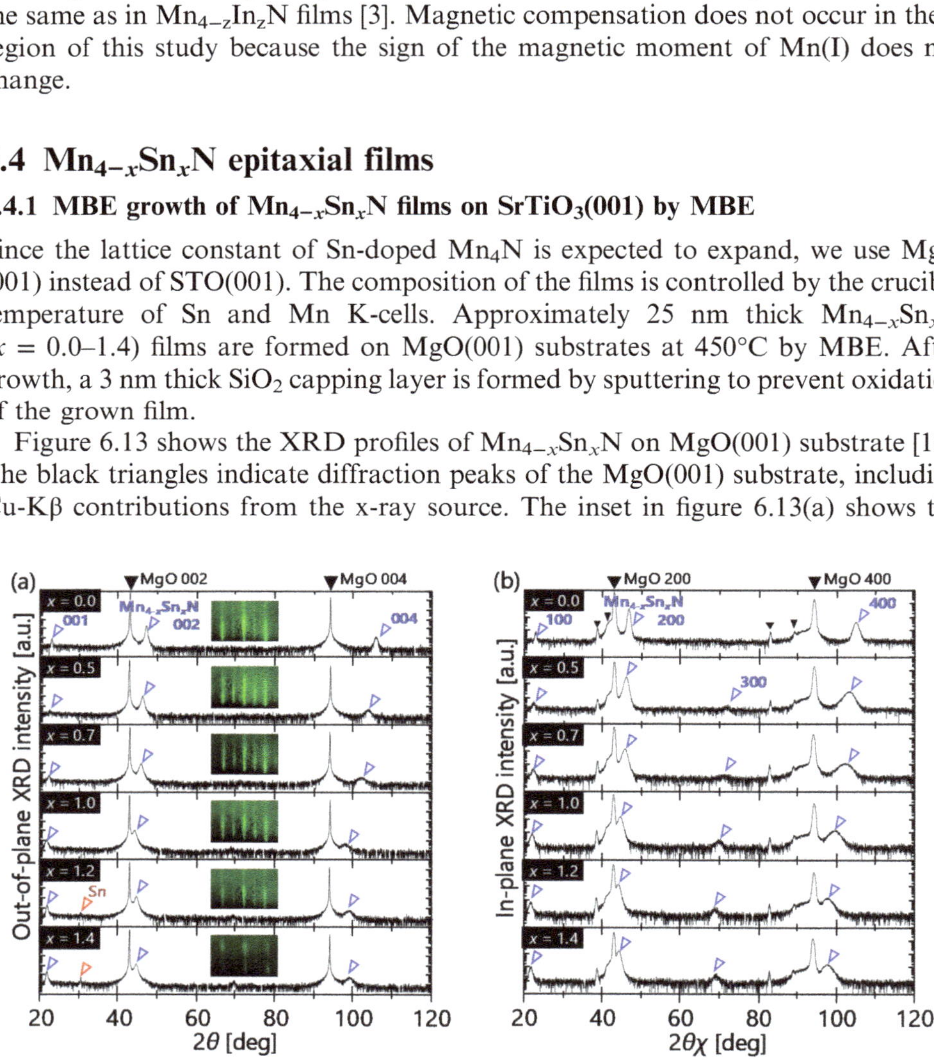

Figure 6.13. (a) Out-of-plane and (b) in-plane XRD patterns of $Mn_{4-x}Sn_xN$ films. RHEED streaks are taken along the MgO[100] azimuth. White arrows indicate the superlattice diffraction lines. Reproduced from [18]. CC BY 4.0.

RHEED images. They show a streaky pattern derived from single-crystalline $Mn_{4-x}Sn_xN$. White arrows in the RHEED image indicate superlattice diffraction lines that are mainly due to nitrogen atoms at the body-centered sites. At $x = 1.4$, the RHEED streaks become blurred and spots appear, suggesting that the crystallinity of the grown thin film is degraded by the excess Sn atoms. In the out-of-plane XRD profiles [figure 6.13(a)], $Mn_{4-x}Sn_xN$ 002 and 004 reflections are observed at all x, while diffraction peaks of segregated metal Sn appeared for $x > 1.0$. In-plane XRD profiles [figure 6.13(b)] show reflections of $Mn_{4-x}Sn_xN$ 100, 200, and 400 for all x. From these results, we can say that we have succeeded in epitaxial growth of $Mn_{4-x}Sn_xN$ ($x \leqslant 1.4$) films. In the XRD profiles, the diffraction peak positions in the in-plane and out-of-plane XRD profiles shift to smaller angles with increasing x, suggesting that the lattice constant of $Mn_{4-x}Sn_xN$ increases with increasing Sn composition. For the peak at $2\theta\chi \approx 70°$, although prohibited by the x-ray extinction rule, it is regarded as a reflection of $Mn_{4-x}Sn_xN$ 300 because it is not easy to explain such a shift of the peak with x due to the presence of Sn clusters and other phases in the Mn-Sn alloys [19, 20]. The 300 reflection in figure 6.13(b) is presumably due to the Sn-induced change in the crystal structure factor.

Figure 6.14 shows the lattice constants of the $Mn_{4-x}Sn_xN$ film calculated from the peak positions of the XRD profile. Error bars indicate errors due to x-ray divergence angle. In the range $x \leqslant 1.0$, the lattice constant increases monotonically because the radius of Sn atoms is larger than that of Mn. However, for $x > 1.0$, there is little change in the lattice constant. This is presumably because the Sn atoms could not replace any more Mn sites in Mn_4N and excess Sn atoms precipitates. The c/a of $Mn_{4-x}Sn_xN$ thin films is less than 1 for $x < 0.9$ but greater than 1 for $x = 1.0$.

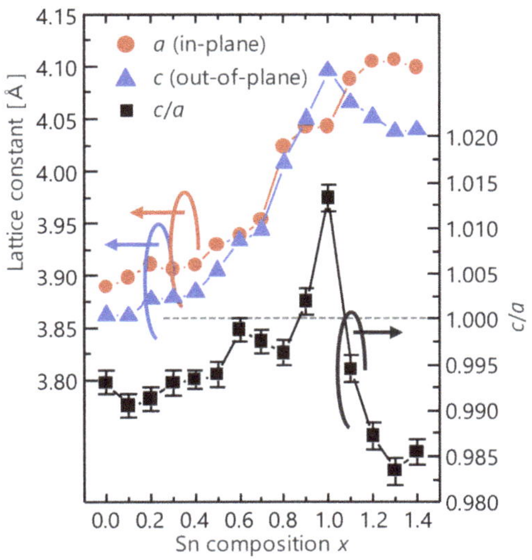

Figure 6.14. Dependences of in-plane lattice constant a, out-of-lattice constant c, and their ratio c/a on x in $Mn_{4-x}Sn_xN$ films on MgO(001). Reproduced from [18]. CC BY 4.0.

6.4.2 Characterization of $Mn_{4-x}Sn_xN$ films

(1) Magneto-transport properties

Hall resistivity and longitudinal resistivity are measured at room temperature by the Van der Pauw method using PPMS (Quantum Design). The ordinary Hall effect contribution appears as the slope of ρ_{xy} in the high field region where the magnetization is saturated. The ordinary Hall coefficient is negative at $x = 0$ [figure 6.15(a)], but changes to positive at $x \geqslant 0.5$ [figures 6.15(b)–(f)]. This means that the majority carriers change from electrons to holes as x is increased. The hysteresis loop in ρ_{xy} is the AHE contribution and is the second term in Eq. (4.1). For $x \leqslant 0.5$ [figures 6.15(a) and (b)], the loop rotates clockwise and exhibits a negative AHE. In contrast, for $x = 0.7$ [figure 6.15(c)], the loop rotates counterclockwise, implying a positive AHE. As x is increased further to $x \geqslant 1.0$ [figures 6.15(d)–(f)], the AHE becomes negative again.

Figure 6.16(a) shows the Sn composition dependence of the anomalous Hall angle θ_{AH} and Hall coefficient R_H of the $Mn_{4-x}Sn_xN$ films at room temperature. θ_{AH} is expressed using longitudinal resistivity ρ_{xx} as,

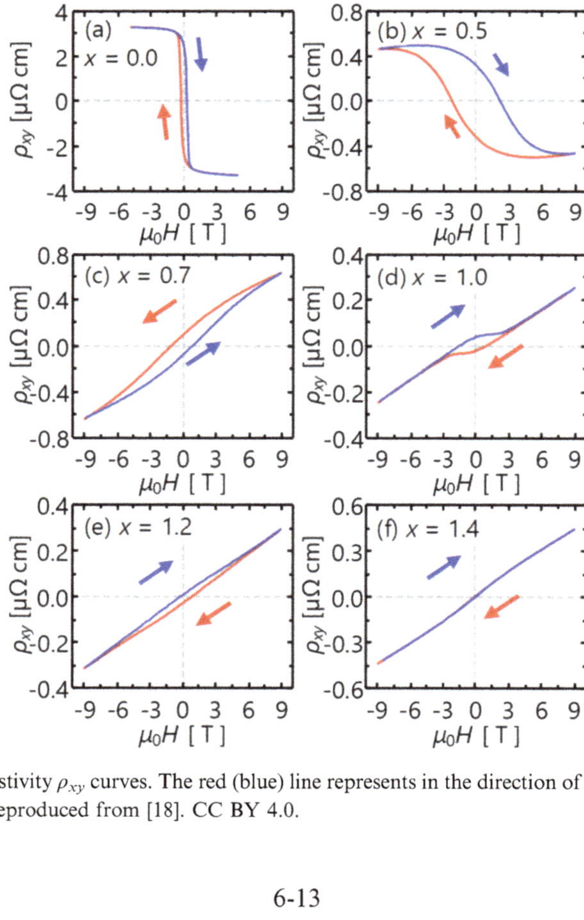

Figure 6.15. Hall resistivity ρ_{xy} curves. The red (blue) line represents in the direction of decreasing (increasing) the magnetic field. Reproduced from [18].

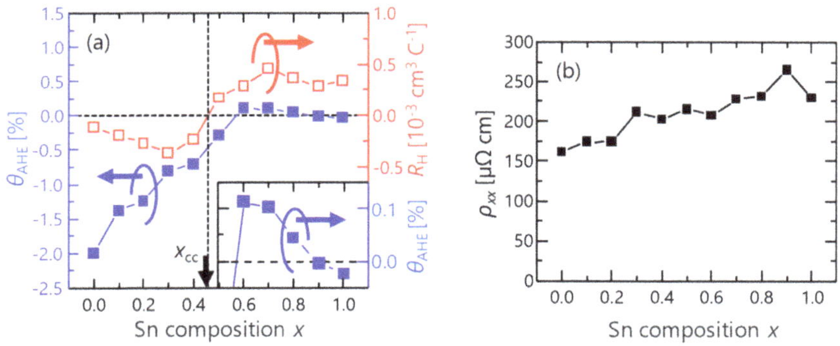

Figure 6.16. Dependences of (a) anomalous Hall angle θ_{AHE}, Hall coefficient R_H, and (b) longitudinal resistivity ρ_{xx} on x in $Mn_{4-x}Sn_xN$ films. Inset in (a) is the enlarged view at $x \geqslant 0.6$. Reproduced from [18]. CC BY 4.0.

$$\theta_{AH} = \frac{\rho_{AHE}}{\rho_{xx}}. \tag{6.1}$$

ρ_{AHE} is obtained by subtracting the ordinary Hall effect contribution from ρ_{xy}. The x dependence of ρ_{xx} is summarized in figure 6.16(b). The composition is limited to the range $x \leqslant 1.0$ because the AHE loop does not close for $x > 1.0$. As shown in figure 6.16(a), the sign of R_H reverses between $x = 0.4$ and 0.5, suggesting that the dominant carrier type changes from electrons ($R_H < 0$) to holes ($R_H > 0$) as x is increased. The sign reversal of θ_{AH} in $Mn_{4-x}Sn_xN$ is observed in all but two Sn compositions except x_{cc}, at which the sign of R_H reverses. This result indicates that it is independent of the carrier type change. Sign reversal of AHE is also observed in $Mn_{4-x}X_xN$ (X = Fe [21], Co [8], Ni [16], In [3], Dy [22], Au [23]). However, there is only one composition in which sign reversal of the AHE occurs. Thus, this is the first report on Mn_4N compounds in which multiple sign reversals occur in the doping element composition dependence of θ_{AH}.

XMCD measurements are a simple method to determine the direction of the magnetic moment relative to the magnetization of Mn(I) and Mn(II) atoms in Mn_4N compounds [24]. We have reported by XMCD measurements that the sign reversal of the AHE loop is accompanied by a reversal of the magnetic moment of the Mn(II) atoms in $Mn_{4-x}X_xN$ (X = Co [8], Ni [16], In [3], Au [13]). However, the magnetization of the $Mn_{4-x}Sn_xN$ epitaxial film does not saturate, even at 9 T, as shown in figure 6.15, which makes XMCD measurements of this sample difficult. Therefore, assuming that the Mn(I) atoms at the corner sites are replaced by Sn atoms, we deduce the magnetic structure of $Mn_{4-x}Sn_xN$ at $x = 0.75$ based on experimental results. The origin of the AHE inversion seems to lie in changes in the magnetic structure, such as magnetic compensation in $Mn_{4-x}Ni_xN$ [17] and $Mn_{4-x}Co_xN$ [8, 12], and ferrimagnetic-ferromagnetic phase transition in $Mn_{4-x}In_xN$ [3]. In these cases, the reversal of the magnetic moment of Mn(II) atoms has been confirmed. $Mn_{4-x}Sn_xN$ can also have a noncollinear magnetic

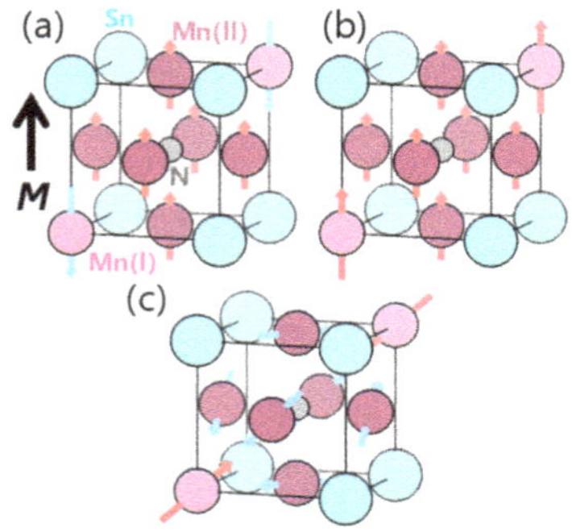

Figure 6.17. Possible magnetic structures of $Mn_{3.25}Sn_{0.75}N$. Sn atoms are assumed to occupy the corner-site Mn(I) atoms: (a) collinear ferrimagnetic structure, (b) collinear ferromagnetic structure, and (c) noncollinear magnetic structure.

structure, as shown in figure 6.17(c). Other possible causes include changes in band structure at the E_F [22, 25] and the sign reversal of spin polarization [26].

6.5 Summary

Chapters 5 and 6 deal with Ni, Co, Cr, In, Au, and Sn-doped Mn_4N films, with magnetic compensation occurring at room temperature only in Ni and Co doped Mn_4N films. Here we discuss the conditions for magnetic compensation in impurity-doped Mn_4N. The difference between these elements is the magnetic moment and the preference of site occupancy. Since the magnitude of $\mu_{Mn(I)}$ is larger than that of $\mu_{Mn(II)}$ in Mn_4N and $\mu_{Mn(I)}$ is parallel to the magnetization, magnetic compensation is considered to occur in at least the following cases:

(i) The added magnetic elements occupy Mn(I) or Mn(II) sites with magnetic moments antiparallel to $\mu_{Mn(I)}$. In this respect, the increase of Ni(I) and Co(I) atoms, whose magnetic moments are antiparallel to $\mu_{Mn(I)}$, only decreases the magnetization and leads to magnetic compensation. In $Mn_{4-x}Cr_xN$, μ_{Cr} aligns antiparallel to $\mu_{Mn(I)}$, reducing magnetization. However, this is limited to a small Cr composition. With respect to site selectivity, Ni and Co have a strong tendency to preferentially occupy I sites, whereas Cr has a weaker tendency. This difference significantly affects the presence or absence of MCP. Considering it this way, it can be stated that magnetic compensation is likely to occur even when nonmagnetic elements are doped in the following case.

(ii) The added nonmagnetic elements preferentially occupy I sites. Such site selectivity is not observed for In, Au, or Sn in the Mn_4N films. The atomic radius of Cr is closer to Mn than to Co. Therefore, Cr atoms are less likely to disrupt the crystallinity of the Mn_4N film, even at compositions higher than Co. This is probably the reason why the RHEED streaks with superlattice diffractions are maintained even at $x = 1.2$ in $Mn_{4-x}Cr_xN$ in figure 6.1. In contrast, for $Mn_{4-x}Co_xN$, such superlattice diffractions disappear at $x = 1.0$. Similar results have been obtained with Ni-doped Mn_4N. There is not enough data on site selectivity to discuss this point further at this time. However, Ni and Co with smaller atomic radii preferentially occupy corner sites, which may be associated with the atomic radius of a doped element.

Bibliography

[1] Komori T, Horiuchi T, Mitarai H, Yasuda T, Amemiya K and Suemasu T 2022 *J. Mag. Mag. Mater.* **564** 170050

[2] Amemiya K *et al* 2013 *J. Phys.: Conf. Ser.* **425** 152015

[3] Yasuda T, Komori T, Mitarai H, Honda S, Ghosh S, Vila L, Attané J P, Amemiya K and Suemasu T 2022 *J. Phys.* D **55** 115003

[4] Ghosh S *et al* 2021 *Nano Lett.* **21** 2580

[5] Kim K-J *et al* 2017 *Nat. Mater.* **16** 1187

[6] Caretta L *et al* 2018 *Nat. Nanotechnol.* **13** 1154

[7] Komori T, Hirose T, Gushi T, Toko K, Hanashima T, Vila L, Attanè J P, Amemiya K and Suemasu T 2020 *J. Appl. Phys.* **127** 043903

[8] Mitarai H *et al* 2020 *Phys. Rev. Mater.* **4** 094401

[9] Stöhr J and König H 1995 *Phys. Rev. Lett.* **75** 3748

[10] Ito K, Toko K, Takeda Y, Saitoh Y, Oguchi T, Suemasu T and Kimura A 2015 *J. Appl. Phys.* **117** 193906

[11] Ito K, Sanai T, Yasutomi Y, Zhu S, Toko K, Takeda Y, Saitoh Y, Kimura A and Suemasu T 2014 *J. Appl. Phys.* **115** 17C712

[12] Ito K *et al* 2020 *Phys. Rev.* B **101** 104401

[13] Horiuchi T, Komoir T, Yasuda T, Hirose T, Toko K, Amemiya K and Suemasu T 2023 *AIP Adv.* **13** 025107

[14] Kawasaki M, Takahashi K, Maeda T, Tsuchiya R, Shinohara M, Ishiyama O, Yonezawa T, Yoshimoto M and Koinuma H 1994 *Science* **266** 1540

[15] Zaefferer S 2007 *Ultramicroscopy* **107** 254

[16] Komori T, Gushi T, Anzai A, Vila L, Attané J P, Pizzini S, Vogel J, Isogami S, Toko K and Suemasu T 2019 *J. Appl. Phys.* **125** 213902

[17] Komori T, Hirose T, Gushi T, Toko K, Hanashima L, Vila J P, Attané K, Amemiya and Suemasu T 2020 *J. Appl. Phys.* **127** 043903

[18] Yasuda T, Komori T, Hirose T, Horiuchi T, Toko K and Suemasu T 2023 *AIP Adv.* **13** 015119

[19] Tomiyoshi S 1982 *J. Phys. Soc. Jpn.* **51** 803

[20] Stange M, Fjellvåg H, Fruseth S and Hauback B C 1997 *J. Alloy Compd* **259** 140

[21] Anzai A, Takata F, Gushi T, Toko K and Suemasu T 2018 *J. Cryst. Growth* **489** 20

[22] Meng M, Wu S X, Zhou W Q, Ren L Z, Wang Y J, Wang G L and Li S W 2015 *J. Appl. Phys.* **118** 053911

[23] Li D, Wang G, Li H, Wu S and Li S 2020 *Mater. Res. Bull.* **122** 110646

[24] Ito K, Honda S and Suemasu T 2022 *Nanotechnology* **33** 062001

[25] Zhang S *et al* 2020 *Nano Lett.* **20** 709

[26] Hao R *et al* 2019 *J. Magn. Magn. Mater.* **478** 187

IOP Publishing

Rare-earth-free Ferrimagnetic Mn_4N Spintronics

Takashi Suemasu

Chapter 7

Ultrafast CIDWM in compensated Mn_4N strips

In this chapter, current-induced DW dynamics in $Mn_{4-x}Ni_xN$ strips with x close to MCP are demonstrated. We see a much faster CIDWM in the $Mn_{4-x}Ni_xN$ strips than those in Mn_4N strips described in chapter 4. First, we briefly revisit how the compensation point could be achieved in $Mn_{4-x}Ni_xN$. Next, we study the magnetization and magnetotransport properties of $Mn_{4-x}Ni_xN$ thin films, and then move on to the CIDWM experiments. Next, we discuss analytical modeling of DW dynamics in antiferromagnetically coupled $Mn_{4-x}Ni_xN$ thin films. Finally, we focus on *ab initio* calculations that can account for experimental results for $Mn_{4-x}Ni_xN$ thin films below and above MCP. The Joule heating effect is also discussed.

7.1 Reaching the compensation point

In some ferrimagnets, ACP and MCP can be done by controlling either the temperature or composition of the material [1–3]. As described earlier, this ability of ferrimagnets has been an active research area in recent years to study CIDWM dynamics [4–6]. Experiments thus far have demonstrated that large SOT-driven DW velocities have been obtained in ferrimagnets containing heavy metals near the MCP and APC [5, 6]. However, the ferrimagnets used in these experiments contain rare-earth and critical elements. Well-known materials are GdCo [5, 7], TbCo [6], GdFeCo [4], and TbFeCo [8].

As we saw in chapter 4, Mn_4N has demonstrated high DW velocities reaching over 900 m s^{-1} [9]. The fact that there is no DMI in the bulk or at the interface has been shown by the asymmetric bubble expansion method [10], leading us to believe that the DW in Mn_4N is driven by STT. To approach the compensation point of this system, Mn can be replaced by a small amount of Ni in Mn_4N. Ni atoms preferentially occupy Mn atoms at corner sites, leading to a decrease in magnetization. In chapter 5, the MCP should be somewhere between $x = 0.1$ and 0.25 in $Mn_{4-x}Ni_xN$ [11]. Figure 7.1 shows the normalized magnetization as a function of temperature for $Mn_{4-x}Ni_xN$ thin films [12]. Magnetization increases in Mn_4N and

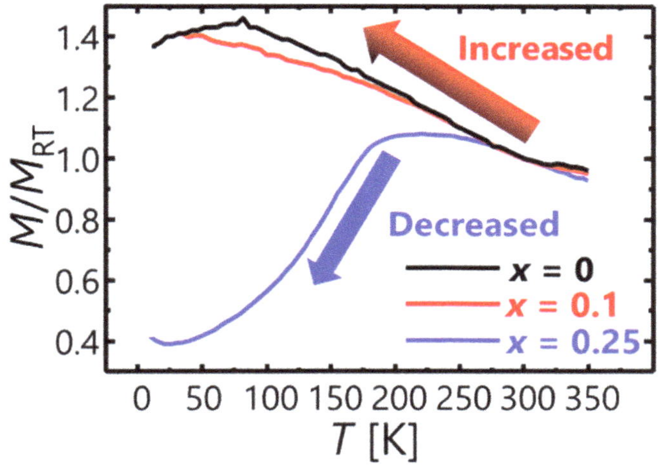

Figure 7.1. Dependences of normalized magnetization for Mn_4N (black), $Mn_{3.9}Ni_{0.1}N$ (red), and $Mn_{3.75}Ni_{0.25}N$ (blue). Measurement is performed with field cooling. Reprinted from [12], with the permission of AIP Publishing.

$Mn_{3.9}Ni_{0.1}N$ thin films, but decreases in $Mn_{3.75}Ni_{0.25}N$ as the temperature decreases to 5 K. It is important to note that the MCP is not reached at any temperature. However, by changing the Ni composition, the MCP can be crossed at room temperature between $Mn_{3.9}Ni_{0.1}N$ and $Mn_{3.75}Ni_{0.25}N$, as described in chapter 5 [11]. Therefore, the Ni composition of these thin films must be fine-tuned to study the movement of DW near the MCP.

7.2 Growth and structural characterization

$Mn_{4-x}Ni_xN$ epitaxial films with a Ni concentration x in the range $x = 0.05 - 0.3$ are fabricated at 450°C on STO(001) substrates by MBE. The detailed growth procedure is given in chapter 5.

Figure 7.2(a) shows a high-resolution cross-sectional STEM image of 30 nm thick $Mn_{3.75}Ni_{0.25}N$ films [13]. A highly ordered crystalline structure with negligible defect density has been confirmed. The smooth interface between the $Mn_{3.75}Ni_{0.25}N$ and STO substrate is also observed in this image, indicating a good lattice matching between $Mn_{3.75}Ni_{0.25}N$ and STO. High-angle annular dark-field STEM (HAADF-STEM) images of the full stack including a SiO_2 capping layer and elemental maps by EDX spectroscopy are shown in figures 7.2(b)–(e). The results show that Ni is uniformly distributed throughout the entire thin film.

7.3 Magnetization and transport measurements

The magnetization curves are measured on these thin films at room temperature using a VSM-SQUID setup to evaluate the saturation magnetization. Figure 7.3 shows the Ni composition x dependence of saturation magnetization [13]. As discussed earlier in chapter 5, the magnetic moment of Mn atoms at I site aligns

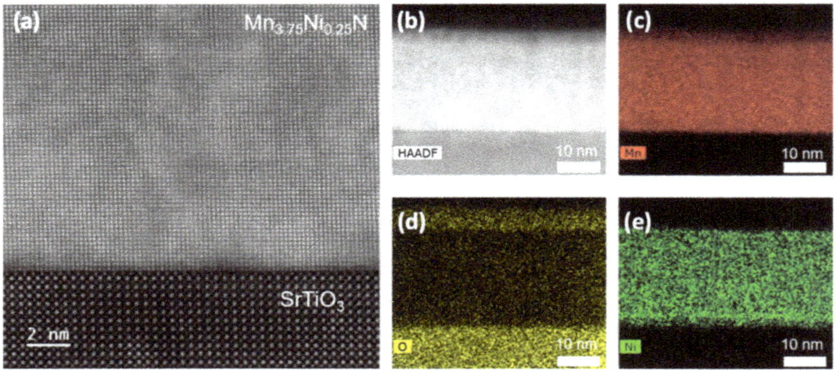

Figure 7.2. (a) STEM image of a 30 nm thick $Mn_{3.75}Ni_{0.25}N$ film on STO. (b) HAADF-STEM image of the film with a 3 nm of SiO_2 capping layer and the associated EDX elemental map of (c) Mn, (d) O, and (e) Ni. Reprinted with permission from [13]. Copyright (2021) American Chemical Society.

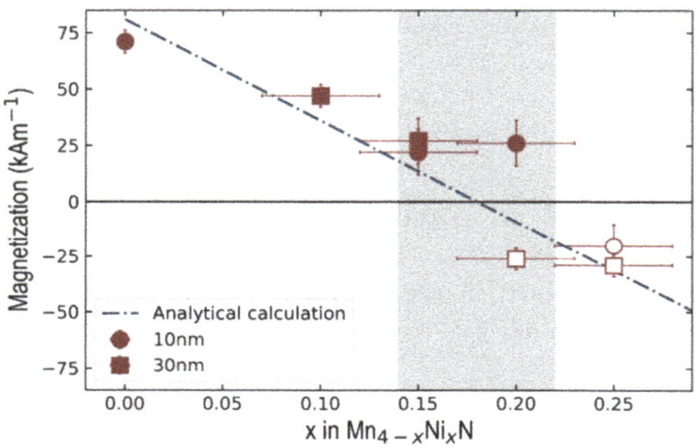

Figure 7.3. Dependence of saturation magnetization on x in $Mn_{4-x}Ni_xN$. The filled (open) symbols show the magnetization below (above) the compensation point. The spread of the observed deviation of x in $Mn_{4-x}Ni_xN$ is in the gray shaded area. The dotted line represents the magnetization obtained analytically. Reprinted with permission from [13]. Copyright (2021) American Chemical Society.

parallel to the magnetization of $Mn_{4-x}Ni_xN$ thin films below the MCP, whereas that of the Mn atoms at site II above the MCP. Thus, the saturation magnetization likewise indicates a thin film with a positive saturation magnetization value below the MCP and a negative saturation magnetization value above the MCP, as shown in figure 7.3.

AHE curves for $Mn_{4-x}Ni_xN$ thin films are measured using the Hall-cross patterning method and the van der Pauw method for the blanket layer. Figure 7.4 shows the AHE curves with various Ni compositions ($x = 0, 0.15, 0.2, 0.25$) [13]. The sharp switching of magnetization in all of the samples indicates that large PMA is retained after patterning. The sign of θ_{AHE} varies between $x = 0.15$ and 0.25, indicating that the MCP is close to the analytically obtained value $x = 0.18$. For the estimation of whether the Ni composition of the thin film is below or above the

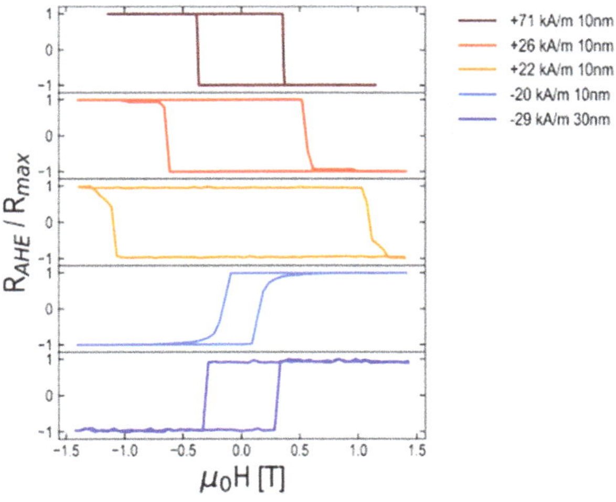

Figure 7.4. AHE curves measured for $Mn_{4-x}Ni_xN$ films at $x = 0$ (71 kA m^{-1}), 0.15 (26 kA m^{-1}), 0.2 (22 kA m^{-1}), 0.25 (−20 kA m^{-1}), and 0.25 (−29 kA kA m^{-1}). The sign of θ_{AH} changes between $x = 0.15$ and 0.2. Reprinted with permission from [13]. Copyright (2021) American Chemical Society.

MCP, a negative θ_{AHE} indicates a composition below the MCP [14–16] and a positive θ_{AHE} [11, 17, 18] indicates a composition above the MCP, as shown in figure 7.4. In agreement with previous results [11, 17], it is observed in figure 7.3 that the saturation magnetization decreases with increasing Ni concentration. The sign change of saturation magnetization is observed between $x = 0.15$ and 0.2. There is also a difference in the saturation magnetization, even for the same Ni composition $x = 0.2$ between the 10 nm and 30 nm thick $Mn_{4-x}Ni_xN$ thin films. This may be the result of a slight deviation from the targeted Ni content or the presence of a dead layer.

The results of XMCD measurements discussed in chapter 4 confirm that the Ni atoms occupy the Mn at I site and the magnetic moment is aligned parallel with that of Mn at II site [11]. Therefore, increasing the Ni composition x in the $Mn_{4-x}Ni_xN$ films is expected to lower the saturation magnetization. Above a certain Ni composition, the sum of the magnetic moments of the Mn atoms at II site and Ni atoms at I site becomes larger than the sum of the magnetic moments of the Mn atoms at I site. This reverses the saturation magnetization, which then points to the magnetic moment of the Mn atoms at II site, as shown in figure 7.5. MCP is also estimated by analytical calculations using magnetic moment values from neutron diffraction studies [19, 20] to have a Ni composition near $x = 0.18$, which is only 3.6% of all of the atoms. It is surprising to note here that such a small amount of Ni leads to the MCP at room temperature.

7.4 Current induced domain wall motion

To demonstrate current-induced DW motion in $Mn_{4-x}Ni_xN$, $Mn_{4-x}Ni_xN$ epitaxial films are processed into wires with a width of 1 μm and a length of 20 μm by

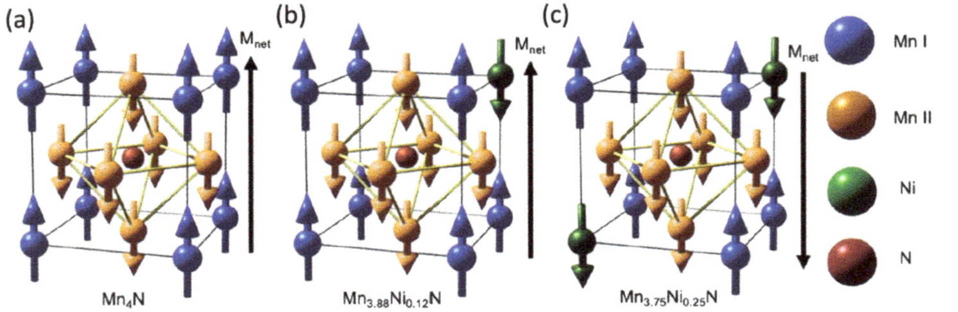

Figure 7.5. Crystal structure of (a) Mn_4N, (b) $Mn_{3.88}Ni_{0.12}N$, and (c) $Mn_{3.75}Ni_{0.25}N$. Reprinted with permission from [13]. Copyright (2021) American Chemical Society.

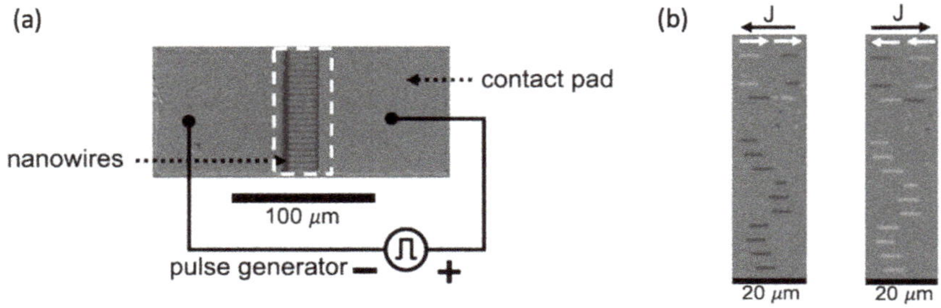

Figure 7.6. (a) Device image of 20 parallel strips of $Mn_{4-x}Ni_xN$ (below MCP) used for the DW motion study. DWs are injected from the contact pads. (b) Differential polar MOKE images of the DW displacement. The white arrows show the direction of the DW motion. Reprinted with permission from [13]. Copyright (2021) American Chemical Society.

standard EB lithography combined with Ar ion milling. An overall view of the device is shown in figure 7.6(a).

An out-of-plane magnetic field pulse, close to the sample's coercive field, is used to nucleate the reverse magnetic domain at the nucleation pad and inject it into the end of the strips. The DW is then displaced by injecting a 1 – 5 ns wide current pulse. Pulse shapes are captured and stored using an oscilloscope. Figure 7.6(b) shows an example of a differential MOKE image with black and white contrast, where the DW moves in the opposite direction of current pulses. To obtain an accurate estimate of DW velocity, DW displacements in different wires are averaged. Thus, the averaged DW velocity is obtained by dividing the averaged displacement by the full width at half maximum of the averaged pulse width, considering the number of current pulses injected. This procedure is performed for all $Mn_{4-x}Ni_xN$ thin films with different x values. For each Ni composition x, the DW displacement is measured as a function of pulsed current density. The average velocities as a function of current density for thin films of different Ni concentrations are shown in figure 7.7. Since saturation magnetization is an important parameter here, a velocity curve for saturation magnetization is shown. In addition, given that there is some

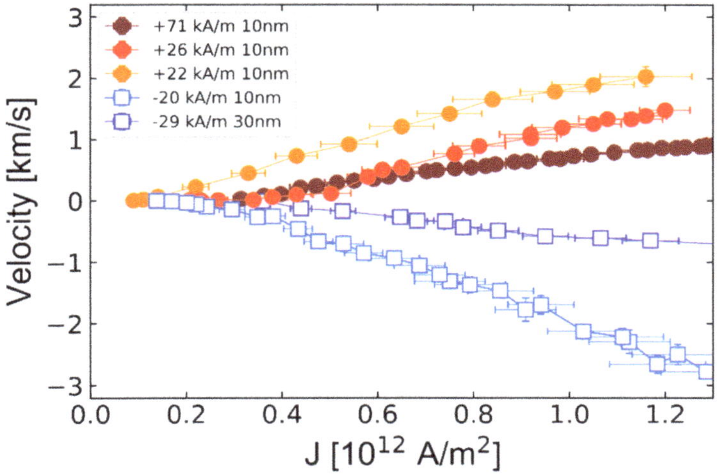

Figure 7.7. Dependences of DW velocity on current density. Here, M_S = 71 kA m^{-1} corresponds to pure Mn_4N [9]. The filled (open) symbols represent Ni concentrations below (above) the MCP in $Mn_{4-x}Ni_xN$. Reproduced with permission from [13]. Copyright (2021) American Chemical Society.

variation in nominal composition, we chose to indicate velocity in terms of saturation magnetization.

DW velocity due to current is indicated as positive(negative) below(above) the compensation point. With increasing Ni concentration, a decrease in saturation magnetization and an increase in DW velocity are observed, with a maximum of 2000 m s^{-1} at a current density of $j = 1.17 \times 10^{12}$ A m^{-2} and below the compensation point of 22 kA m^{-1}. Below the compensation value, a general trend of increasing DW mobility (dv/dj) is observed as one approaches the MCP (i.e., as saturation magnetization decreases). Up to this point, the direction of motion of DW is in the direction of electron flow.

Interestingly, when the Ni composition x exceeds the MCP, the direction of DW motion is reversed and the DW moves in the opposite direction to the electron flow. For a Ni composition of x = 0.25 (M_S = −20 kA m^{-1}), very large DW velocities approaching 3000 m s^{-1} are obtained at a current density of $j = 1.26 \times 10^{12}$ A m^{-2}. Above the MCP, the DW mobility begins to decrease with increasing saturation magnetization, eventually reaching values similar to those observed for Mn_4N. The increase and subsequent decrease in DW velocity near the compensation point are very similar to those observed for the DW motion in other ferrimagnetic materials [4, 6, 7]. Figure 7.7 shows that the threshold current density at which DW begins to move is very low in these thin films. The lowest threshold current density is close to $j = 0.1 \times 10^{12}$ A m^{-2} and is observed at x = 0.2 (M_S = 22 kA m^{-1}). DW velocities obtained from different thin films show a similar trend, starting with a thermal assist region, followed by a linear region corresponding to the precession of the DW motion driven by STT [21]. For Ni compositions x = 0.2 (M_S = 22 kA m^{-1}) and

0.25 (M_S = −20 kA m^{-1}), near the compensation point, the thermal assist region appears to expand and the linear onset is very slow compared to pure Mn_4N films. Threshold current densities are also smaller than Mn_4N in both samples, indicating that the pinning in these films is very small. These threshold current density values are comparable to other ferrimagnets, such as 10^{12} A m^{-2} in TbFeCo [22] and GdCo [7]. However, it is much lower than ferrimagnetic insulators such as TmYIG [23] and BiYIG [24]. The DW velocities obtained in this experiment are the highest among those reported using STT at room temperature without the assistance of an in-plane magnetic field.

7.5 Analytical modeling of the two sub-lattice system

The collective coordinate q–φ model [4, 25] is employed to explain the ultrahigh DW velocity, as well as the reversal of the DW motion. The q–φ is expanded into a ferrimagnetic system using the effective parameters of the two sublattices '1' and '2'. The two sublattices of Mn_4N are composed of the same atom (Mn). Since there have been no precise measurements on the individual gyromagnetic factors γ and the amount of Ni atoms is as small as around 5% at most, we assume $\gamma_1 = \gamma_2$ for simplicity. This means that the MCP and ACP are coincident in $Mn_{4-x}Ni_xN$. In addition, we assume that the damping factors of both sublattices (α_1, α_2) are equal ($\alpha_1 = \alpha_2$) due to the strong antiferromagnetic coupling in this system. Taking these into account, the DW velocity, at the asymptotic limit of the precessional regime, which is well above the critical current density (j_C), is given by,

$$v = \frac{L_S + \beta L_\alpha}{L_S^2 + L_\alpha^2} L_S \boldsymbol{u} \tag{7.1}$$

Here, $L_S \boldsymbol{u} = \frac{Ph}{2e} je_J$. je_J is the current density, P ($P_1 = P_2$) is the effective spin polarization of the two sublattices, α is the Gilbert damping parameter, β is the non-adiabatic contribution to the STT, $L_S = (M_1 - M_2)/\gamma$ is the angular momentum density and $L_\alpha = (\alpha/\gamma)(M_1 + M_2)$.

Figure 7.8 plots the experimental DW velocity at $j = 1 \times 10^{12}$ A m^{-2} in figure 7.7 versus saturation magnetization, $M_S = \gamma L_S = (M_1 - M_2)$. The best fit is obtained with $P = 0.65$, $\alpha = 0.013$, and $\beta = 0.002$, as shown by the red line in figure 7.8. Considering the q–φ model, the reversal of the direction of DW motion is expected to occur at $L_S = -\beta L_\alpha$, just below the ACP. This result justifies our assumption that $\gamma_1 = \gamma_2$, and thus the MCP and the ACP in this system coincide. These results indicate that the large STT-driven DW velocities are attributed to the increased mobility as one approaches ACP, along with the large P. The reversal of the DW motion is also predicted by the analytical q–φ model, as discussed later. The sign of the spin polarization P is assumed to be unchanged by the angular momentum density L_S. Thus, the reversal of the DW motion direction is considered to be associated with a change of the net spin polarization relative to the angular momentum after crossing the ACP. To put it differently, the spin polarization P becomes effectively negative after crossing the ACP.

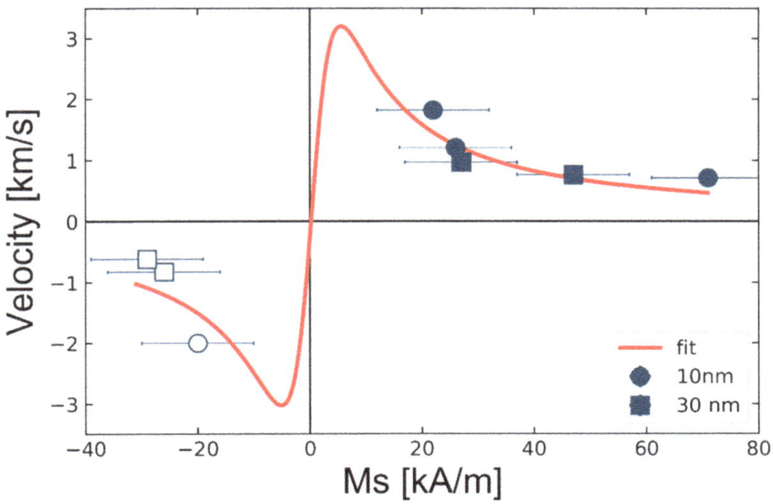

Figure 7.8. DW velocity against net magnetization, measured at $j = 1 \times 10^{12}$ A m^{-2} (blue circles and squares). The best fit is obtained with $P = 0.65$, $\alpha = 0.013$, and $\beta = 0.002$ using the q–φ model from equation (7.1). Reprinted with permission from [13]. Copyright (2021) American Chemical Society.

7.6 First-principles calculations and analysis

First-principles band structure calculations are performed to verify the assumption that the spin polarization remains in the same direction after crossing the ACP (or MCP) in $Mn_{4-x}Ni_xN$. These first-principles calculations are performed in the framework of density functional theory (DFT) using the spin-polarized relativistic Korringa–Kohn–Rostoker (SPR-KKR) and VASP [26–34]. The detailed calculation procedure is given in reference [13]. The calculated total magnetic moment is found to be 1 μ_B (3.3 μ_B for Mn(I), −0.8 μ_B for Mn(II)) along the [001] axis. This is in good agreement with those reported [35, 36]. Mn(I) is replaced by Ni in $1 \times 1 \times 4$ and $1 \times 1 \times 8$ unite cells to model Ni concentrations of $x = 0.125$ and 0.25, respectively. To validate the supercell approach, the effect of Ni substitution in $Mn_{4-x}Ni_xN$ is calculated using the coherent potential approximation (CPA) implemented in the SPR-KKR code [32–34].

Figure 7.9(a) shows the total magnetic moment calculated as a function of Ni composition x. A similar trend of decreasing magnetization with increasing x followed by MCP is observed at $x = 0.15$ for the SPR-KKR approach (green) and at $x = 0.17$ for the VASP approach (orange). These calculated magnetic moments reproduce the experimental data. The spin polarization and transport properties can be estimated by studying the s orbitals of materials responsible for electron transport. The PDOS of s orbitals for Mn(I) and Mn(II) atoms is shown in figures 7.9(b) and (c), respectively. The calculated polarizations are 'upward' at the E_F (toward the magnetic moment of Mn(I), also taken here as the global quantization axis) for both Mn(I) and Mn(II) atoms, with the polarization of Mn(II) being one order of magnitude larger than that of Mn(I). This is evidence that electron transport in Mn_4N is not through the Mn(I) atoms but through the Mn(II) atoms.

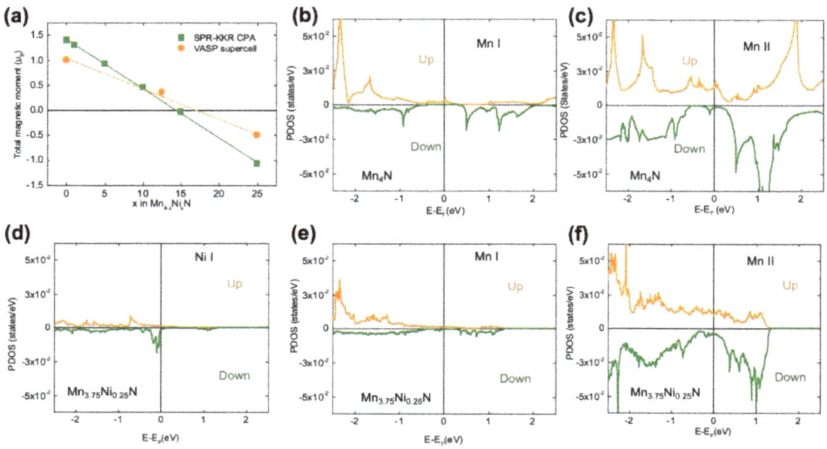

Figure 7.9. (a) Net magnetic moment as a function of the Ni concentration showing that the MCP is at $x = 0.15$ by SPR-KKR CPA (green) and at $x = 0.17$ by VASP (orange). PDOS of *s* orbitals for (b) Mn(I) and (c) Mn(II) at $x = 0.25$ (Mn_4N). PDOS of *s* orbitals for (d)Ni(I), (e)Mn(I), and (f) Mn (II) atoms at $x = 0.25$ (above MCP). Here, Up represents the majority spin. Reprinted with permission from [13]. Copyright (2021) American Chemical Society.

Figures 7.9(d)–(f) show the PDOS of *s* orbitals in $Mn_{3.75}Ni_{0.25}N$, wherein the Ni composition is above the MCP. A closer look at the E_F shows that for Ni, Mn(I), and Mn(II) atoms, the polarization here is also toward the magnetic moment of Mn (I). PDOS is one order of magnitude higher for Mn(II) than for Mn(I) or Ni(I). Thus, electrons with spins parallel to the global quantization axis dominate the transport in $Mn_{3.75}Ni_{0.25}N$. The *ab initio* calculations show that the spin polarization direction stays the same as that of Mn(I) before and after ACP. Thus, these calculations can explain the experimental results in agreement with the predictions made by the q–φ model (figures 7.10).

7.7 Analytical modeling of Joule heating in Mn_4N wires

In this section, the effect of Joule heating in nanowires is investigated when pulse currents flow. The purpose of this study is to find the temperature of the nanowires and whether or not the Curie temperature (T_C) is reached with the current density and pulse width used in the experiment. We adopt the simple analytical formula given by You *et al* [37, 38]. This model takes into account that the nanowires are heated by the current pulse and the heat is dissipated by the substrate. The calculations are based on the following assumptions:

i) The length L of wire is infinite with respect to its thickness.
ii) Wire temperature is uniform throughout the wire.
iii) From the substrate, the heat source is at the surface of the nanowire ($z = 0$) and heat flows only through the substrate, as shown in figure 7.11.
iv) The temperature of the wire is uniform and equal to the temperature of the interface.

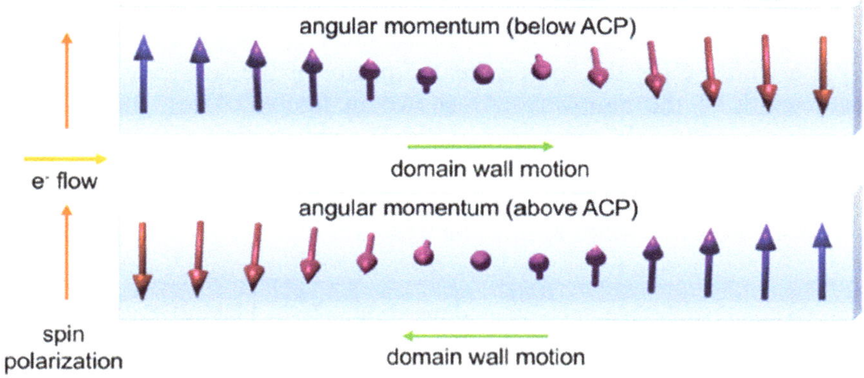

Figure 7.10. Schematic of the Bloch DW with the individual net magnetic moments below (top) and above (bottom) the ACP. The orange arrow represents the direction of spin polarization, which remains in the same direction regardless of Ni composition. The direction of the electron flow is shown by the yellow arrow. Since the spin polarization direction remains the same after the ACP, while the net angular momentum changes its direction, the STT drives the domain walls in the opposite direction above the ACP. Reprinted with permission from [13]. Copyright (2021) American Chemical Society.

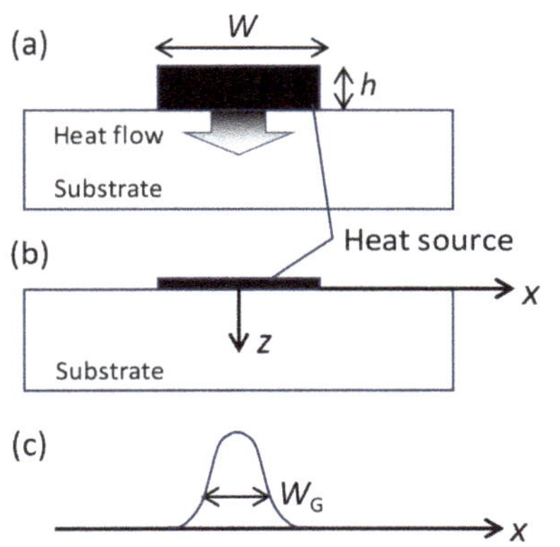

Figure 7.11. (a) Illustration of a nanowire with width W and height h. The nanowire behaves as the heat source, and the heat flows towards the substrate. (b) Diagram showing approximate scale of substrate and the nanowire, and (c) Gaussian profile of the power density in the nanowire.

The Joule heating for a current density j is expressed as $RI^2 = L(Whj)^2/(Wh\sigma_W) = L(Wh)j^2/\sigma_W$, where σ_W is the electrical conductivity, W is the width, and h is the height of the nanowire. Considering the aforementioned assumptions, the overall temperature profile is given as,

$$T(t) = \frac{Whj^2}{\pi\mu_S\sigma_W\rho_S C_S}\left(\operatorname{arcsinh}\left(\frac{2\sqrt{\mu_S t}}{w_G}\right) - \theta(t - t_p)\operatorname{arcsinh}\left(\frac{2\sqrt{\mu_S(t - t_p)}}{w_G}\right)\right). \quad (7.2)$$

Here, μ_S is the diffusivity of the substrate given as $\mu_S = \kappa_S/\rho_S$ with κ_S being the thermal conductivity and ρ_S the density of the substrate. Meanwhile, t is the time, t_p is the pulse width, C_S is the specific heat capacity of the substrate, and w_G is the Gaussian width of the nanowire. As shown in figure 7.11(c), the width of the nanowire is reduced to the Gaussian width. The reason why Gaussian profiles are used here is that they are easy to calculate. $w_G = \alpha W$, here α is a parameter to adjust. The α parameter is estimated to be around 0.5 when the t is much smaller than W/μ_S [37]. At $t = t_p$, we can further reduce the equation to,

$$T(t) = \frac{Whj^2}{\pi\mu_S\sigma_W\rho_S C_S}\left(\operatorname{arcsinh}\left(\frac{2\sqrt{\mu_S t}}{w_G}\right)\right). \tag{7.3}$$

Here, we calculate and compare the Joule heating effect using the above equation.

In this section, we compare the Mn_4N films on STO and $Mn_{4-x}Ni_xN$ films on STO by calculating the Joule heating effect using the above equation. For these calculations, the properties of STO are (specific heat capacity) $C_S = 536.31\ \mathrm{J\ K^{-1}}$ at 300 K [39], (density) $\rho_S = 5120\ \mathrm{kg\ m^{-3}}$, and (thermal conductivity) $\kappa_S = 11\ \mathrm{W\ m^{-1}\ K^{-1}}$ at 300 K [40–42]. For Mn_4N, the resistivity is set to 180 μΩ cm (which is converted to conductivity) [9] and $w_G = \alpha W$ with $\alpha = 0.5$ and $W = 1$ μm for the wire width.

Figure 7.12 shows the temperature rise in the Mn_4N nanowire [43]. Figure 7.12(a) plots the temperature variation versus current density with various current pulse widths. The green horizontal dotted line shows the T_C of Mn_4N at 450 K. The 0 K here is taken as 300 K. Similarly, figure 7.12(b) shows the temperature variation versus current pulse width with a current density varying from $j = 0.2 \times 10^{12}\ \mathrm{A\ m^{-2}}$ to $2 \times 10^{12}\ \mathrm{A\ m^{-2}}$. The curve obtained here suggests that the Joule heating of the nanowires does not reach T_C in the CIDWM experiment in Mn_4N.

Figure 7.13 shows the temperature rise in the $Mn_{3.85}Ni_{0.15}N$ nanowire [43]. In these calculations, the resistivity of the $Mn_{3.85}Ni_{0.15}N$ thin film is set to 225 μΩ cm and the other parameters are kept the same as those used in figure 7.11. Overall, these calculations are consistent with the experimental results, suggesting that the nanowire heating does not reach T_C at the current densities and pulse widths used in

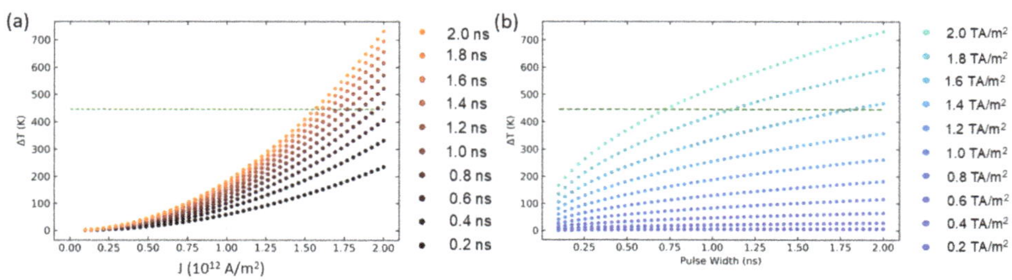

Figure 7.12. Temperature rise in the Mn_4N wire with a width of 1 μm. (a) Change in temperature against current density with various current pulse widths. (b) Change in temperature against current pulse width with various current densities. The green line represents the T_C of Mn_4N. Reproduced from [43]. CC BY 4.0.

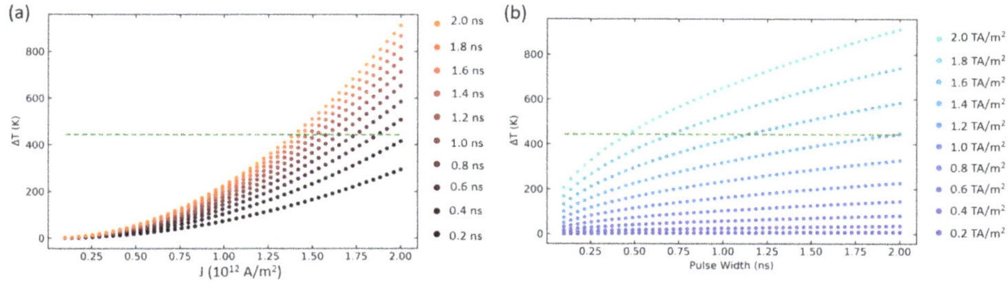

Figure 7.13. Temperature rise in the $Mn_{3.85}Ni_{0.15}N$ wire with a width of 1 μm. (a) Change in temperature against current density with various current pulse widths. (b) Change in temperature against current pulse width with various current densities. The green line represents the T_C of Mn_4N. Reproduced from [43]. CC BY 4.0.

the $Mn_{4-x}Ni_xN$ films. Even though it does not take convection effects into account, heat dissipation from the atmosphere, and increased resistivity of the wires due to Joule heating, heat dissipation through the atmosphere should not be negligible, and therefore the results presented here may be overestimates. In addition, considering that resistivity increases with heating, less current flows through the wire, which in principle should reduce heating.

Figure 7.14 shows the temperature rise in (a)(b) Mn_4N and (c)(d) $Mn_{3.85}Ni_{0.15}N$ nanowires when the nanowire width is varied with a pulse width of 1.5 ns in (a) and (c), and the current density $j = 1.5 \times 10^{12}$ Am^{-2} in (b) and (d) [43]. A pulse width of 1.5 ns and a current density of $j = 1.5 \times 10^{12}$ Am^{-2} are chosen to be very high to allow estimation of heating under extreme conditions. These curves show less heating in Mn_4N than in $Mn_{3.85}Ni_{0.15}N$ due to the lower resistivity of Mn_4N. Narrowing the width from 1 μm to 0.3 μm results in only a slight change in heating. Only at 0.1 μm is there a significant reduction in heating. Next, let us see how wire thickness affects the heat generation of Mn_4N and $Mn_{3.85}Ni_{0.15}N$.

Figure 7.15 shows the temperature rise in (a)(b) Mn_4N and (c)(d) $Mn_{3.85}Ni_{0.15}N$ nanowires [43]. The wire width is set to $W = 1$ μm. These graphs show a clear thickness dependence of temperature rise. Joule heating studies indicate that while the heating is within the experimental parameters, the T_C is not exceeded. It also shows how varying the thickness of the thin film can reduce the heating effects of these systems. Therefore, to increase the current density and reach even higher magnetic DW velocities, the thickness of the thin film should be reduced. Another way to achieve the same thing is to narrow the width of the nanowires to 0.1 μm. However, it becomes very difficult to observe the domains with a MOKE microscope.

7.8 Summary

The current-induced DW dynamics are investigated for $Mn_{4-x}Ni_xN$ with Ni composition set close to MCP and/or ACP. STEM reveals that the interface between the STO substrate and $Mn_{3.75}Ni_{0.25}N$ is smooth. Ni is found to be uniformly and homogeneously present by EDX analysis. The magnetization of

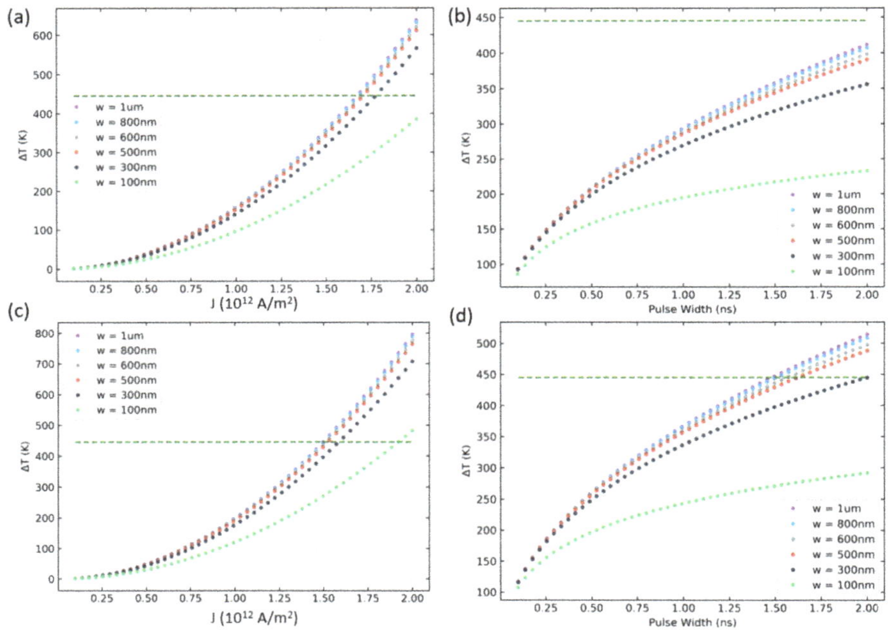

Figure 7.14. Temperature rise in (a) and (b) Mn_4N, and (c) and (d) $Mn_{3.85}Ni_{0.15}N$ wires. (a) and (c) Change in temperature against current density for various wire widths. (b) and (d) Change in temperature against current pulse width for various wire widths when the current density is 1.5×10^{12} Am^{-2}. The green line represents the T_C of Mn_4N. Reproduced from [43]. CC BY 4.0.

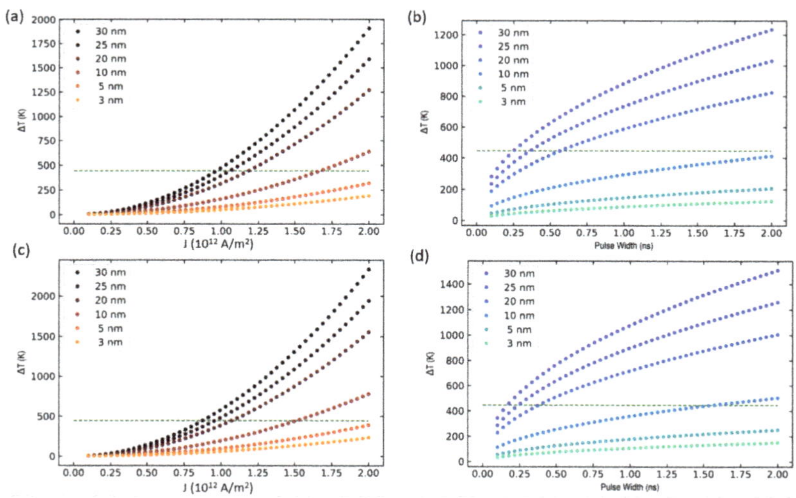

Figure 7.15. Temperature rise in (a) and (b) Mn_4N and (c) and (d) $Mn_{3.85}Ni_{0.15}N$ wires. (a) and (c) Change in temperature against current density for various film thicknesses. (b) and (d) Change in temperature against current pulse width for various film thicknesses when the current density is 1.5×10^{12} A m^{-2}. The green line represents the T_C of Mn_4N. Reproduced from [43]. CC BY 4.0.

the thin films decreases with increasing Ni composition and increases above a certain Ni composition.

The AHE angle switches from negative to positive with increasing Ni concentration, confirming the crossing of MCP. From the AHE curves, M_S measurements, and analytical calculations of saturation magnetization, the MCP is estimated to be around $x = 0.18$. The DW velocity increases as the Ni composition approaches the MCP. Velocities of more than 2000 m s^{-1} are obtained below MCP. After crossing the MCP, the direction of DW motion is reversed and very high DW velocities approaching 3000 m s^{-1} are demonstrated above the MCP. The analytical model can predict such high DW wall velocities using two sub-lattice model with effective parameters. The following two assumptions are taken for the analytical model:

(i) Since both sublattices contain Mn atoms and the fraction of Ni in the system is very low, the MCP and ACP are expected to coincide with an effective gyromagnetic ratio $\gamma = \gamma_1 = \gamma_2$.
(ii) The spin polarization direction must remain unchanged below and above the MCP.

The best fitting of the q–φ model is obtained, which agrees very well with the experimentally obtained DW velocity and confirms the first assumption that the MCP and ACP coincide in $Mn_{4-x}Ni_xN$. Ab initio calculations are performed on $Mn_{4-x}Ni_xN$ thin films using the VASP and SPR-KKR approaches. These approaches confirm the presence of MCP between $x = 0.15$ and 0.17. They are close to the experimentally obtained MCP at around $x = 0.18$. The PDOS of s orbitals indicates that the electron transport is due to Mn(II) atoms and the DOS of Mn(II) at the E_F is one order of magnitude higher than that of Mn(I). The spin polarization direction is also observed in the global quantization axis [001] (parallel to the magnetic moment of Mn(I)). From the PDOS of s orbitals in $Mn_{3.75}Ni_{0.25}N$, it is observed that Mn(II) atoms have one order of magnitude higher DOS at the E_F than Mn(I) and Ni(I). Interestingly, the spin polarization direction also remains oriented toward the global quantization axis [001]. This supports the second hypothesis. The results of *ab initio* calculations and the predictions of the q–φ model explain the high DW velocities and reversal of the DW direction in these $Mn_{4-x}Ni_xN$ thin films.

Bibliography

[1] Stanciu C D *et al* 2006 *Phys. Rev.* B **73** 220402
[2] Binder M *et al* 2006 *Phys. Rev.* B **74** 134404
[3] Je S-G *et al* 2018 *Appl. Phys. Lett.* **112** 062401
[4] Kim K-J *et al* 2017 *Nat. Mater.* **16** 1187
[5] Caretta L *et al* 2018 *Nat. Nanotechnol.* **13** 1154
[6] Siddiqui S A, Han J, Finley J T, Ross C A and Liu L 2018 *Phys. Rev. Lett.* **121** 057701
[7] Cai K *et al* 2020 *Nat. Electron.* **3** 37
[8] Awano H 2015 *J. Mag. Mag. Mater.* **383** 50
[9] Gushi T *et al* 2019 *Nano Lett.* **19** 8716
[10] Pizzini S *et al* 2014 *Phys. Rev. Lett.* **113** 047203

[11] Komori T, Hirose T, Gushi T, Toko K, Hanashima T, Vila L, Attané J P, Amemiya K and Suemasu T 2020 *J. Appl. Phys.* **127** 043903
[12] Komori T, Gushi T, Anzai A, Vila L, Attané J P, Pizzini S, Vogel J, Isogami S, Toko K and Suemasu T 2019 *J. Appl. Phys.* **125** 213902
[13] Ghosh S *et al* 2021 *Nano Lett.* **21** 2580
[14] Gushi T *et al* 2018 *Jpn. J. Appl. Phys.* **57** 120310
[15] Kabara K and Tsunoda M 2015 *J. Appl. Phys.* **117** 17B512
[16] Meng M, Ren S X, Zhou L Z, Wang Y J, Wang G L and Li S W 2015 *Appl. Phys. Lett.* **106** 032407
[17] Komori T, Anzai A, Gushi T, Toko K and Suemasu T 2019 *J. Cryst. Growth* **507** 163
[18] Mitarai H *et al* 2020 *Phys. Rev. Mater.* **4** 094401
[19] Fruchart D, Givord D, Convert P, l'Heritier P and Senateur J P 1979 *J. Phys.* F **9** 2431
[20] Takei W J, Shirane G and Frazer B C 1960 *Phys. Rev.* **119** 122
[21] Metaxas P J *et al* 2007 *Phys. Rev. Lett.* **99** 217208
[22] Ngo D-T, Ikeda K and Awano H 2011 *Appl. Phys. Express* **4** 093002
[23] Avci C O *et al* 2019 *Nat. Nanotechnol.* **14** 561
[24] Caretta L *et al* 2020 *Science* **370** 1438
[25] Haltz E, Krishnia S, Berges L, Mougin A and Sampaio J 2021 *Phys. Rev.* B **103** 014444
[26] Kresse G and Hafner J 1993 *Phys. Rev.* B **47** 558
[27] Kresse G and Furthmüller J 1996 *Comput. Mater. Sci.* **6** 15
[28] Kresse G and Furthmüller J 1996 *Phys. Rev.* B **54** 11169
[29] Blöchl P E 1994 *Phys. Rev.* B **50** 17953
[30] Perdew J P, Burke K and Ernzerhof M 1996 *Phys. Rev. Lett.* **77** 3865
[31] Kresse G and Joubert D 1999 *Phys. Rev.* B **59** 1758
[32] Ebert H, Ködderitzsch D and Minár J 2011 *Rep. Prog. Phys.* **74** 096501
[33] Ebert H, Braun J, Ködderitzsch D and Mankovsky S 2016 *Phys. Rev.* B **93** 075145
[34] Ebert H *The Munich SPR-KKR package* https://www.ebert.cup.uni-muenchen.de/old/index.php?option=com_content&view=article&id=8&catid=4&Itemid=7&lang=en
[35] Meinert M and Phys J 2016 *Condensed Matter* **28** 056006
[36] Isogami S, Masuda K and Miura Y 2020 *Phys. Rev. Mater.* **4** 014406
[37] You C-Y, Sung I M and Joe B-K 2006 *Appl. Phys. Lett.* **89** 222513
[38] Fangohr H, Chernyshenko D S, Franchin M, Fischbacher T and Meier G J 2011 *Phys. Rev.* B **84** 054437
[39] Jacob K and Rajitha G 2011 *J. Chem. Thermodyn.* **43** 51
[40] Muta H, Kurosaki K and Yamanaka S 2005 *J. Alloy Compd* **392** 306
[41] Popuri S R *et al* 2014 *RSC Adv.* **4** 33720
[42] Martelli V, Jiménez J L, Continentino M, Baggio-Saitovitch E and Behnia K 2018 *Phys. Rev. Lett.* **120** 125901
[43] Ghosh S 2022 Manipulation of domain walls in rare-earth-free anti-perovskite ferrimangetic nitrides *PhD Thesis* University of Tsukuba

Takashi Suemasu

Chapter 8

Properties of Mn_4N-based mixed crystals sorted by doped element

Various mixed crystals based on Mn_4N have been made by mixing with a third element. In this chapter, we will briefly introduce the types of studies that have been conducted, while avoiding duplication of those that have been already described in the previous chapters.

8.1 $Mn_{4-x}Al_xN_y$

Polycrystalline Mn_3AlN_y (y = 1.0, 1.1, and 1,2) was prepared by a solid-state reaction (SSR) method and its magnetic, electrical, and thermoelectric properties were measured [1]. Calculations on the electronic structure and T_C for Mn_3AlN have also been performed [2, 3]. Table 8.1 summarizes some of the previous studies [1–3].

8.1.1 Magnetic, electrical, and thermoelectric properties in Mn_3AlN_y (y = 1.0, 1.1, 1.2)

Table 8.2 summaries the studies in reference [1], wherein polycrystalline Mn_3AlN_y samples were prepared by the SSR method. It is important to note that (1) the coercivity of each sample is small, (2) the slope of the electrical resistivity and Seebeck coefficient change around the T_C, and (3) negative thermal expansion (NTE) was not observed. The coercivity was small for all samples with y = 1.0, 1.1, and 1.2, indicating that they are candidates for soft magnetic materials. XRD measurements of the y = 1.0 sample were performed in the temperature range 20 K–850 K, and it is reported that the positions of the diffraction peaks did not change during the ferromagnetic-paramagnetic transition.

Detailed electrical properties were also measured for each sample. It was found that ρ_{xx} is proportional to T^2 at low temperatures below 30 K, confirming Fermi fluid-like behavior. If the electrical properties exhibit Fermi fluid-like behavior in the cryogenic region, then ρ_{xx} is expressed as,

doi:10.1088/978-0-7503-5477-6ch8

Table 8.1. Some previous studies of $Mn_{4-x}Al_xN_y$. Polycrystalline powders are formed by solid state reaction. Temperature dependences of magnetization (M–T), longitudinal resistivity (ρ_{xx}), specific heat (C_p), Seebeck coefficient (S), thermal conductivity (κ), and volume (V) are investigated. Energy-momentum (E–k) relation, spin-resolved DOS, and T_C are also studied.

Exp./Cal.	x	Properties	Reference	Year
Powder/SSR	1	M–T, ρ_{xx}–T, C_p–T, S, κ	[1]	2011
Cal.	1	E–k, spin-resolved DOS, V–T	[2]	2008
Cal.	1	E–k, spin-resolved DOS, T_C–y	[3]	2013

Table 8.2. Lattice constants at room temperature, T_C, and KW ratio in Mn_3AlN_y (y = 1.0, 1.1, 1.2) samples [1].

y	a [Å]	T_C [K]	H_C [Oe]	KW ratio b/γ^2 [μW cm K^{-2}]
1.0	3.8545	818	72	107.7
1.1	3.8598	802	75	19.3
1.2	3.8611	774	80	4.6

$$\rho_{xx} = \rho(0) + bT^2. \tag{8.1}$$

It is reported that $b = 0.1353$ μΩ cm K^{-2}. Furthermore, the specific heat measurement is also performed in this study, and is expressed as,

$$\frac{C_p(T)}{T} = \gamma + \beta T^2 + \delta T^4. \tag{8.2}$$

The Sommerfeld electronic specific heat coefficient γ is calculated to be $\gamma = 35.45$ mJ mol^{-1} K^{-2}. The Kadowaki–Woods ratio (KW ratio) [4] was calculated to be $b\gamma^{-2} = 107.7$ μΩ cm K^{-2}, suggesting a strongly correlated system. Since a KW ratio larger than about 10 μΩ cm K^{-2} can be roughly classified as a heavy Fermion system (HFS), the samples with $y = 1.0$ and 1.1 are considered to be HFS. The Debye temperature is calculated to be about 614 K. In addition, the Seebeck coefficient and thermal conductivity are only measured for Mn_3AlN, and the electron and lattice contributions κ_e and κ_L, respectively, are determined. The results show that the lattice contribution is dominant (figures 8.1 and 8.2).

8.1.2 First-principles calculation of magnetic properties and electronic structure

First-principles studies are reported in references [2, 3]. In reference [3], magnetic properties have also been calculated for compositions below $y = 1.0$, and it is reported that the T_C is above room temperature for all the Al compositions. The calculation of T_C is based on Mohn and Wolfarth's model [5] as,

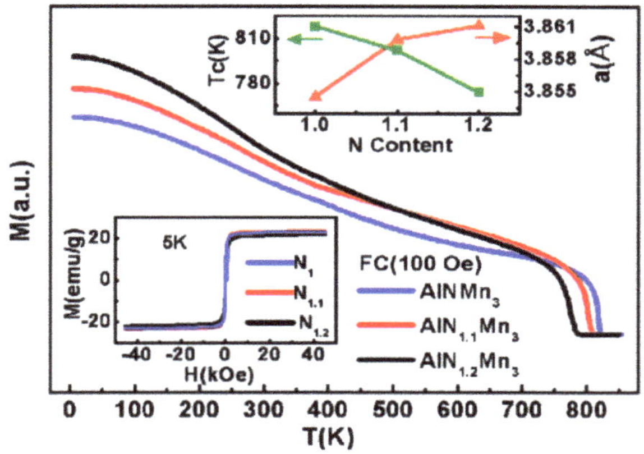

Figure 8.1. Temperature dependent magnetization *M(T)* curve under field-cool process at 100 Oe from 5 K to 850 K. Top inset: the refined lattice parameter (*a*) and the Curie temperature (T_C) for Mn_3AlN_y. Bottom inset: isotherm magnetization curves *M(H)* at 5 K for Mn_3AlN_y. Reprinted from [1], with the permission of AIP Publishing.

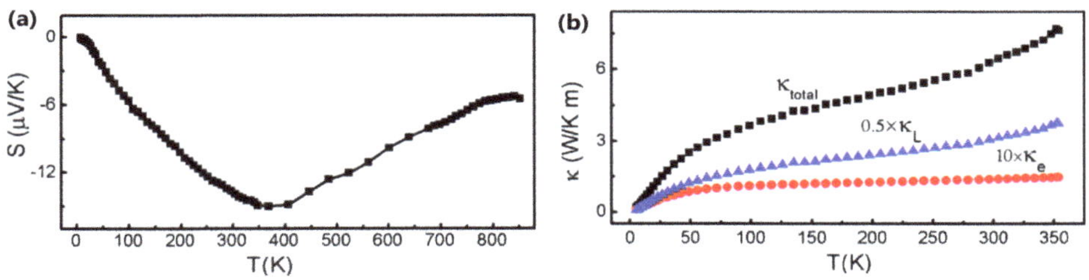

Figure 8.2. (a) Temperature dependence of Seebeck coefficient *S*(*T*) at zero field. (b) Thermal conductivity as a function of temperature at zero field. Blue triangle plots show thermal conductivity from lattice contribution, and red circle plots show that from electron conduction. Reprinted from [1], with the permission of AIP Publishing.

$$\frac{T_C^2}{T_S^2} + \frac{T_C}{T_{SF}} - 1 = 0. \tag{8.3}$$

Here, T_S is the Stoner temperature and T_{SF} is the spin fluctuation temperature, which are given by equation (8.4) [6, 7] and (8.5) [5], respectively, as,

$$T_S = \frac{\Delta E_{ex}}{4k_B}. \tag{8.4}$$

$$T_{SF} = \frac{m^2}{10k_B\chi_0}. \tag{8.5}$$

Here, ΔE_{ex} is the energy splitting due to exchange interactions, m is the magnetic moment per Mn atom, and χ_0 is the magnetic susceptibility. χ_0 is obtained by the method of Wolfarth [8] and Gersdolf [9] as,

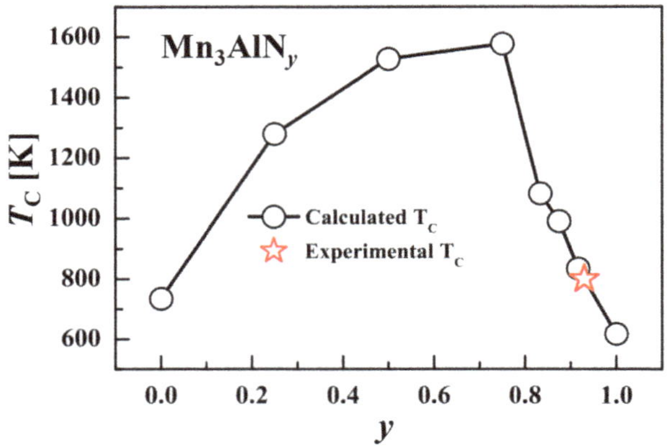

Figure 8.3. Theoretically estimated and experimental T_C of Mn_3AlN_y. Reprinted from [3], with the permission of AIP Publishing.

$$\chi_0^{-1} = \frac{1}{2\mu_B}\left[\frac{1}{2N_{\mathrm{Mn}}^{\uparrow}(E_\mathrm{F})} + \frac{1}{2N_{\mathrm{Mn}}^{\downarrow}(E_\mathrm{F})} - I\right] \tag{8.6}$$

It is also known that $\Delta E_{\mathrm{ex}}$is given by equation (8.7) [10],

$$\Delta E_{\mathrm{ex}} = Im. \tag{8.7}$$

Here, $N_{\mathrm{Mn}}^{\uparrow(\downarrow)}(E_\mathrm{F})$ is the DOS of the upper and lower spins at the E_F, and I is the exchange integral in the DFT calculation. The results of the T_C calculations are shown in figure 8.3. All of the T_C's obtained by the DFT calculations are well above room temperature, which is promising for a ferrimagnetic nitride material. They are also in good agreement with experimental values for samples with $y = 0.92$–0.95. However, there have been no reports on the preparation, crystallinity evaluation, and magnetic properties of the samples with a small amount of doping.

8.2 $Mn_{4-x}Cr_xN_y$

Table 8.3 shows some of the previous studies [11, 12]. The magnetic and magneto-thermal properties of $Mn_{4-x}Cr_xN$ in bulk forms were investigated in 1959 [11]. According to that study, T_C increases with the amount of Cr, while the M_S decreases. However, it was also taken into account that with each increase in Cr composition, the amount of denitrification is $1 - x/4$.

Table 8.3. Some previous reports of $Mn_{4-x}Cr_xN$.

Exp./Cal.	x	Properties	References	Year	Sub.
Powder	0–1.4	M–H, M–T	[11]	1959	–
Thin film/MBE	0–1.5	AHE, XMCD	[12]	2022	STO

8.3 $Mn_{4-x}Fe_xN$

Table 8.4 summarises some of the previous studies [11, 13–18]. In addition to the electronic structure, magnetic anisotropy and charge density have been calculated by first-principles calculations. Bulk samples were investigated up to $x = 0.718$ in 1959. These results are also shown.

8.3.1 Magnetic properties of bulk samples

Bulk $Mn_{4-x}Fe_xN$ samples were studied in 1959 by Juza *et al* [11] and in 1970 by Madar *et al* [13]. In 1970, T_C was reported to be $T_C = 756$ K for $x = 0$ and $T_C = 573$ K for $x = 1$ [13]. Meanwhile, the magnetic moment was reported to be 1.17 μ_B (at 0 K) at $x = 0$ and 2.15 μ_B (at 0 K) at $x = 1$ [19].

8.3.2 Change from in-plane magnetic anisotropy for Fe_4N to perpendicular magnetic anisotropy for Mn_4N in $Mn_{4-x}Fe_xN$ thin films on MgO(001)

Figure 8.4(a) shows the Fe composition x dependence of the lattice constants a, c, and the ratio c/a in $Mn_{4-x}Fe_xN$ (x = 0, 1, 2, 3, 4) thin films on MgO(001). Figures 8.4(b) and (c) show the M–H curves when a perpendicular or in-plane magnetic field is applied. In-plane magnetic anisotropy (IMA) is confirmed for Fe_4N and Fe_3MnN with $c/a \geqslant 1$, while for Mn_4N and Mn_3FeN with $c/a < 1$ the out-of-plane magnetic field required for saturation is smaller than that for in-plane. In addition, the hysteresis loop is open when the out-of-plane field is applied, indicating that they are PMA. However, for Mn_2Fe_2N ($c/a < 1$), the in-plane magnetic field required for saturation is reported to be smaller, indicating the IMA. Figure 8.5 shows the calculated spin-resolved DOS around the E_F for these compounds. The sign of the spin polarization of electrical conductivity can be determined by considering the signs of the anisotropic magneto resistance (AMR) ratio and the spin polarization of the DOS at E_F [20]. Based on the results shown in figure 8.4, and experimentally obtained AMR ratios [15], it can be stated that minority spin transport is dominant in Fe_4N, Fe_3MnN, and Fe_2Mn_2N, while majority spin transport is dominant in $FeMn_3N$ and Mn_4N [15].

Table 8.4. Some previous reports of $Mn_{4-x}Fe_xN$.

Exp./Cal.	x	Properties	References	Year	Sub.
Powder	0–0.72	M–H, M–T	[11]	1959	–
Powder	0–1	M–T	[13]	1970	–
Film/MBE	0–4	M–H, c/a, M_S–x	[14]	2018	MgO
Film/MBE	0–4	Spin-resolved DOS, AMR	[15]	2018	MgO
Thin film/MBE	0–4	ρ_{xx}–T, ρ_{AH}–T	[16]	2018	MgO
Cal.	0–4	Spin-resolved DOS, E–k, Charge density	[17]	2006	–
Cal.	1,3	PD, spin-resolved DOS, magnetic anisotropic energy	[18]	2018	–

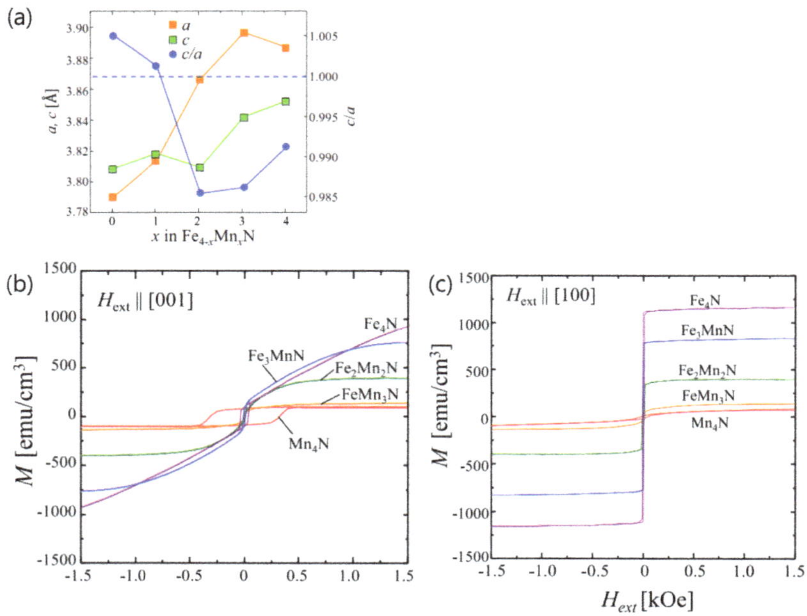

Figure 8.4. (a) In-plane lattice constant a, and out-of-plane lattice constant c of $Mn_{4-x}Fe_xN$. (b and c) Magnetization (M) vs. external magnetic field (H) curves of $Mn_{4-x}Fe_xN$ measured at RT with H_{ext}. Reprinted from [14], Copyright (2018), with permission from Elsevier. (b) Perpendicular to sample surface, the [001] direction of $Mn_{4-x}Fe_xN$, and (c) parallel to sample surface, the [100] direction of $Mn_{4-x}Fe_xN$. Reprinted from [14], Copyright (2018), with permission from Elsevier.

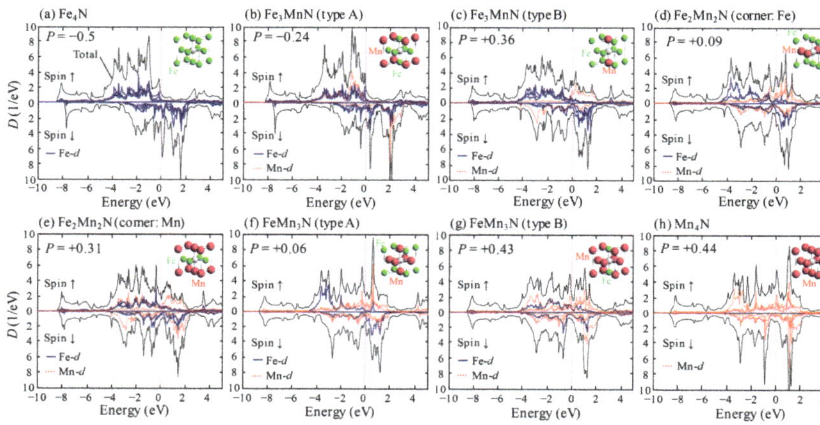

Figure 8.5. Calculated spin-resolved DOS and total DOS of $Mn_{4-x}Fe_xN$. Calculated values of spin polarization at EF are presented. Reprinted from [15], with the permission of AIP Publishing.

8.3.3 First-principles calculation for $Mn_{4-x}Fe_xN$

$Mn_{4-x}Fe_xN$ is unique in that not only Mn but also Fe can have a large magnetic moment, and this assumption is taken into account in first-principles calculations. Most of the studies have assumed a collinear magnetic structure and calculated the electronic structure, charge density, and magnetic anisotropy energy. Figure 8.6 shows one example.

8.4 $Mn_{4-x}Co_xN$

Some of the previous studies are summarized in table 8.5 [13, 21–24]. The magnetic properties of polycrystalline $Mn_{4-x}Co_xN$ were reported by Madar in 1970 [13].

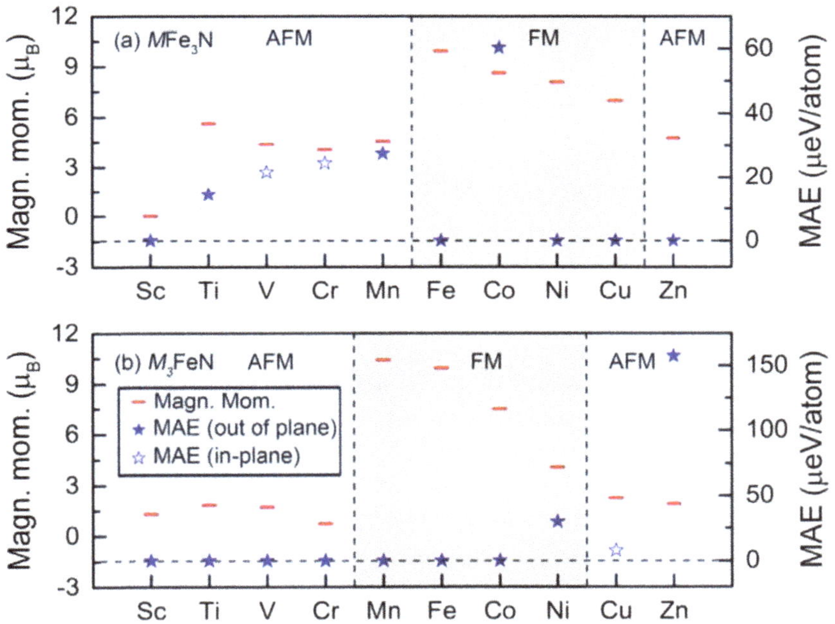

Figure 8.6. Distribution diagrams of the total magnetic moment and magnetic anisotropy energy in MFe_3N (a) and M_3FeN (b). The magnetic coupling is displayed by different color areas. Reprinted from [18], Copyright (2018), with permission from Elsevier.

Table 8.5. List of previous studies of $Mn_{4-x}Co_xN$

Exp./Cal.	x	Properties	References	Year	Sub.
Powder	0–0.7	M–T (x = 0.3–0.6), phase diagram	[13]	1970	–
Film/MBE	0, 0.2	M–H, c/a, magnetic torque	[21]	2016	STO
Film/MBE	0–1.3	M–H, M_S–x, ρ_{AH}–H, XMCD	[22]	2020	STO
Film/MBE	0–4	M–H, M_S–x, c/a, XMCD	[23]	2020	STO
Powder/SSR	0.61	M–H, NPD	[24]	2019	–

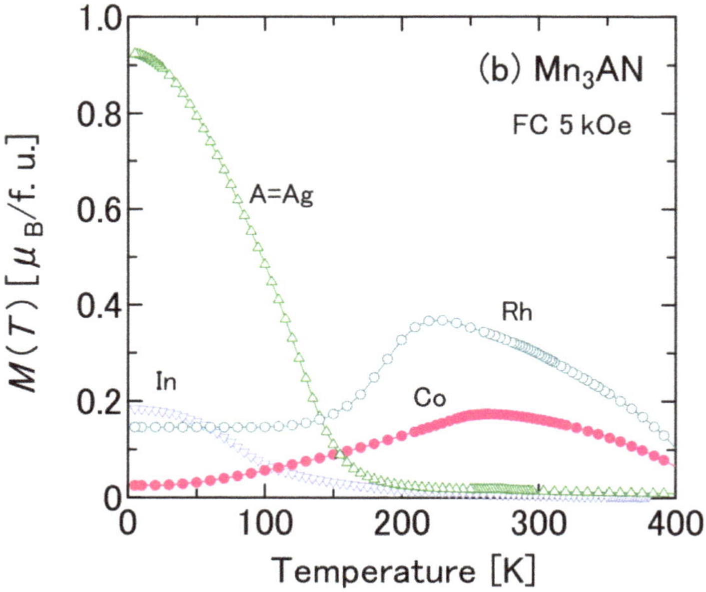

Figure 8.7. Temperature dependence of magnetization in polycrystalline Mn_3CoN. Reproduced from [25]. CC BY 4.0.

Polycrystalline $Mn_3CoN_{0.74}$ was reported to have $T_N = 252$ K [25]. Ito *et al* [21, 23] and Mitarai *et al* [22] reported the preparation of thin film samples on STO(001) substrates, and XMCD showed that the magnetization compensation at room temperature between $x = 0.2$ and 0.8, and between 0.8 and 1.3 [22]. In a study by Sun *et al* (chapter 2), an exchange bias in the M–H curve was observed in polycrystalline $Mn_{3.39}Co_{0.61}N$. They also reported that the neutron powder diffraction method revealed the mechanism of magnetic ordering and magnetization reversal in the samples [24].

8.4.1 Crystalline quality of thin films and magnetic and magneto-transport properties

In reference [21]. $Mn_{4-x}Co_xN$ thin films ($x = 0$, 0.2) were fabricated on STO(001) by MBE. Analyzes using x-ray reciprocal lattice mapping measurements confirmed that $c/a < 1$ for both samples. The M_S was measured by VSM and found to be 80 kA m^{-1} ($x = 0$) and 60 kA m^{-1} ($x = 0.2$) at room temperature. Magnetic torque measurements were also performed on both samples, as shown in figure 8.8. The sawtooth-like curves show that the magnetization was not saturated. Since the magnetization direction was not parallel to H, K_u values were underestimated when we derived them from the maximum amplitude of the torque curves. Thus, we used the 45 degrees-torque method to estimate K_u from unsaturated magnetic torque curves. Figure 8.8(c) shows the $(T/H)^2$–T plots using the averaging-out absolute values at $\theta = 45°$, 135°, 225°, and 315°. The M_S and K_u^{eff} values were obtained from the intersections of the fitting lines with the vertical and horizontal axes, respectively.

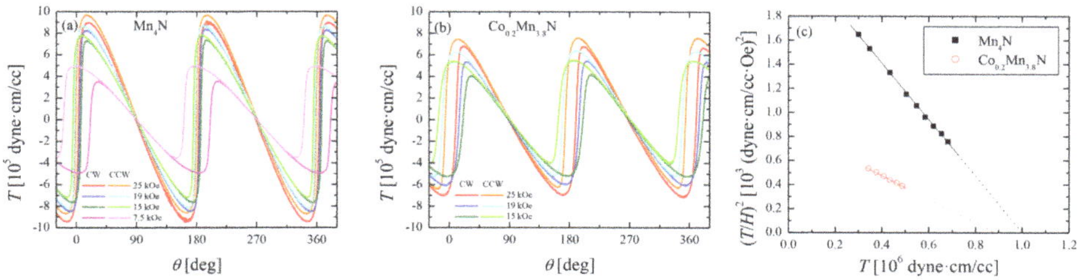

Figure 8.8. Magnetic torque curves of (a) Mn_4N and (b) $Mn_{3.8}Co_{0.2}N$ films measured at room temperature under various magnetic fields. (c) $(T/H)^2$ versus T plots. Reproduced from [21]. CC BY 4.0.

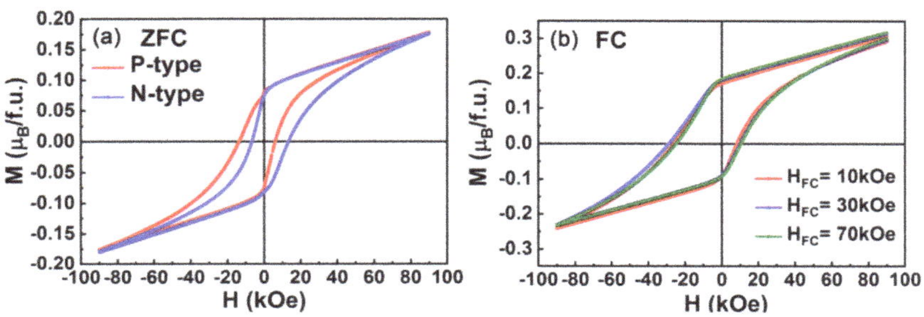

Figure 8.9. (a) Zero-field-cooled (ZFC) $M(H)$ loops of positive type (P-type) and negative type (N-type) measured at 10 K. (b) Field-cooled (FC) $M(H)$ loops of P-type after field cooling in $H_{FC} = 10$, 30, and 70 kOe, respectively. Reprinted figure with permission from [24], Copyright (2019) by the American Physical Society.

Using equation $K_u = K_u^{eff} + 2\pi K M_S^2$, the K_u values were calculated to be 1.0×10^5 J m^{-3} and 8.9×10^4 J m^{-3}, respectively. For those who have interest in the magnetic properties, magneto-transport properties, and XMCD spectra of $Mn_{4-x}Co_xN$ (x=0–4), please visit chapter 5.

8.4.2 Discovery of exchange bias and observation of magnetic order in $Mn_{3.39}Co_{0.61}N$ by NPD

Exchange bias has been observed in $Mn_{3.61}Co_{0.39}N$ produced by an SSR method, as shown in figure 8.9 [24]. In addition, the exchange bias behavior of the hysteresis curve in the M–H curve was observed due to the ferromagnetic coupling between the ferromagnetic phase in the (111) plane and the antiferromagnetic phase of Γ_{5g} type by neutron powder diffraction (NPD) method in figure 8.10. In particular, the temperature dependence of the magnetic moments of the ferromagnetic phase and the antiferromagnetic phase of the Γ_{5g} type was measured, wherein the magnetic moments are larger in the temperature range below 80 K.

Figure 8.11 shows a schematic diagram of the mechanism of magnetization reversal with exchange bias. After applying a magnetic field to the extent of saturation in the positive direction and then demagnetizing, the magnetic moments of Mn and Co at the corner sites rotate in accordance with the applied magnetic field. Based on this argument, it is known that an exchange bias occurs if K_{AFM} (magnetic anisotropy energy of the antiferromagnetic phase) $\geqslant J_{INT}$ (interface

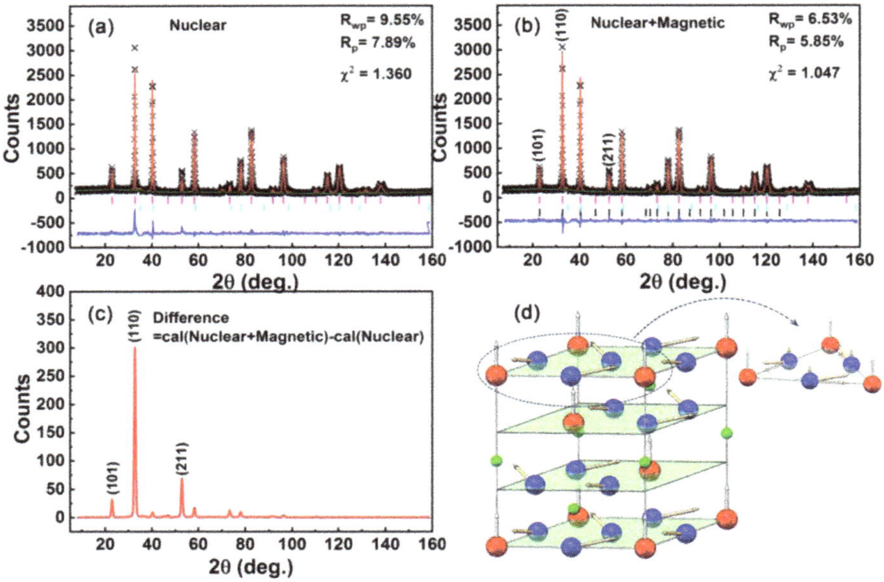

Figure 8.10. NPD pattern recorded at 5 K. In (a) and (b), the crosses signify the experimental intensities; the solid green and red lines show the calculated pattern of background and peaks, respectively; while the blue line means the difference between experimental and calculated intensities. The vertical bars with the colors of pink and cyan show the Bragg reflections of $Mn_{3.39}Co_{0.61}N$ and MnO, respectively. The black vertical bars in (b) show Bragg reflections in the $Mn_{3.39}Co_{0.61}N$ magnetic structure. (c) indicates the difference between the calculated data of (a) and (b), which is generated by magnetic contribution of $Mn_{3.39}Co_{0.61}N$. (d) shows the magnetic structure of $Mn_{3.39}Co_{0.61}N$, in which the white arrows in the [111] direction indicate FM spins on the corner atoms in the cubic structure, and the golden arrows indicate canted Γ_{5g} AFM spin structure. Reprinted figure with permission from [24], Copyright (2019) by the American Physical Society.

coupling constant) from 'Exchange bias' [26]. However, if $J_{INT} \gg K_{AFM}$, then the spins of the ferromagnetic and antiferromagnetic phases must rotate simultaneously and no exchange bias behavior occurs. Therefore, in $Mn_{0.61}Co_{0.39}N$, K_{AFM} (2.8 meV) > J_{INT} (J_{Mn-Co} =1.17 meV), and an exchange bias occurs (figure 8.11).

8.5 $Mn_{4-x}Ni_xN$

Some of the previous studies are summarized in table 8.6 [27–49]. As can be seen from this table, a lot of studies have been done on this material. Magnetic and magnetothermal properties of bulk samples were investigated in the 1950s–1970s, such as the T_C and M_S in 1959 [11] by Juza *et al* and the *M–T* curve for $x = 0–1$ in 1970 [13] by Madar, revealing their magnetic properties and T_C. In addition, in 1980 Madar measured the T_C and M_S of $Mn_{4-x}Ni_xN$ samples. Furthermore, magnetic susceptibility measurements were conducted by l'Htritier in the $x = 0–1$ region in 1980 [50]. In 1978, Fruchart *et al* analyzed the magnetic structure of Mn_3NiN [27] and its temperature dependence in the low-temperature region below 266 K. The Γ_{5g}-AFM structure was found to be stable in the temperature range below 163 K, while the $\Gamma_{4g} + \Gamma_{5g}$-AFM structure was stable in the temperature range of 163–266 K.

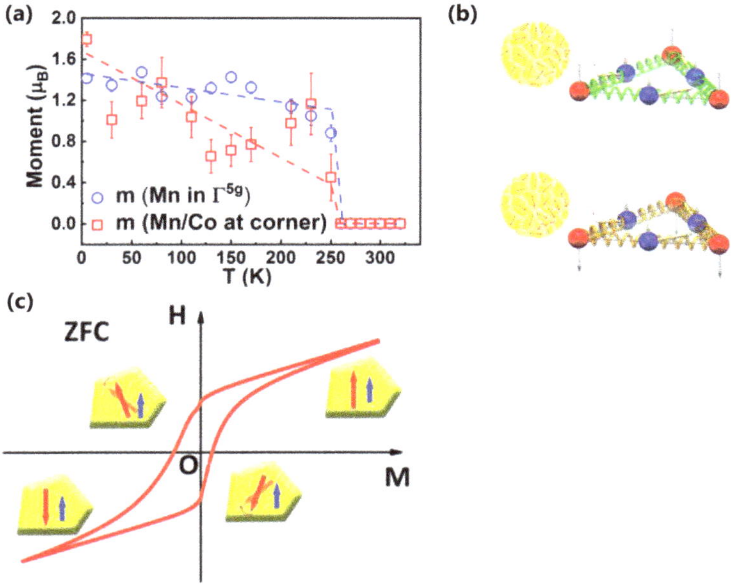

Figure 8.11. (a) Temperature dependence of FM and AFM moments from 5 to 300 K. The lines are guides to the eyes. (b) Schematics of the exchange coupling interaction between canted Γ_{5g} AFM spins and FM spins at $+H_{max}$ and $-H_{max}$, respectively. (c) Schematic of spin rotation in the magnetization process of the ZFC hysteresis loop, the red arrows indicates FM spins for Co/Mn at the corner, while the blue arrows indicate the FM component along the [111] direction for Mn spins at the face center. Reprinted figure with permission from [24], Copyright (2019) by the American Physical Society.

However, there has been no active research since the late-1980s, when reports of calculations on piezomagnetic effects and spin waves were published. Mn_3NiN has experienced a resurgence in research since the 2010s, and detailed identification of its magnetic structure using nuclear magnetic resonance and neutron powder diffraction methods has been attempted. At the same time, the discovery of Inver alloy-like properties with no thermal expansion and properties as an L-TCR material with almost constant electrical resistivity at different temperatures has led to further research. In addition, Mn_3NiN thin films began to appear in the 2010s, and their magnetic properties and their temperature dependencies have been investigated.

The remarkable series of subsequent studies are those of Boldrin *et al* and Johnson *et al,* particularly the studies on the piezomagnetic effect (PME) [35] and the Barocaloric effect (BCE) [36]. It should be noted that these studies have greatly cleared up the prospect of the relationship between magnetic ordering and strain in Mn_3NiN. They have also investigated the relationship between magneto-transport properties and strain by depositing Mn_3NiN on oxide substrates [37]. The discovery of the contribution of the magnetic octupole by anomalous Nernst effect (ANE) measurements was reported in 2022 [40].

As described in chapter 5, magnetic compensation between $x = 0.1$ and 0.25 was also observed at room temperature in $Mn_{4-x}Ni_xN$ thin films. CIDWM has also been investigated in this material, as shown in chapter 7, and a speed of 3,000 m s^{-1} has

Table 8.6. Some previous studies of $Mn_{4-x}Ni_xN$.

Exp./Cal.	x	Properties	References	Year	Sub.
Powder	0–0.6	M–H, M–T	[11]	1959	–
Powder/SSR	1	M–T	[27]	1978	–
Powder /SSR	1	DSC, NPD	[28]	2013	–
Powder /SSR	~1	M–H, M–T, LTE, NPD, MVE	[29]	2015	–
Powder /SSR	1	TCR, DCT, a–, M–, ρ–T	[30]	2010	–
Powder /SSR	0.67–1	Near zero TCR, χ-T, NPD	[31]	2016	–
Film/SD	1	ρ_{xx}–T, M–T	[32]	2011	Quartz
Film/SD	1	M–T, χ–T, M–H	[33]	2015	Quartz
Powder/SSR	0.5, 1	NTE, M–H, M–T, LTE	[34]	2017	–
Film/PLD	1	PME, T_N–, M–ε_{xx}, M–H, M–T	[35]	2018	BTO, STO, LSAT
Powder/SSR	1	BCE, T_N-p	[36]	2018	–
Film/PLD	1	M–T, Biaxial strain	[37]	2019	BTO, STO, LSAT
Film/PLD	1	σ_{AH}, AHE, σ_{AH}–T	[38]	2019	BTO, STO, LSAT
Film/PLD	1	σ_{AH}–ε_{xx}, ρ_{xy}–E	[39]	2021	BTO
Film/PLD	1	ANE, octupole-AFM domain	[40]	2022	STO
Film/MBE	0–4	M–H, M_S–x	[41]	2019	MgO
Film/MBE	0–0.5	M–H, AHE	[42]	2019	MgO STO
Film/MBE	0.1, 0.25	XMCD	[43]	2020	STO
Film/MBE	0.25	Domain wall motion	[44]	2021	STO
Cal.	1	Magnetostriction, PzM	[45]	1989	–
Cal.	1	Spin wave spectrum	[46]	1990	–
Cal.	1	Calculation of ANE	[47]	2020	–
Cal.	1	AHC vs strain (Kagome lattice)	[48]	2022	–
Cal.	1	Spin-phonon coupling	[49]	2022	–

been reported for the STTs alone [44]. In terms of first-principles calculations, anomalous Hall conductivity (AHC) and anomalous Nernst coefficient (ANC) have been calculated focusing on the Kagome structure in the (111) plane, and Mn_3NiN is reported to be a candidate material for spin-thermoelectric applications. Recently, spin-phonon coupling has also been calculated. Mn_3NiN is expected to be utilized in spintronics devices in various aspects.

8.5.1 Piezomagnetic effect (PME) and Barocaloric effect (BCE) in Mn_3NiN

It is known that in the magnetostriction effect, the magnetization responds to the square of the electric field, while in PME the magnetization responds to the first

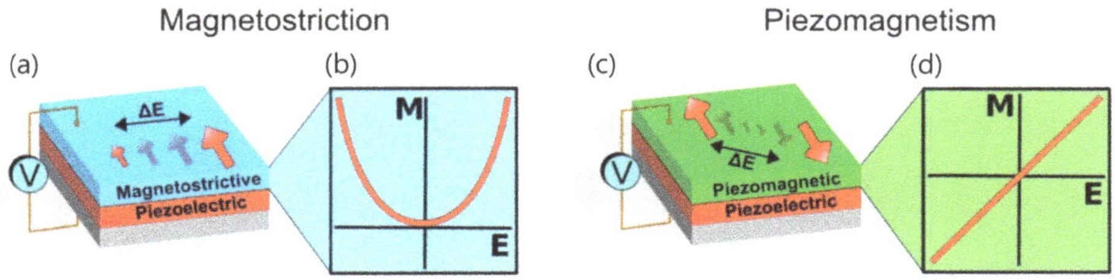

Figure 8.12. Current artificial multiferroic devices that control magnetization with electric field rely on magnetostriction in multiferroic laminates as shown in (a). Magnetostriction varies quadratically with strain and is only significant in materials that possess large spin-orbit coupling. Device operation requires an external magnetic field to enable magnetization reversal (b). In contrast, piezomagnetism (c) is linear in strain allowing sensitive control of magnetization at small strains and, most importantly, magnetization reversal without the need for an external magnetic field (d). Reproduced with permission from [35]. Copyright (2018) American Chemical Society.

order of the electric field. In 2018, Boldrin *et al* [35] used the PLD method to fabricate Mn_3NiN thin films on BTO(001), STO(001), and LSAT(001) substrates, and the relationship between strain and magnetic properties was investigated. It is reported that T_N increases when an in-plane biaxial tensile strain is applied, while T_N decreases when a biaxial compressive strain is applied. The remanent magnetization M_{rem} at 100 K is reported to be large regardless of the direction of strain. BTO is known to undergo a phase transition from a tetragonal to an orthorhombic structure at 275 K and from an orthorhombic to a rhombohedral structure at 187 K. While the magnetization of Mn_3NiN does not change much in the structural phase transition of BTO at 275 K, a large change in magnetization is reported in the structural phase transition at 187 K. The latter phase transition has been shown to induce in-plane uniaxial strain on Mn_3NiN, which may result in PME. A BCE was also reported in 2019 [38]. Near the AFM-PM transition point, the maximum adiabatic entropy change due to pressure is $\Delta S = 35$ J K^{-1} kg^{-1} ($P = 0.28$ GPa) and the adiabatic temperature change is $\Delta T = 5.8$ K ($P = 0.56$ GPa) (figures 8.12–8.16).

8.5.2 First-principles calculation of spin-phonon coupling in Mn_3NiN

Spin-phonon coupling in Mn_3NiN is discussed in reference [49] based on first-principles calculation as shown in figure 8.17.

8.6 $Mn_{4-x}Cu_xN$

Some of the previous studies on $Mn_{4-x}Cu_xN$ are summarized in table 8.7 [13, 25, 51–70]. The magnetostriction effect, ferromagnetic shape memory effect (FSME), magneto-volume effect (MVE), and negative thermal expansion (NTE) of Mn_3CuN have been actively investigated and reported since the early-2010s. ESR measurements have revealed short-range antiferromagnetic ordering. Recently, thin film samples have been prepared and their electrical and magneto-transport properties have been reported as summarized in table 8.7. In this subsection, we focus on the FSME and magneto-transport properties of Mn_3CuN, which are not covered in chapter 2 (figure 8.18).

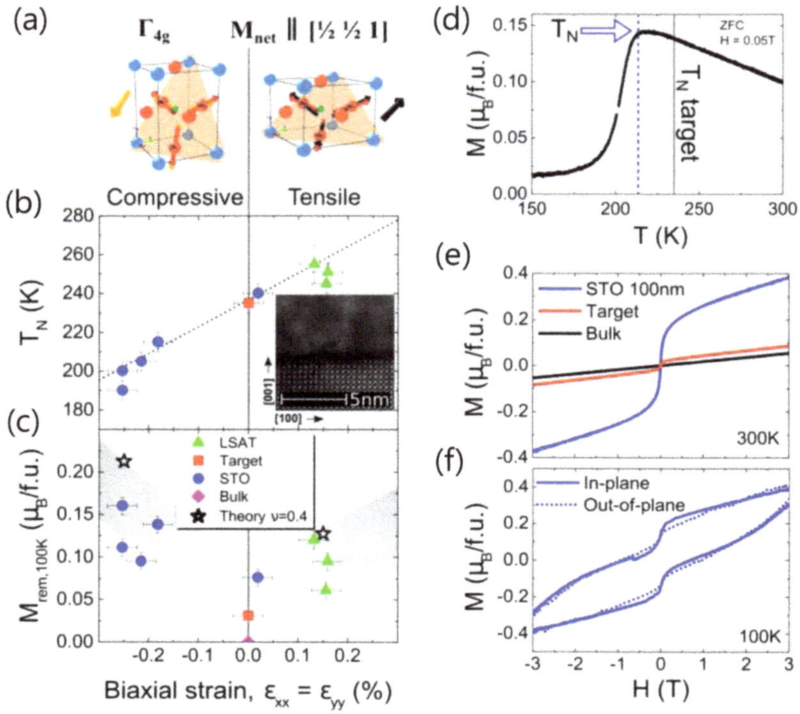

Figure 8.13. (a) Crystal and magnetic structure of Mn_3NiN. Red arrows in both panels indicate local magnetic moments on Mn sites according to G_{4g} representation in unstrained (cubic) lattice with $M_{net} = 0$. Yellow (black) arrows indicate magnetic moments under compressive (tensile) strain. M_{net} is parallel or antiparallel to M_{Mn3} with compressive and tensile strain, respectively. All magnetic moments are in the (111) plane indicated in orange. Canting and changes of size are not to scale. (b) Néel temperature as a function of biaxial strain, $\varepsilon_{xx} = \varepsilon_{yy}$, for all Mn_3NiN films (inset) HRTEM image of a Mn_3NiN film on STO. (c) Remnant magnetization collected with field applied in-plane for all samples. Bulk and target values are presented for comparison. The theoretical data points are calculated according to the method set out in 27 assuming a Poisson ratio $\nu = 0.4$, where $\varepsilon_{xx}/\varepsilon_{zz} = -2\nu/(1\text{-}\nu)$, in agreement with the value found with TEM. (d) Magnetization as a function of temperature for a 100nm Mn_3NiN film grown on STO showing the shift of T_N. (e) $M(H)$ loop at 300K for the same film in (d) as well as the target and bulk samples. (f) $M(H)$ loop at 100K for the same film in (d) with the field applied in- or out-of-plane. Reproduced with permission from [35]. Copyright (2018) American Chemical Society.

8.6.1 Magnetostriction and ferromagnetic shape memory effect (FSME) in Mn_3CuN

FM-PM phase transition occurs in Mn_3CuN at 143 K. A large magnetostriction effect of 0.2 % has been reported in FM-Mn_3CuN [51]. FSME has also been studied [52, 53]. The origin of the magnetostriction effect and FSME has been discussed in terms of the electronic structure of Mn_3AN, where a narrow band near the E_F, which is a hybrid of Mn-3*d* and N-2*p* orbitals, overlaps the conduction band, and its degree of overlap is known to depend on the valence electron number of dopant A. In this case, Cu has the lowest valence electron number n_v among dopant elements. It has been pointed out that this can cause cubic-to-tetragonal transformation, and

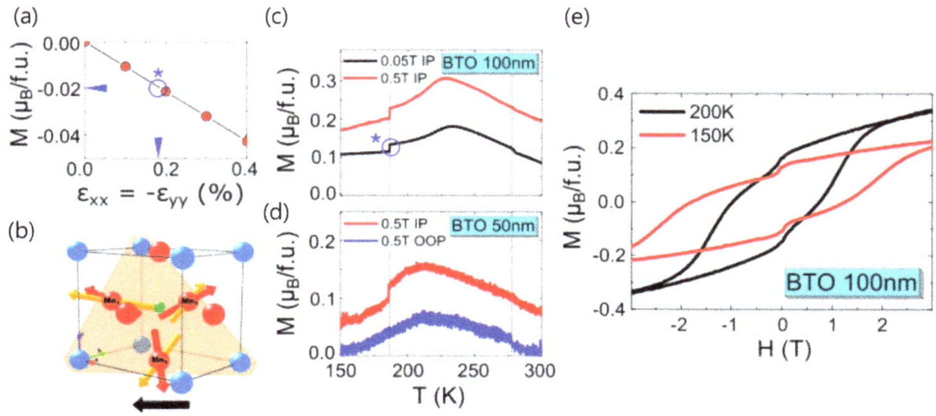

Figure 8.14. (a) Predicted variation in M with uniaxial strain. The blue circle, star symbol, and horizontal arrow indicate the experimentally measured change in M from (c) and the vertical arrow indicates the inferred strain. (b) Predicted lattice and magnetic structure under uniaxial strain. The net moment (black) is wholly in the plane of the film. (c and d) Experimental $M(T)$ of 100 and 50 nm Mn_3NiN films on BTO showing a large jump in M at the orthorhombic to rhombohedral transition at 187 K in BTO when measured with the applied field out-of-plane, which produces uniaxial strain. (e) The corresponding $M(H)$ loops in the AFM state of Mn_3NiN above and below the 187 K BTO transition. Reproduced with permission from [35]. Copyright (2018) American Chemical Society.

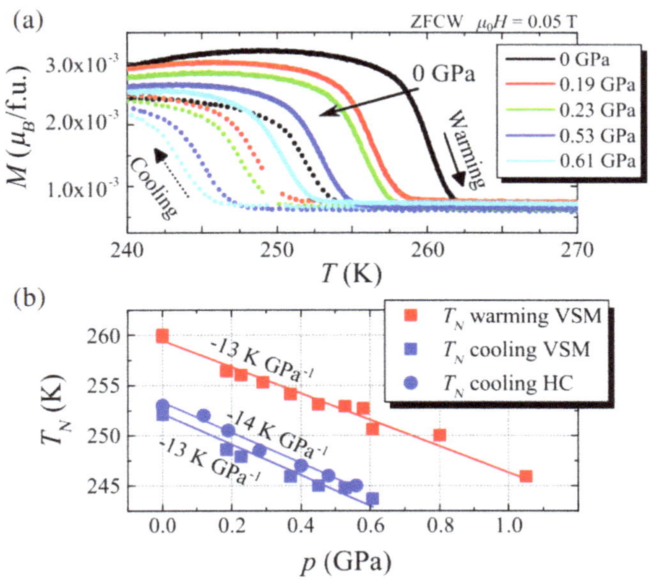

Figure 8.15. (a) Magnetization under an applied field of $\mu_0 H = 0.05$ T as a function of the temperature and measured at different pressures after zero-field cooling. (b) Pressure dependence of T_N taken from the derivative of the $M(T)$ VSM data and the peak of the HC data. Solid lines are linear fits to the data. Reproduced with permission from [36]. CC BY 4.0.

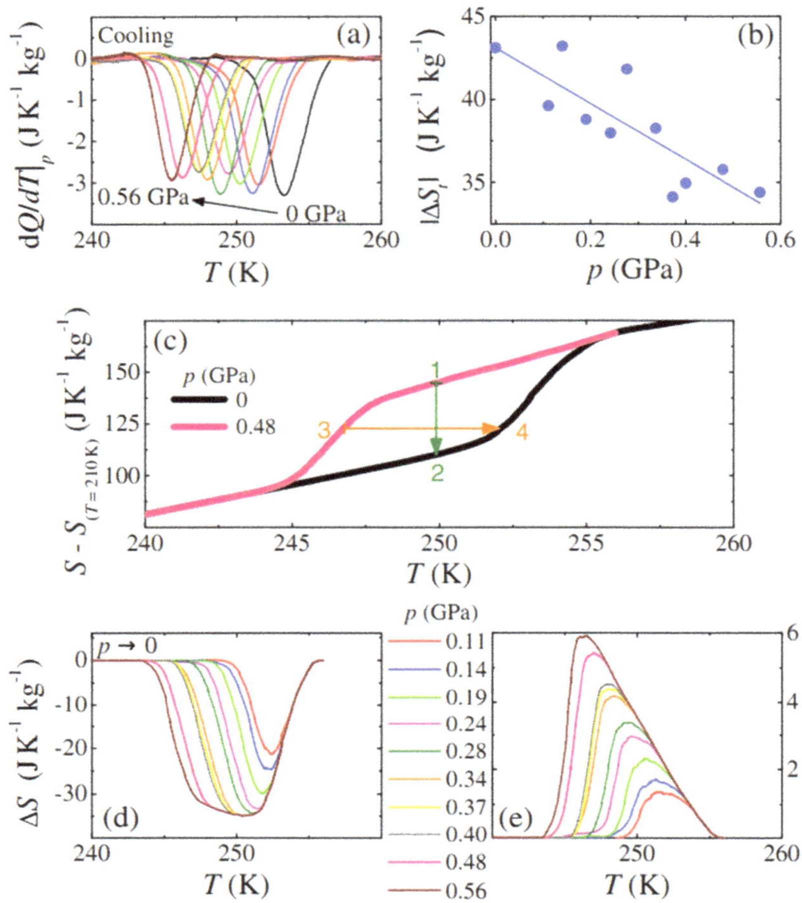

Figure 8.16. (a) Temperature-dependent heat flow under pressure $[(dQ)/(dT)]\,|_p$ on cooling through the transition at different values of increasing pressure p after baseline subtraction. The full legend is the same as in (d) and (e). (b) Entropy changes cooling through T_N obtained from the data in (a). (c) Isobaric entropy curves $S - S_{210\ K}$ on cooling at 0- and 0.48-GPa applied pressure. The green arrow from 1 to 2 and orange arrow from 3 to 4 indicate the isothermal entropy and adiabatic temperature changes, respectively, upon removal of pressure. (d and e) Isothermal entropy and adiabatic temperature changes obtained from the data in (a) as explained in (c). Reproduced from [36]. CC BY 4.0.

the band Jahn–Teller effect is known to be the physical background of martensitic transformation in ferromagnetic shape memory alloys. Linear thermal expansion $\Delta L(T)/L$ and strain of Mn_3CuN under parallel ($H_{||}$) and perpendicular ($H_{\perp}$) magnetic fields are shown in figure 8.19.

8.6.2 Magneto-transport properties in Mn_3CuN thin films

The magneto-transport properties of Mn_3CuN thin films deposited by sputtering on MAO(001), MgO(001), and STO(001) substrates have also been evaluated, and it has been reported that the T_C decreases for sample formed on MAO, MgO, and

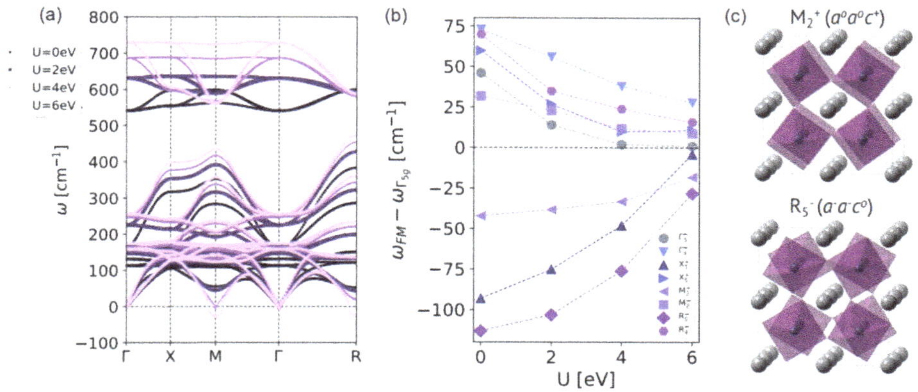

Figure 8.17. (a) Phonon dispersion for the Γ_{5g} ordering in the Mn_3NiN computed at different U values in which, the lattice volume was fixed to the experimentally obtained value. The latter is in order to explore the effect of the correlation into the vibrational modes. Here, the unstable modes are shown at negative frequency values by notation. (b) Obtained change in the frequencies, $\Delta\omega = \omega_{FM} - \omega_{\Gamma 5g}$, when the magnetic transitions in induced. In the latter, the extracted phonon frequencies of the lowest vibrational modes at the high-symmetry points in the Brillouin zone. (c) Representation of the structural distortions associated with the M_2^+ and R_5^- modes where the octahedral rotation and tilting can be appreciated, respectively. Reprinted from [49], Copyright (2022), with permission from Elsevier.

STO, in that order, figure 8.20 [61]. The inset of figure 8.20(a) shows the relationship between T_C and lattice constants of Mn_3CuN thin films deposited on each substrate. Hall effect measurements have also been performed for Mn_3CuN/MgO samples in various temperature ranges, and it is reported that the normal Hall coefficient is negative in the temperature range 5–150 K, while it is found to be positive above 200 K. It is also reported that the hysteresis loop of the anomalous Hall effect opens in the temperature range below 150 K because of the FM-PM transition.

For the normal Hall effect, the carrier density and mobility were measured, and it was reported that the hole density decreased drastically in the temperature region below T_C, while the electron carrier density slightly increased. In addition, it was reported that the mobility of holes slightly increased while the mobility of electrons rapidly increased in the temperature region below T_C. It is also characteristic that the hole conductivity decreases rapidly below T_C and increases at lower temperatures, which dominates the overall conductivity of Mn_3CuN. As mentioned earlier, a structural phase transition occurs simultaneously with the magnetic phase transition in Mn_3CuN. During the structural phase transition, the hole carrier density near the E_F decreases, which explains the carrier behavior in Mn_3CuN.

8.7 $Mn_{4-x}Zn_xN$

Some of the previous studies on $Mn_{4-x}Zn_xN$ are summarized in table 8.8 [13, 71–77]. Mn_3ZnN has been developed largely as an NTE material. However, since phase separation was confirmed by NPD, the development of phase-separation memory has been aimed at and resistive switching has been studied.

Table 8.7. Some previous studies of $Mn_{4-x}Cu_xN$.

Exp./Cal.	x	Properties	Reference	Year	Sub.
Powder	0–1	M–H, magnetic phase diagram	[13]	1970	–
Powder	0–0.3	M–T, a–x, magnetic moment per atom	[51]	2022	–
Powder/SSR	1	FSME, magnetostriction	[52]	2008	–
Powder/SSR	1	FSME, N-deficient	[53]	2010	–
Powder/SSR	1	FSME, LTE, T_C-(1-δ)	[54]	2013	–
Powder/SSR	1	Critical behavior	[55]	2013	–
Powder/SSR	~1	$Mn_3Cu_{1-x}A_xN$, LTE, MVE	[25]	2014	–
Film/MBE	0.4	XPS, M–H	[56]	2014	MgO
Film/PLD	1	RMS, XPS, M–T	[57]	2020	MgO
Powder/SSR	~1	LTE, C_p–T (spin glass)	[58]	2010	–
Film/SD	1	EDX, XPS, χ–T, ρ_{xx}–T	[59]	2010	Quartz
Film/SD	1	ρ_{xx}–T, M–T, M–H	[60]	2013	YSZ MgO
Film/PLD	1	SEM/TEM, J–V, ρ_{xx}–T etc	[61]	2017	n-Si
Film/SD	1	ρ_{xx}–T, M–T, ρ_{AH}–H, n, μ–T	[62]	2017	MgO STO MAO
Powder/SSR	1	XMCD	[63]	2011	–
Powder/SSR	0.5–1	M–T, ρ_{xx}–T, TCR	[64]	2012	–
Powder/SSR	0.6–0.9	M–T, χ–T, ESR, SR-AFM	[65]	2015	–
Cal.	1	Ge-doped DOS, μ, charge	[66]	2010	–
Cal.	~1	Ge-doped elastic properties	[67]	2011	–
Cal.	1	XMCD spectra calculation	[68]	2014	–
Cal.	1	TE, Spin-resolved DOS, σ_{xx}	[69]	2018	–
Cal.	1	Ground-state phase, J	[70]	2019	–

In the temperature range of 140 K–177 K, the coexistence of two phases, C1 and C2, in Mn_3ZnN has been shown by the NPD method as described in chapter 2. The temperature dependence of electrical resistivity was investigated during this phase separation, and a jump in electrical resistivity near the phase-separation temperature (~190 K in this study) can be observed in figure 8.22(b) [74]. It is important to note that the cause of the resistivity jump is neither metal-semiconductor transition nor AFM transition but phase separation. It is also known that most resistive switching phenomena originate from either strong electronic correlation or magnetic ordering. To determine whether either of these is dominant, a Kadowaki–Woods ratio was investigated for Mn_3ZnN and was reported to be 2.026×10^{-4} μW cm K^{-2}. This result indicates that electronic correlations are dominant in Mn_3ZnN.

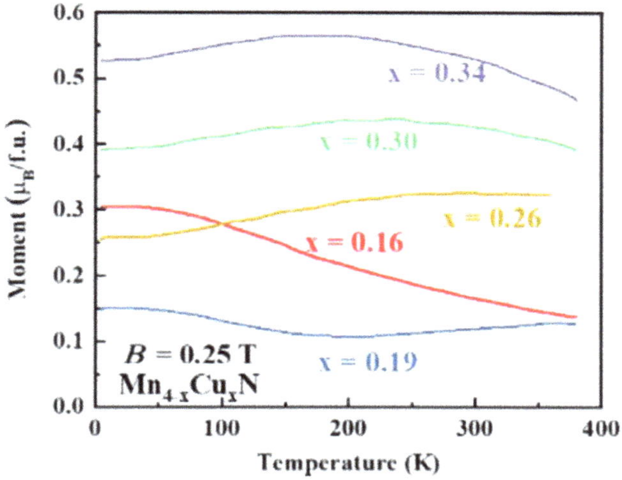

Figure 8.18. Temperature dependence of (a) magnetization and (b) magnetic susceptibility in $Mn_{4-x}Cu_xN$. Reproduced from [51]. CC BY 4.0.

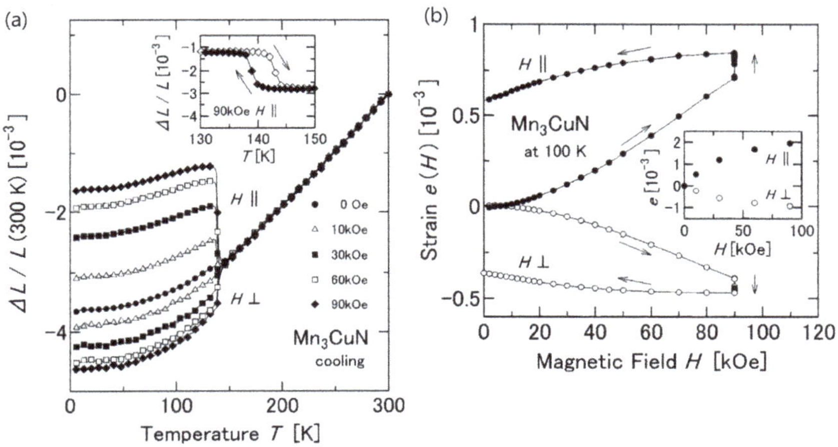

Figure 8.19. (a) Linear thermal expansion $\Delta L(T)/L$ of Mn_3CuN parallel ($H_{||}$) and perpendicular ($H_{\perp}$) to the field up to 90 kOe in a cooling process. Inset: temperature hysteresis at the transition. (b) Field-induced strain $e(H)$ of Mn_3CuN at 100 K. The field was applied at the rate of 12 Oe/s up to 90 kOe, fixed for 3 h, then removed at the same rate. Even under the constant field the strain gradually increases. Inset: $e(H)$ estimated from the thermal expansion data shown in (a). Reprinted from [52], with the permission of AIP Publishing.

8.8 $Mn_{4-x}Ga_xN$

Table 8.9 summarizes some of the previous studies [13, 51, 71, 78–95]. As described in chapter 2, Mn_3GaN has a long history of development as an NTE material. Magnetic ordering has also been studied most intensively in Mn_3AN materials, and phenomena such as spin-glass structures [78, 79], BCE [81], and ferromagnetic ordering [87] have been reported. Recently, SOT switching has been studied

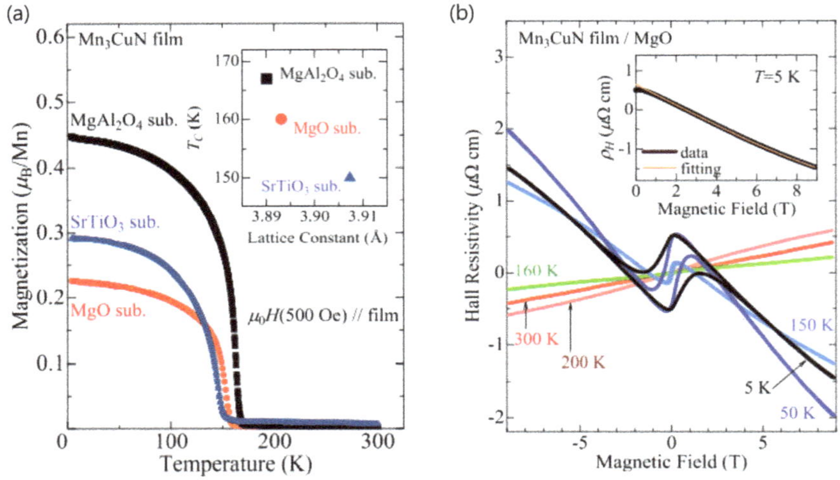

Figure 8.20. (a) Temperature dependence of magnetization of Mn_3CuN films prepared on different substrates and annealed with Ti powders. A magnetic field of 500 Oe was applied parallel to the film surface. The inset shows the Curie temperature (T_C) as a function of out-of-plane lattice constant. (b) Magnetic field dependence of Hall resistivity (ρ_H) of the Mn_3CuN film grown on MgO substrate. The inset shows the result of fitting the two-carrier model to the 5 K data at $m_0H \geqslant 5$ T. Reproduced with permission from [62] (figure 8.21), Copyright (2017) by the American Physical Society.

Table 8.8. Some previous studies of $Mn_{4-x}Zn_xN$.

Exp./Cal.	x	Properties	reference	Year
Powder	1	NPD, M–T, a–T	[13]	1970
Powder/SSR	1	Heat capacity	[71]	1983
Powder	1	LTE, M–T, C-doped	[72]	2012
Powder/SSR	1	Phase-separation, NPD, χ–T	[73]	2012
Powder	1	Resistive switching, $1/\chi$–, ρ_{xx}–, a–T	[74]	2012
Powder/SSR	1	TEM, EDX, XPS	[75]	2013
Cal.	1	NTE, magnetic structure	[76]	2010
Cal.	1	Ising model, M–J, M–H	[77]	2021

extensively, including ST-FMR measurements [90]. Small-doped materials have also attracted attention as candidates for room temperature magnetization compensation materials [91] (figures 8.23–8.27).

8.9 $Mn_{4-x}Ge_xN$

Table 8.10 shows two previous studies [13, 51]. In reference [51], the possible magnetic compensation at room temperature has been discussed (figure 8.28).

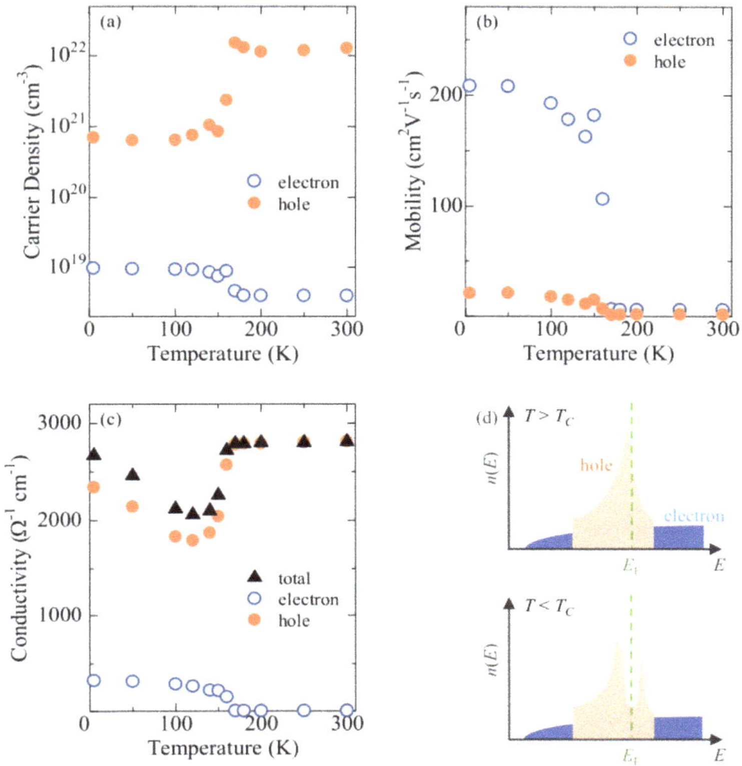

Figure 8.21. Temperature dependence of (a) carrier concentration and (b) carrier mobility extracted from the Hall coefficient data. (c) The electrical conductivity calculated from the carrier concentration and mobility plotted in (a) and (b). (d) Schematic illustration of the electronic structure deduced from the Hall resistivity data. Reprinted figure with permission from [62], Copyright (2017) by the American Physical Society.

8.10 $Mn_{4-x}Rh_xN$

From interest in giant NTE, magnetostructural correlations have been investigated systematically using sintered polycrystalline samples of stoichiometric $Mn_{4-x}Rh_xN$ and solid solution $Mn_3Cu_{1-x}Rh_xN$ [96]. Table 8.11 shows two of previous studies. In reference [51], the possible magnetic compensation at room temperature has been discussed (figure 8.29).

8.11 $Mn_{4-x}Pd_xN$

Various magnetic properties of $Mn_{4-x}Pd_xN$ have been studied, such as magneto-volume effect and thermal expansion effect, as shown in table 8.12 [21, 96–98] (figures 8.30 and 8.31).

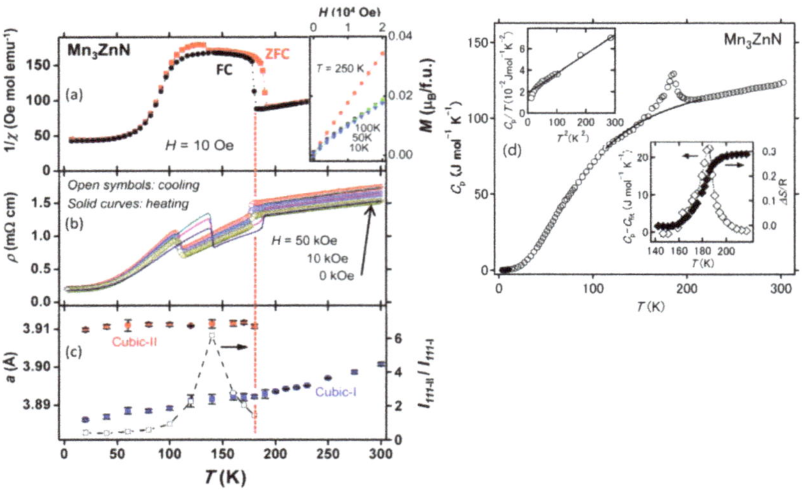

Figure 8.22. (a) T dependence of the inverse c of Mn_3ZnN, (b) ρ in various magnetic fields, and (c) the cubic lattice parameter and the intensity ratio between the 111 peaks for the Cubic–I and –II phases. (d) Temperature dependence of C_p of Mn_3ZnN. The left-hand and right-hand insets show a linear fit to the C_p/T vs. T^2 curve and an estimation of DS, respectively. Reprinted from [74], with the permission of AIP Publishing.

8.12 $Mn_{4-x}Ag_xN$

There have been many measurements on NTE coefficient and low-temperature coefficient of resistance (TCR) for $Mn_{4-x}Ag_xN$, including $Mn_3Ag_{1-x}Cu_xN$. This point is also introduced in chapter 2. Here, their magnetic properties are described. Table 8.13 shows some of the previous studies [99–106] (figures 8.32 and 8.33).

8.13 $Mn_{4-x}In_xN$

Some of the previous studies are summarized in table 8.14 [107–111]. Mekata evaluated the magnetic and magnetothermal properties of $Mn_{4-x}In_xN$ samples [107]. Recently, magnetic properties of polycrystalline Mn_3InN were investigated in 2012, and magnetic, magneto-transport, and magneto-optical properties of $Mn_{4-x}In_xN$ epitaxial films were investigated in 2022 (figure 6.7).

8.14 $Mn_{4-x}Sn_xN$

Some of the previous studies are summarized in table 8.15 [34, 107, 112–115]. The magnetic and magnetothermal properties of polycrystalline $Mn_{4-x}Sn_xN$ were reported by Mekata [107]. Boldrin *et al* also performed LTE measurements on $Mn_{3.5}Sn_{0.5}N$ and Mn_3SnN [35]. In 2020, the magneto-transport properties on Mn_3SnN were measured. SOT-switching was reported in 2021 [113]. Recent studies of $Mn_{4-x}Sn_xN$ epitaxial films are described in chapter 6.

Table 8.9. Some previous studies of $Mn_{4-x}Ga_xN$.

Exp./Cal.	x	Properties	reference	Year	Sub.
Powder	0–1	M–T, magnetic phase diagram	[13]	1970	
Powder	0–0.4	M–T, a–T, magnetic moment per atom	[51]	2022	
Powder/SSR	1	C_p	[71]	1983	
Powder/SSR	0.75–1	Spin-glass, phase diagram	[78]	1986	
Powder/SSR	1	Spin-glass, M–T, χ–T	[79]	2008	
Powder/SSR	0.9–1	M–T, M–H phase diagram	[80]	2010	
Powder/SSR	1	BCE	[81]	2015	
Powder	0.5–0.7	AFM-robust FM, NPD	[82]	2018	
Powder/SSR	1	3D-ising arrot plot	[83]	2020	
Film/MBE	1	TEM, M–H, M–T	[84]	2006	(Ga,Mn)N
Film/SD	1	ρ_{xx}–T, magnetocapacitance	[85]	2013	STO LSAT BSTO
Film/SD	1	Mn_3GaN/Co_3FeN, AMR	[86]	2015	MgO
Film/SD	~1	FM-phase c/a, PMA, M–H	[87]	2018	MgO
Film/SD	1	SOT, Pt, AFM-, FM-Mn_3GaN	[88]	2019	MgO
Film/SD	1	SOT, Pt, Ta, AFM-Mn_3GaN	[89]	2021	MgO
Film/SD	1	Spin polarization, FMR	[90]	2020	LSAT
Cal.	1	SP, Spin-DOS, M–a	[91]	2005	
Cal.	1	Frustrated-triangular	[92]	2010	
Cal.	1	Ferroelectric/piezomagnetic hetero structure, ME effect	[93]	2011	
Cal.	1	Spin-DOS, C_p–T, χ–T	[94]	2021	
Cal.	1	AHC	[95]	2020	

8.15 $Mn_{4-x}Sb_xN$

Table 8.16 shows some of previous reports [108, 116, 117]. Mn_3SbN undergoes a phase transition from a high-temperature paramagnetic state to a low-temperature ferromagnetic state at $T_C = 360$ K, accompanied by a cubic-to-tetragonal structural deformation [116, 117]. It has also been discovered that Mn_3SbN in the tetragonally distorted FM phase exhibits large magnetostriction of up to 450 ppm at room temperature from the linear thermal expansion under a magnetic field of 90 kOe [118] (figure 8.35).

8.16 $Mn_{4-x}Pt_xN$

Table 8.17 summarizes some of previous reports [99, 118, 119]. In reference [119], Mn_3PtN_x ($x = 0$, 0.25, 0.5, 0.75, and 1.0) compounds were prepared by SSR, and the structure, magnetic properties, and thermal expansion behaviors with different nitrogen content were systematically investigated. The crystal structure changes at

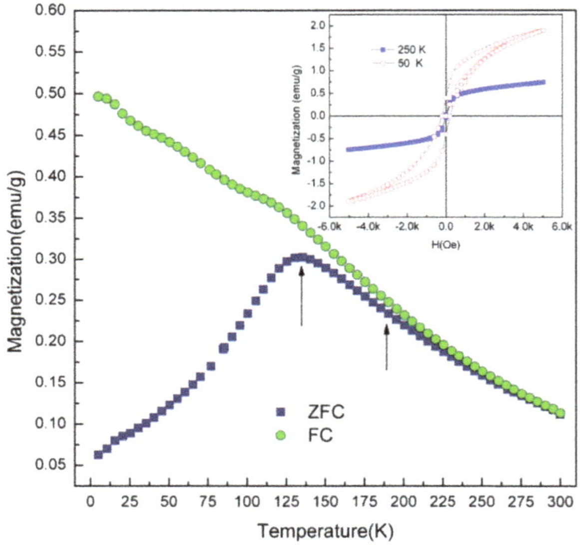

Figure 8.23. Temperature dependence of FC and ZFC dc magnetization at 100 Oe. Arrows indicate the onset of the weak and strong irreversibility. Inset: hysteresis loops of Mn_3GaN measured at T= 50 and 250 K, respectively. Reprinted from [79], with the permission of AIP Publishing.

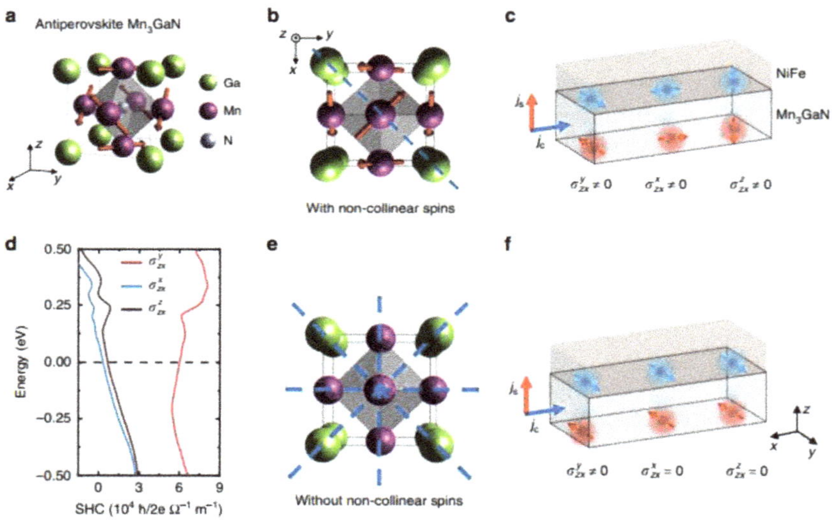

Figure 8.24. The concept of the unconventional spin-Hall effect in Mn_3GaN. (a) The crystallographic unit cell of antiperovskite Mn_3GaN with the antiferromagnetic Γ_{5g} spin structure where Mn spins (arrows) form a Kagome-type lattice in the (111) plane. x, y, and z correspond to the cubic [101], [010], and [001] axes, respectively. (b) Spin structure of Mn_3GaN projected onto the (001) plane. The blue-dashed line corresponds to the (110) mirror plane. (c) Schematic illustrations of the Py/Mn_3GaN bilayer and the allowed spin-Hall spin polarization in the low-symmetry state (a). This indicates non-zero spin-Hall conductivities $\sigma_{zx}{}^{y}$, $\sigma_{zx}{}^{x}$, and $\sigma_{zx}{}^{z}$, which correspond to spin polarizations along y, x, and z direction, respectively (with the charge current along x and spin flow along z). (d) Calculated spin-Hall conductivities $\sigma_{zx}{}^{y}$, $\sigma_{zx}{}^{x}$, and $\sigma_{zx}{}^{z}$ for Mn_3GaN in the antiferromagnetic phase as a function of Fermi energy. (e) Crystal structure of Mn_3GaN without non-collinear spin structure (i.e., above the antiferromagnetic transition temperature T_N) in the (001) plane, which gives rise to a high-symmetry state. (f) Allowed spin polarization in the high-symmetry state, where only the conventional spin-Hall conductivity $\sigma_{zx}{}^{y}$ is non-zero. Reproduced from [90]. CC BY 4.0.

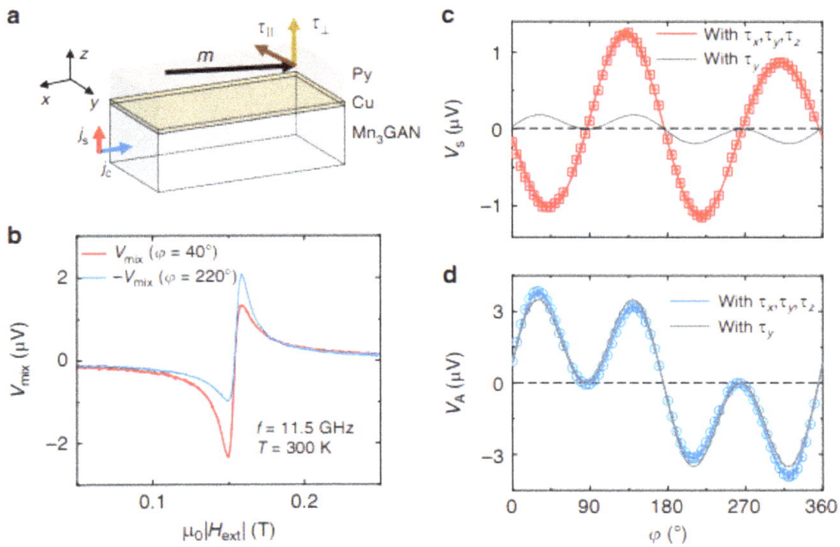

Figure 8.25. ST-FMR measurements. (a) Schematic of the ST-FMR geometry for the Py(10 nm)/Cu(2 nm)/Mn_3GaN(20 nm) structures. (b) ST-FMR spectra for the sample at 300 K (antiferromagnetic phase) with Py magnetization oriented at 40 and 220 degrees relative to the current axis. (c and d) Symmetric (c) and antisymmetric (d) ST-FMR components for the direction (x-axis). The applied microwave frequency and power are 11.5 GHz and 15 dBm, respectively. The error bars indicate fitting uncertainties. Reproduced from [90]. CC BY 4.0.

room temperature from cubic to hexagonal, and then back to cubic as the N content gradually decreases. It was found from the isothermal magnetization curves that all of the obtained Mn_3PtN_x show AFM characteristics. In addition, anomalies were observed in thermal expansion curves around the magnetic transition from AFM to PM. This result indicates the presence of strong spin-lattice coupling (figure 8.36).

8.17 $Mn_{4-x}Au_xN$

Table 8.18 summarizes some of previous reports. In Refs. [120, 121], the sign change of AHE was observed in $Mn_{4-x}Au_xN$ epitaxial films. As described in chapter 6, this is related to the reversal of the magnetic moment of Mn atoms at face-centered site by Au doping (figure 8.37).

8.18 $Mn_{4-x}Gd_xN$

Table 8.19 summarizes one of previous reports [122]. It has been confirmed that Gd doping causes suppression of the AHE due to heavy Fermion system [122]. Samples with x = 0, 0.08, 0.1, 0.14, 0.24, 0.3, 0.4, and 0.48 were prepared on MgO(001) substrate by MBE. The magnetic properties and the temperature dependence of ρ_{xx} and ρ_{AH} are reported in detail in this study. The magnetic properties are considered to be due to competition between Kondo coupling by localized 4f electrons and

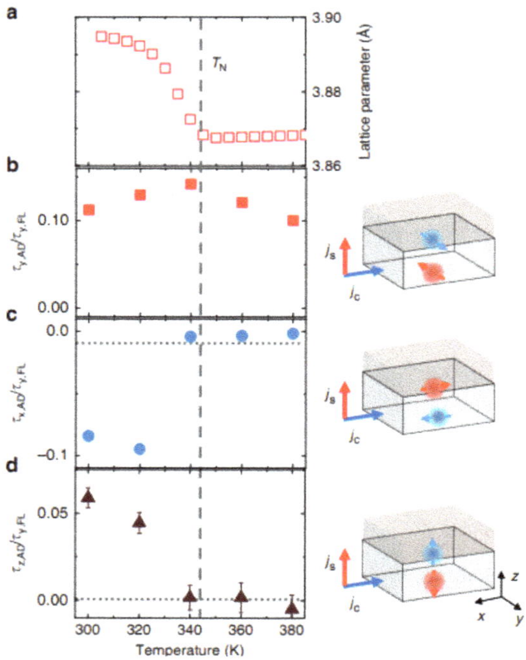

Figure 8.26. Temperature dependence of spin-orbit torques. (a) Out-of-plane lattice parameter of a 30 nm Mn_3GaN/LSAT sample as a function of temperature; the lattice parameters anomaly indicates the Mn_3GaN Néel temperature T_N of ∼345 K. (b–d) The torque ratios $\tau_{y,AD} = \tau_{y,FL}$, $\tau_{x,AD} = \tau_{y,FL}$ and $\tau_{z,AD} = \tau_{y,FL}$ as a function of the temperature for the Py(10 nm)/Cu(2 nm)/Mn_3GaN(20 nm) device measured at the microwave frequency of 7 GHz. The schematics on the right-hand panel show the geometry of the spin-Hall effect with different spin polarizations. The error bars indicate fitting uncertainties. Some of them are smaller than the size of the symbols. Reproduced from [90]. CC BY 4.0.

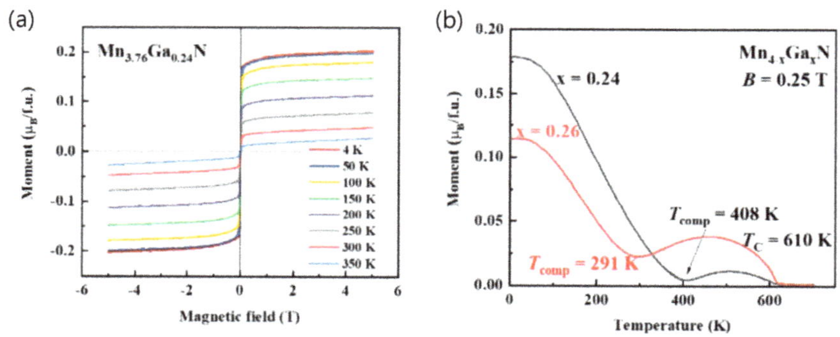

Figure 8.27. (a) *M–H* curves for $Mn_{4-x}Ga_xN$ ($x = 0.24$) at different temperatures and (b) thermomagnetic scans in $Mn_{4-x}Ga_xN$ ($x = 0.24$ and 0.26). Reproduced from [51]. CC BY 4.0.

Table 8.10. Some previous studies of $Mn_{4-x}Ge_xN$.

Exp./Cal.	x	Properties	References	Year
Powder	0–0.6	*M–H*, magnetic phase diagram	[13]	1970
Powder	0–0.4	*M–T, a–T*, magnetic moment per atom	[51]	2022

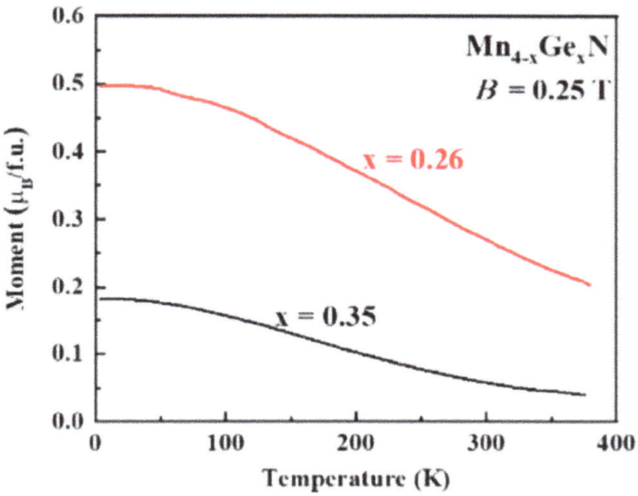

Figure 8.28. *M–T* curves in $Mn_{4-x}Ge_xN$ (x = 0.26 and 0.35). Reproduced from [51]. CC BY 4.0.

Table 8.11. Some previous studies of $Mn_{4-x}Rh_xN$.

Exp./Cal.	x	Properties	Reference	Year
Powder	1	*M–T*, NPD, *a–T*	[96]	1967
Powder	$Mn_{4-x}Rh_xN$ (x = 0.5,1) $Mn_3Cu_{1-x}Rh_xN$ (x = 0–1)	NTE, *M–T*	[50]	2014

conduction electrons and RKKY interactions at low temperatures, but it is unclear to what extent both interactions contribute to the room temperature magnetic properties. However, it has been reported that the in-plane coercivity increases with increasing composition. It is speculated that this is due to the antiferromagnetic coupling between the 4*f* electrons of Gd and the 3*d* electrons of Mn.

The temperature dependence of ρ_{xx} has been investigated, showing Fermi liquid-like behavior in the temperature range 5–20 K and a significantly larger electronic

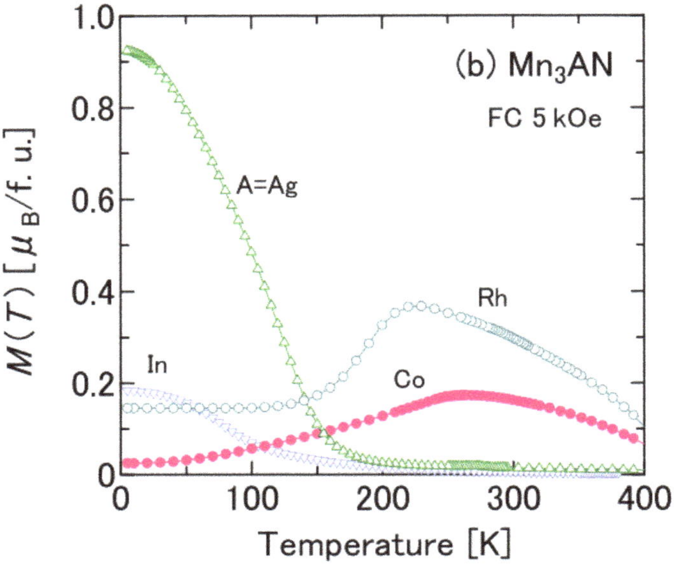

Figure 8.29. Temperature dependence of magnetization in Mn_3RhN. Reproduced from [50]. CC BY 4.0.

Table 8.12. Some previous studies of $Mn_{4-x}Pd_xN$.

Exp./Cal.	x	Properties	Reference	Year	Sub.
Powder	1	a–T	[96]	1967	–
Powder	0.5,1 $Mn_3Cu_{1-x}Pd_xN$ (x = 0–1)	NTE, M–T	[50]	2014	–
Film/SD	0–0.36	ρ_{xx}–T, M–H, M–T, MST	[97]	2018	Si
Powder	1	ρ_{xx}–T, C–T, M–T, LTE, TE	[98]	2021	–

specific heat coefficient than that of common metals. Kondo coupling by the localized magnetic moment and conduction electrons is dominant in the temperature range of 50–110 K. Above this temperature range, it is found that weak metal conducting occurs due to the decrease in the mean free path of conduction electrons caused by disordering. A detailed analysis is also performed for ρ_{AH}, and the methodology follows the method of Hou *et al* [123], as shown in equation (8.8). where ρ_{xx0} indicates the residual resistivity and ρ_{xx} indicates the longitudinal electrical resistivity at finite temperature.

$$\rho_{AH} = \alpha\rho_{xx0} + (\beta_0 + \beta_1 - \gamma)\rho_{xx0}^2 + (\gamma - 2\beta_1)\rho_{xx0}\rho_{xx} + \beta_1\rho_{xx}^2 \tag{8.8}$$

The first term involving α indicates the contribution of skew scattering, while the second and third terms indicate the contribution of static and dynamic disorder

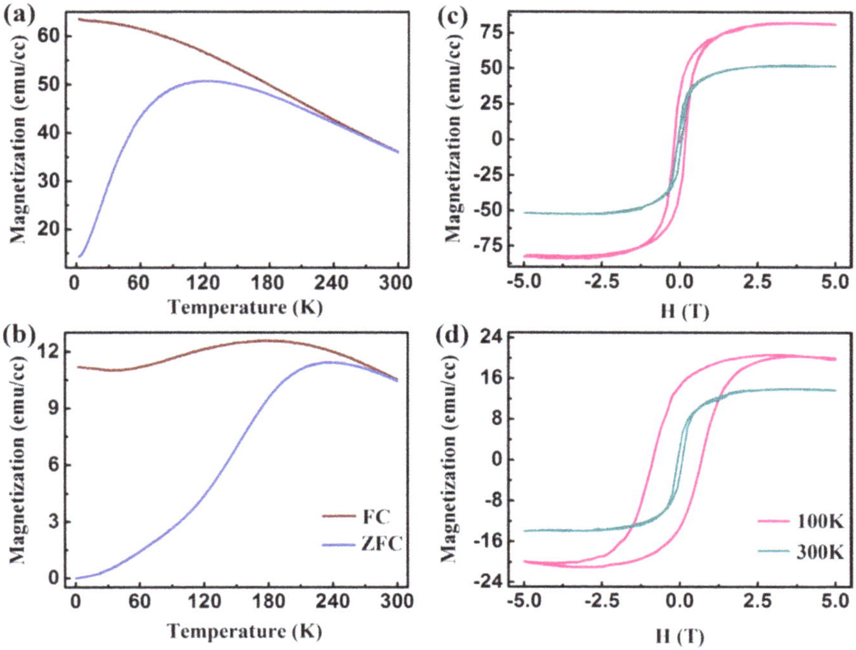

Figure 8.30. (a) and (b) Temperature dependence of magnetization, $M(T)$, for the two $Mn_{4-x}Pd_xN$ samples with $x = 0.0$ and $x = 0.17$ under zero-field cooling and field cooling at $H = 5000$ Oe, respectively; (c) and (d) the corresponding magnetic hysteresis loops, $M(H)$, measured at 300K and 100K with H lying parallel to the samples. Reproduced from [97].

scatterings, respectively. Although α is smaller than pure Mn_4N in the equation, the coefficients of the second and third terms are larger than those of pure Mn_4N. It can be concluded that the contribution of skew scattering is small, although multiple scattering mechanisms are involved. From this point of view, we conclude that the suppression of the AHE due to skew scattering occurs when Gd is introduced to Mn_4N in this paper (figures 8.38–8.39).

8.19 $Mn_{4-x}Dy_xN$

Table 8.20 summarizes some of the previous reports. The magneto-transport properties of $Mn_{4-x}Dy_xN$ have been studied, especially the mechanism of the AHE [124, 125]. Meng *et al* reported on the relationship between the anomalous Hall effect and carriers (especially carrier density) [124], and then presented a study in which they analyzed the AHE [125]. In the first report, the carrier density was calculated according to equation (8.9), and the relevance to the anomalous Hall conductivity was mentioned.

$$\boldsymbol{J}_{\mathrm{AH}} = 2ne^2\lambda \boldsymbol{E} \times \boldsymbol{S}. \tag{8.9}$$

Here, $\boldsymbol{J}_{\mathrm{AH}}$ is the anomalous Hall current density, n is the carrier concentration, e is the elemental charge, λ is the spin-orbit interaction coefficient, $\boldsymbol{E}$ is the Hall electric

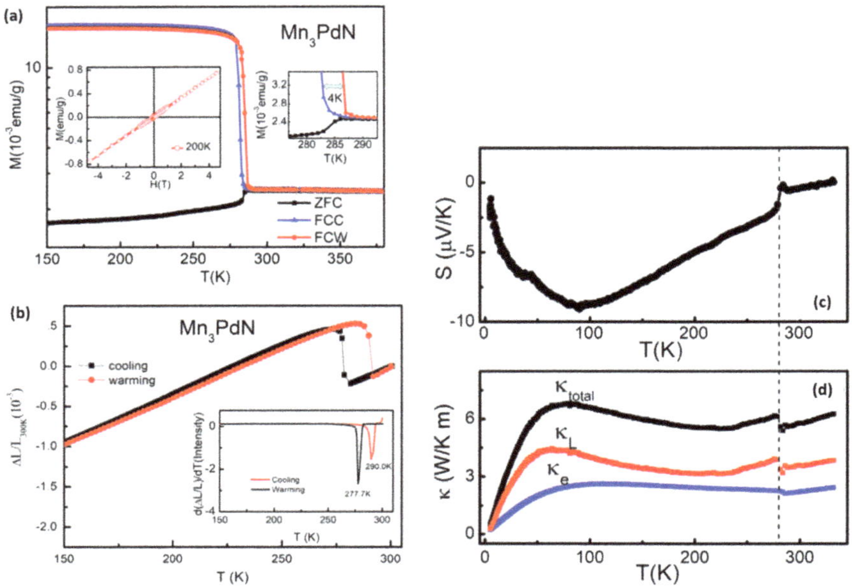

Figure 8.31. (a) The temperature dependent of the magnetization of Mn_3PdN ranging from 150 K to 380 K under ZFC, FCC and FCW modes. The left-hand inset shows the *M–H* curve measured at 200 K. The right-hand inset shows the enlarged view at the transition temperature region. (b) The temperature dependent of the linear thermal expansion $\Delta L/L_{300K}$ for Mn_3PdN. The inset shows the $d(\Delta L/L_{300K})/dT$ curve. (c) The temperature-dependent Seebeck coefficient of Mn_3PdN. (d) The thermal conductivity as a function of temperature *T* from 5 to 330 K. Reprinted from [98], Copyright (2021), with permission from Elsevier.

Table 8.13. Some previous studies of $Mn_{4-x}Ag_xN$.

Exp./Cal.	x	Properties	References	Year	Sub.
Powder	1	NPD, χ–T, M–T	[99]	1976	–
Powder	0.76–1	NTE, TCR, NPD, M–T, LTE	[100]	2015	–
Film/SD	0–0.87	TCR, ρ_{xx}–T, M–T, M–H	[101]	2016	Si
Powder	~1	Cu-doped TCR	[102]	2011	–
Powder	~1	Cu-doped TCR, M–T, MR	[103]	2012	–
Powder	~1	Cu-doped TCR, surface morphology	[104]	2022	Alumina
Powder	~1	TCR, Sn,Zn-doped, M–T, Cp–T	[105]	2014	–
Calc.	1	spin-rotation, PME NC-AFM	[106]	1992	–

field, and $\boldsymbol{S}$ is the spin angular momentum of an electron. The sign of the AHE of pure Mn_4N is negative, but at $x = 0.08$ it flips to positive and at $x = 0.4$ it flips again to negative [124]. In this article, the change in band structure at the E_F due to Dy doping is considered to be the origin for this. In particular, it was considered that the Hole-like band was located near the E_F at $x = 0.08$, while at $x = 0.4$ the Fermi

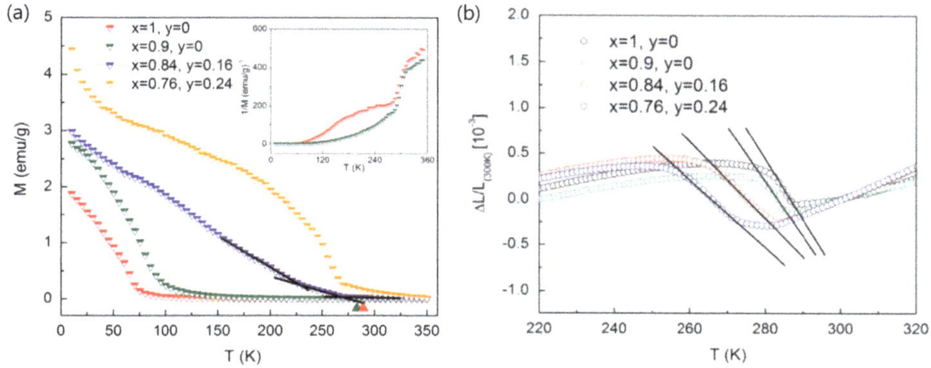

Figure 8.32. (a) Temperature dependence of magnetization for $Mn_{3+y}Ag_xN_y$ compounds. The black-solid lines are the guidance for magnetic transitions for $x = 0.84$. The inset shows temperature dependence of the reciprocal of magnetization. (b) Linear thermal expansion coefficient for $Mn_{3+y}Ag_xN$ compounds. Reprinted from [100], Copyright (2015), with permission from Elsevier.

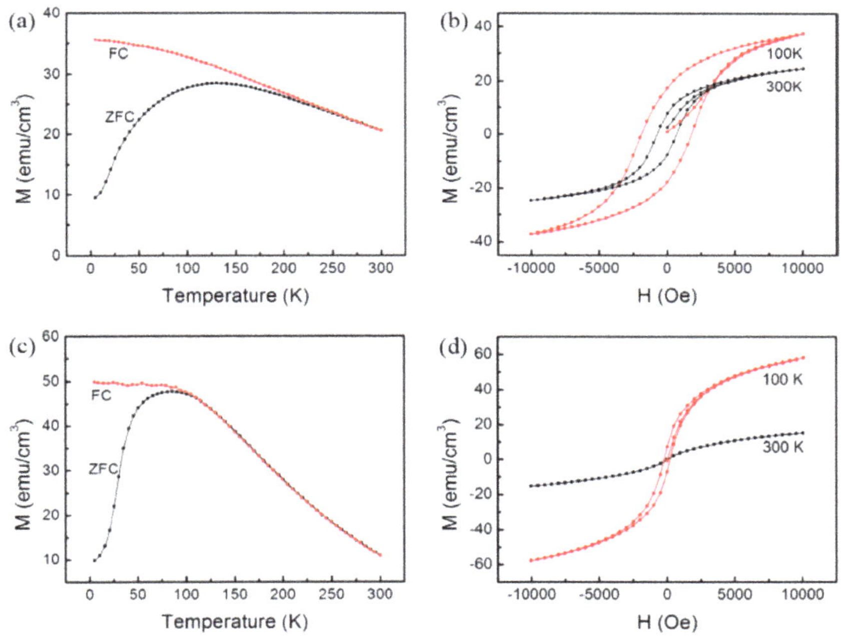

Figure 8.33. Temperature dependence of ZFC and FC dc-magnetizations at 5000 Oe and the hysteresis loops measured at $T = 300$ K and 100 K for the $Mn_{4-x}Ag_xN$ deposits with (a), (b) $x = 0.0$; and (c), (d) $x = 0.87$, respectively. Reproduced with permission from reference [101]. Reproduced from [101]. © IOP Publishing Ltd. All rights reserved.

Table 8.14. Some previous studies of $Mn_{4-x}In_xN$.

Exp./ Cal.	x	Properties	reference	Year	Sub.
Powder	0–0.4	M–x(compensation), M–T	[107]	1962	–
Powder	1	Magnetic structure	[108]	1973	–
Powder	1	χ–T, M–H, AFM + WFM(75–150 K), AFM(175–300K) PM (>300 K)	[109]	2012	–
This film	0–1.4	c/a, FWHM, AHE, R_H, H_C–x sign reversal in θ_{AH}	[110]	2022	MgO (001)
	0–0.12				STO(001)
Thin film	0–0.41	XMCD(0.15, 0.27, 0.41) FIM-FM transition	[111]	2022	MgO (001)

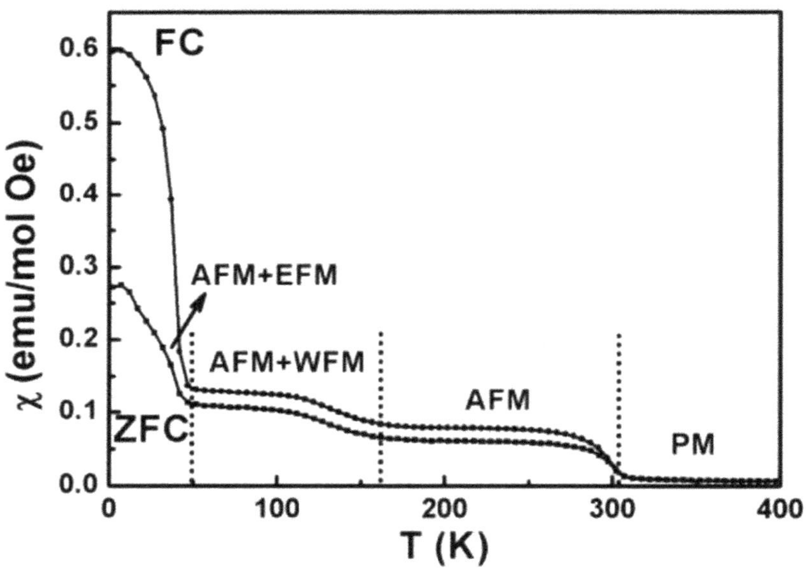

Figure 8.34. Temperature dependence of magnetization in Mn_3InN. Reproduced with permission from reference [109]. Reproduced from [109]. © The Author(s). Published by IOP Publishing Ltd. CC BY 4.0.

surface is reconstructed and the electron-like band is again located near the E_F. Such a sign reversal due to Fermi surface reconstruction has been confirmed by angle-resolved photoemission spectroscopy for $TaSe_2$ and $Cu_{0.2}NbS_2$ [126], and the authors cite the same paper to explain this phenomenon.

However, what is observed in these materials is only the normal Hall effect and it is difficult to directly identify the cause of the sign reversal of the AHE from the Fermi surface reconstruction. Although the carrier density has been measured, its

Table 8.15. Some previous studies of $Mn_{4-x}Sn_xN$.

Exp./Cal.	x	Properties	Reference	Year	Sub.
Powder	0–0.5	*M–x*(compensation), *M–T*	[107]	1962	–
Cal.	1	Study of magnetic order	[114]	1977	–
Powder	0.5, 1	NTE, *M–H*, *M–T*, LTE	[34]	2017	–
Film/SD	1	Poisson ratio, AHE, strain	[112]	2020	MgO
Film/SD	1	SOT-switching octupole	[113]	2021	MgO
Film/MBE	0–1.4	AHE, anomalous Hall angle	[115]	2022	MgO

Table 8.16. Some previous studies of $Mn_{4-x}Sb_xN$.

Exp./Cal.	x	Measurement and data	Reference	Year
Powder	1	*M–T*, Phase transition	[116]	1971
Powder	1	*M–T*, Phase transition	[108]	1973
Powder	1	*strain–H*, *M–H*, LTE, FSME	[117]	2012

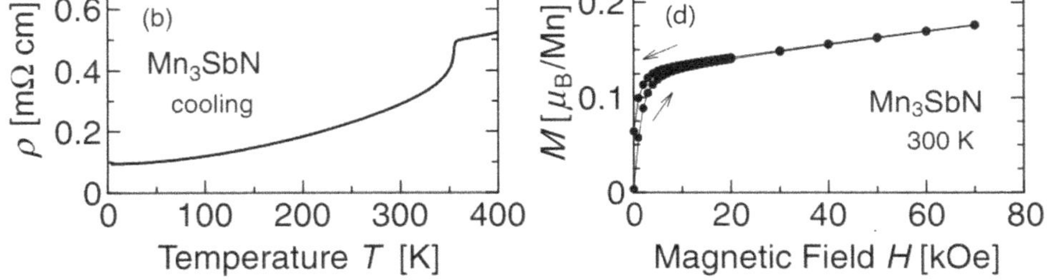

Figure 8.35. Temperature dependence of magnetization *M*(*T*) (a) and electrical resistivity *r*(*T*) (b) for Mn_3SbN, both of which were measured upon cooling. (c) Field-induced strain *e*(*H*) and (d) field-induced magnetization *M*(*H*) of Mn_3SbN at 300 K. For both measurements, the field was increased at a rate of 1 kOe/min up to 70 kOe, fixed for 2 h, and then removed at the same rate. Reproduced from [117].

Table 8.17. Some previous studies of $Mn_{4-x}Pt_xN_y$.

Exp./Cal.	x	Properties	References	Year
Powder	1	χ–T, NPD, a–T	[118]	1971
Powder	1	M–T	[99]	1976
Powder	$x = 1$, $y = 0$–1	χ–T, M–H, LTE, a–T	[119]	2018

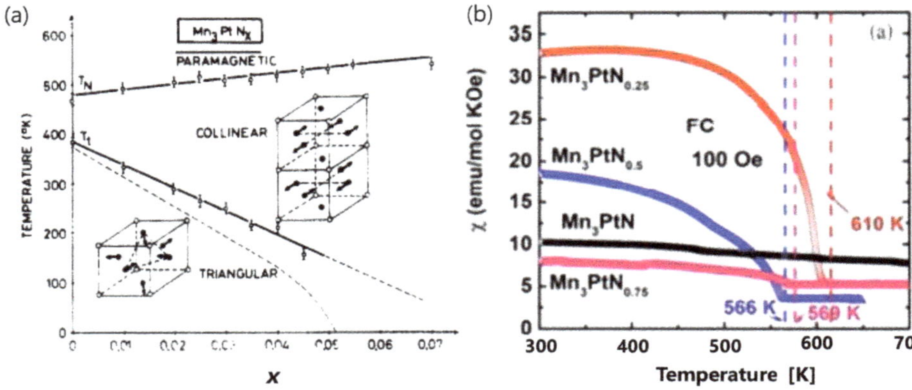

Figure 8.36. (a) Magnetic phase diagram in Mn_3PtN_y. Reprinted from [116], Copyright (2012), with permission from Elsevier.(b) Temperature dependence of magnetic susceptibility for Mn_3PtN_y ($y = 0$, 0.25, 0.5, 0.75, 1.0). Reproduced from [119]. CC BY 4.0.

Table 8.18. Some previous studies of $Mn_{4-x}Au_xN$.

Exp./Cal.	x	Properties	References	Year	Sub.
Film/PAMBE	0, 0.5	XPS, AHE, ρ_{xx}–T, ρ_{AH}–T	[120]	2020	MgO
Film/MBE	0–0.3	AHE, XMCD	[121]	2023	STO

sign is not shown. Furthermore, the spin-resolved DOS or spin-resolved E-k diagram for $Mn_{4-x}Dy_xN$ is not known. Therefore, it is not clear whether the sign reversal of the AHE is due to the change of carrier type or the sign reversal of the spin polarization. However, it was experimentally demonstrated that the carrier density is proportional to the anomalous Hall conductivity σ_{AH}, given by,

$$\sigma_{AH} = \frac{\rho_{AH}}{\rho_{xx}^2 + \rho_{AH}^2}. \tag{8.10}$$

A detailed discussion of the AHE for 80 nm-thick $Mn_{3.5}Dy_{0.5}N$/MgO(001) thin films was provided in reference [125]. The analysis method follows that of Hou *et al* [123],

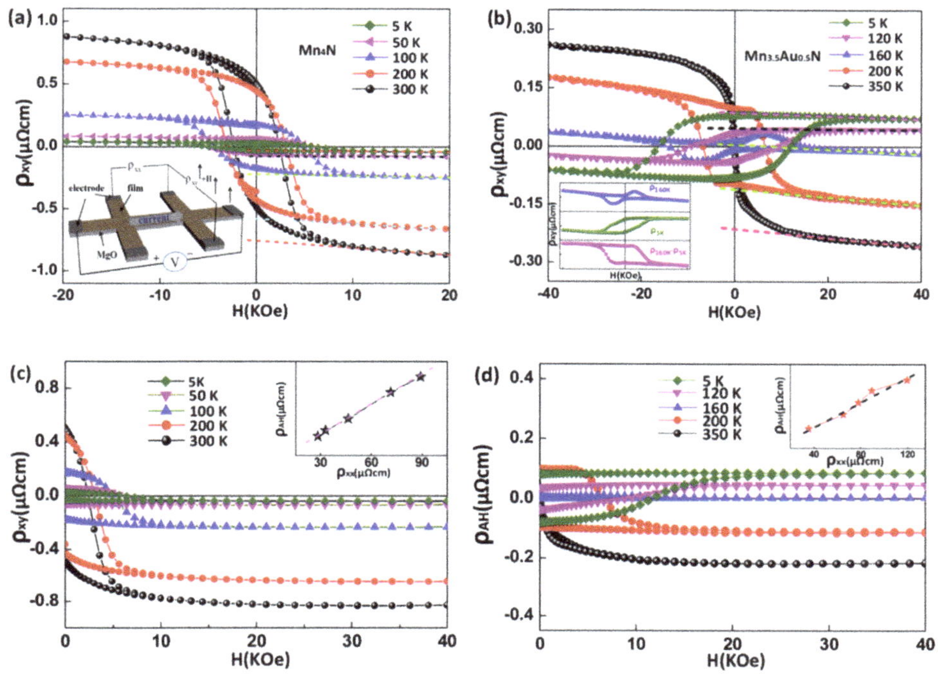

Figure 8.37. (a) Magnetic field dependence of ρxy for Mn_4N films; the inset is the schematic diagram of electrical measurement. (b) Magnetic field dependence of ρ_{xy} for $Mn_{3.5}Au_{0.5}N$; the inset is difference curve of ρxy between 160 K and 5 K. Reprinted from [12], Copyright (2018), with permission from Elsevier.

Table 8.19. Previous study of $Mn_{4-x}Gd_xN$.

Exp./Cal.	x	Properties	Reference	Year	Sub.
Film/MBE	0–0.48	M–T, χ–T, ρ_{AH}–T, MR	[122]	2018	MgO

as in the case of the $Mn_{4-x}Gd_xN$ thin film [122]. In $Mn_{4-x}Dy_xN$, the skew scattering is significantly reduced with Dy doping and multiple scattering mechanisms contribute to the skew scattering. The crystallinity was evaluated using RHEED, XRD, and cross-sectional TEM, and it was confirmed that high-quality crystals were successfully produced. The Néel temperature has also been measured, with a value of $T_N = 255$ K. This is lower than the T_N of pure Mn_4N and is attributed to the reduction of Mn–N–Mn ferromagnetic coupling due to the introduction of Dy (figures 8.40–8.43).

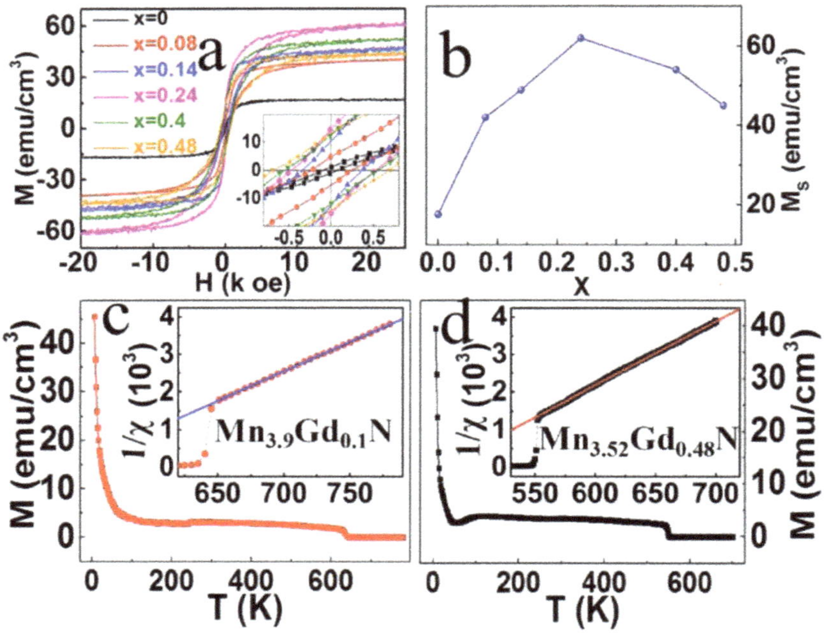

Figure 8.38. (a) M–H curves of samples with different x at room temperature. The inset shows the enlarged pattern around zero field. (b) Saturation magnetization (M_S) as a function of x. (c) and (d) ZFC curves of $Mn_{3.9}Gd_{0.1}N$ and $Mn_{3.52}Gd_{0.48}N$ between 5 K and 700 K. Red and blue solid lines in the inset of (c) and (d) are the fitting results using the Curie–Weiss law. Reprinted from [122], with the permission of AIP Publishing.

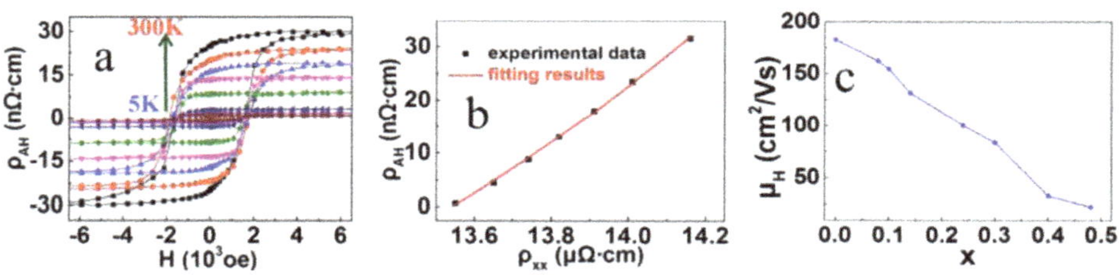

Figure 8.39. (a) ρ_{AH}–H curves of $Mn_{3.86}Gd_{0.14}N$ at different temperatures. ρ_{AH} against ρ_{xx} for $Mn_{3.86}Gd_{0.14}N$ is shown in (b), black dots are experimental results, and the red line is the fitting result using equation (8.8). (c) The Hall mobility of different samples at 15 K as a function of x. Reprinted from [122], with the permission of AIP Publishing.

Table 8.20. Some previous studies of $Mn_{4-x}Dy_xN$.

Exp./Cal.	x	Properties	Reference	Year	Sub.
Film/MBE	0–0.4	M–H, ρ_{xx}–T, ρ_{xy}–H	[124]	2015	MgO
Film/MBE	0.5	M–T, M–H, TEM, ρ_{AH}–T	[125]	2016	MgO

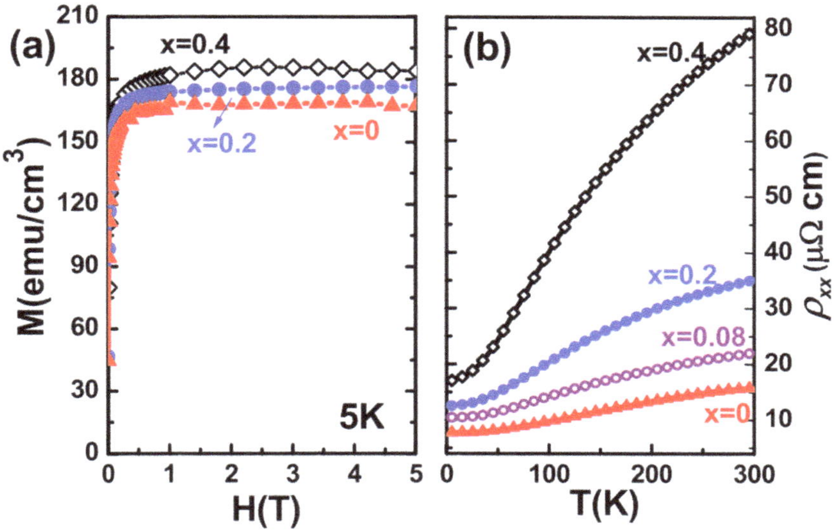

Figure 8.40. (a) Magnetic field H dependence of magnetization M in $Mn_{4-x}Dy_xN$ at 5 K under out-of-plane magnetic fields. (b) Longitudinal resistivity ρ_{xx} as a function of temperature. Reprinted from [124], with the permission of AIP Publishing.

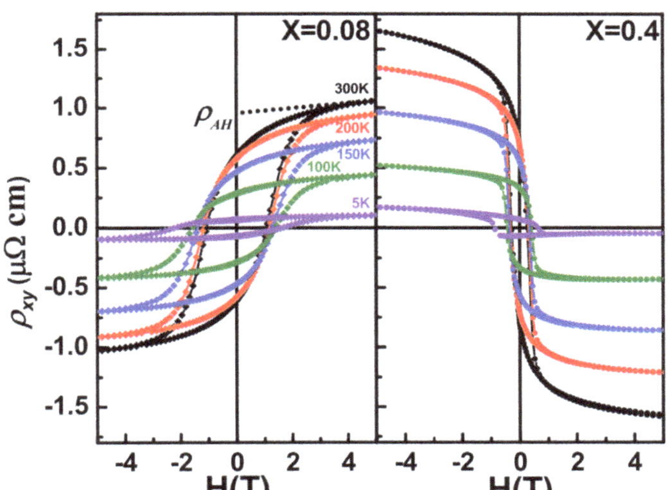

Figure 8.41. Representative Hall loops of $Mn_{4-x}Dy_xN$ with $x = 0.08$ and 0.4 as a function of H with different temperature. Reprinted from [124], with the permission of AIP Publishing.

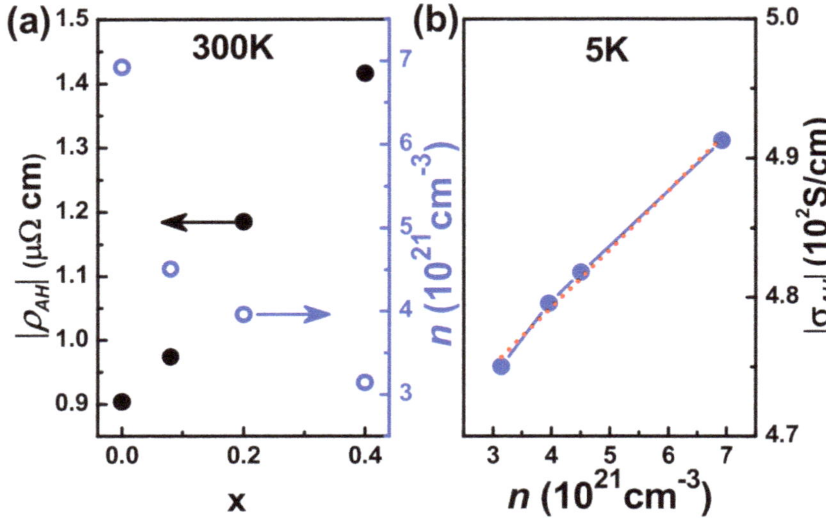

Figure 8.42. (a) | ρ_{AH} | (black-solid circles) of $Mn_{4-x}Dy_xN$ versus x at 300 K; carrier density n (open blue circles) versus x determined from R_0 at 300 K. (b) | σ_{AH} | versus n at 5 K; the red dotted line shows a | σ_{AH} | ∝ n correlation. Reprinted from [124], with the permission of AIP Publishing.

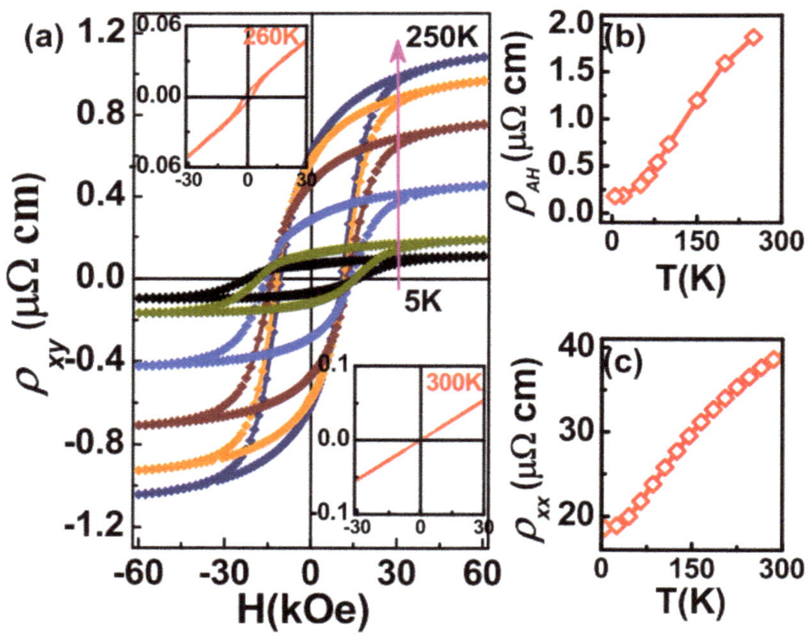

Figure 8.43. (a) ρ_{xy}–H curves at different temperatures for $Mn_{3.5}Dy_{0.5}N$. The arrow indicates the increase of temperature from 5 K to 250 K. Upper and lower insets: ρ_{xy} at 260 K and 300 K, respectively; (b) ρ_{AH} and (c) ρ_{xx} as a function of temperature, respectively. Reprinted from [125], with the permission of AIP Publishing.

Bibliography

[1] Lin J C, Wang B S, Tong P, Lu W J, Zhang L, Zhu X B, Yang Z R, Song W H, Dai J M and Sun Y P 2011 *Appl. Phys. Lett.* **98** 092507
[2] Ouyang Y, Chen H, Tong M, Du Y, Feng Y and Zhong X 2008 *Comput. Mater. Sci.* **44** 97
[3] Shao D F, Lu W J, Lin J C, Tong P, Jian H B and Sun Y P 2013 *J. Appl. Phys.* **113** 023905
[4] Kadowaki K and Woods S B 1986 *Solid State Commun.* **58** 507
[5] Mohn P and Wohlfarth E P 1987 *J. Phys. F: Met. Phys.* **17** 2421
[6] Qi Q, Skomski R and Coey J M D 1994 *J. Phys. Condens. Matter* **6** 3245
[7] Ibach H and Luth H 1993 *Solid State Physics: An Introduction to Principles of Materials Science* (Berlin: Springer)
[8] Wohlfarth E P 1962 *Phys. Lett.* **3** 17
[9] Gersdorf R 1962 *J. Phys. Radium* **23** 726
[10] Mohn P 2006 *Magnetism in the Solid State: An Introduction* (Berlin: Springer)
[11] Juza R, Deneke K and Puff H 1959 *Z. Elektrochem* **63** 551
[12] Komori T, Horiuchi T, Mitarai H, Yasuda T, Amemiya K and Suemasu T 2022 *J. Mag. Mag. Mater.* **564** 170050
[13] Madar R 1970 *Thesis* University of Paris (France), Paris, No. 686, Centre de Documentation du CNRS
[14] Anzai A, Takata F, Gushi T, Toko K and Suemasu T 2018 *J. Cryst. Growth* **489** 20
[15] Anzai A, Gushi T, Komori T, Honda S, Isogami S and Suemasu T 2018 *J. Appl. Phys.* **124** 123905
[16] Isogami S, Anzai A, Gushi T, Komori T and Suemasu T 2018 *Jpn. J. Appl. Phys.* **57** 120305
[17] Chen L 2006 *J. Appl. Phys.* **100** 113717
[18] Li Z R, Mi W B and Bai H L 2018 *Comput. Mater. Sci.* **142** 145
[19] Wijn H P J 1988 *Alloys and Compounds of d-Elements with Main Group Elements. Part 2* (Berlin: Springer)
[20] Kokado S, Tsunoda M, Harigaya K and Sakuma A 2012 *J. Phys. Soc. Jpn.* **81** 024705
[21] Ito K, Yasutomi Y, Kabara K, Gushi T, Higashikozono S, Toko K, Tsunoda M and Suemasu T 2016 *AIP Adv.* **6** 056201
[22] Mitarai H *et al* 2020 *Phys. Rev. Mater.* **4** 094401
[23] Ito K *et al* 2020 *Phys. Rev.* **B 101** 104401
[24] Sun Y *et al* 2019 *Phys. Rev. Mater.* **3** 024409
[25] Takenaka K, Ichigo M, Hamada T, Ozawa A, Shibayama T, Inagaki T and Asano K 2014 *Sci. Technol. Adv. Mater.* **15** 015009
[26] Nogués J and Sculler I K 1999 *J. Magn. Magn. Mater.* **192** 203
[27] Fruchart D and Bertaut E F 1978 *J. Phys. Soc. Jpn.* **44** 781
[28] Wu M, Wang C, Sun Y, Chu L, Yan J, Chen D, Huang Q and Lynn J W 2013 *J. Appl. Phys.* **114** 123902
[29] Deng S *et al* 2015 *Chem. Mater.* **27** 2495
[30] Sun Y, Wang C, Chu L, Wen Y, Niea M and Liu F 2010 *Scr. Mater.* **62** 686
[31] Deng S, Sun Y, Wang L, Wu H, Shi K, Hu P, Huang Q and Wang C 2016 *Appl. Phys. Lett.* **108** 041908
[32] Na Y, Wang C, Chu L, Ding L, Yan J, Xue Y, Xie W and Chen X 2011 *Mater. Lett.* **65** 3447
[33] Na Y, Wang C and Xiang J 2015 *Mater. Lett.* **152** 213
[34] Boldrin D and Cohen L F 2017 *J. Alloy Compd* **699** 887

[35] Boldrin D *et al* 2018 *ACS Appl. Mater. Interfaces* **10** 18863
[36] Boldrin D *et al* 2018 *Phys. Rev.* X **8** 041035
[37] Boldrin D *et al* 2019 *Adv. Funct. Mater.* **29** 1902502
[38] Boldrin D *et al* 2019 *Phys. Rev. Mater.* **3** 094409
[39] Johnson F, Boldrin D, Zemen J, Pesquera D, Kim J, Moya X, Zhang H, Singh H K, Samathrakis I and Cohen L F 2021 *Appl. Phys. Lett.* **119** 222401
[40] Johnson F *et al* 2022 *Appl. Phys. Lett.* **120** 232402
[41] Komori T, Anzai A, Gushi T, Toko K and Suemasu T 2019 *J. Cryst. Growth* **507** 163
[42] Komori T, Gushi T, Anzai A, Vila L, Attané J P, Pizzini S, Vogel J, Isogami S, Toko K and Suemasu T 2019 *J. Appl. Phys.* **125** 213902
[43] Komori T, Hirose T, Gushi T, Toko K, Hanashima T, Vila L, Attanè J P, Amemiya K and Suemasu T 2020 *J. Appl. Phys.* **127** 043903
[44] Ghosh S *et al* 2021 *Nano Lett.* **21** 2580
[45] Gomonaj E V 1989 *Phase Transit.* **18** 93
[46] Gomonay H 1990 *J. Magn. Magn. Mater.* **86** 301
[47] Zhou X, Hanke J P, Feng W, Blügel S, Mokrousov Y and Yao Y 2020 *Phys. Rev. Mater.* **4** 024408
[48] Torres-Amaris D, Bautista-Hernandez A, González-Hernández R, Romero A H and Garcia-Castro A C 2022**15** arXiv:2204.07607v1 [cond-mat.str-el]
[49] Florez-Gomez L, Ibarra-Hernandez W and Garcia-Castro A C 2022 *J. Mag. Mater. Mag.* **562** 169813
[50] Ph. l'Htritier *Thesis* University of Grenoble (France) 1980 at Centre de Documentation du CNRS, Paris
[51] Zhang R, He Y, Fruchart D, Coey J M D and Gercsi Z 2022 *Acta Mater.* **234** 118021
[52] Asano K, Koyama K and Takenaka K 2008 *Appl. Phys. Lett.* **92** 161909
[53] Takenaka K, Shibayama T, Asano K and Koyama K 2010 *J. Phys. Soc. Jpn.* **79** 7
[54] Takenaka K, Hamada T, Shibayama T and Asano K 2013 *J. Alloy Compd* **5775** 5291
[55] Yin Y, Han J C, Yuan Q, Ling L S and Song B 2013 *J. Magn. Magn. Mater.* **346** 203
[56] Yu F, Ren L, Meng M, Wang Y, Yang M, Wu S and Li S 2014 *J. Appl. Phys.* **115** 133911
[57] Kawaguchi T, Suzuki J, Sakamoto N, Suzuki H and Wakiya N 2021 *J. Cerm. Soc. Jpn.* **129** 377
[58] Huang R, Li L, Wua Z, Chua X, Xua X and Qianc L 2010 *Solid State Commun.* **150** 1617
[59] Sun Y, Wang C, Na Y, Chu L, Wen Y and Nie M 2010 *Mater. Res. Bull.* **45** 1230
[60] Aoyama M, Takenaka K and Ikuta H 2013 *J. Alloy Compd* **577** S314
[61] Shi K, Wang C, Sun Y, Wang L, Deng S, Hu P, Lu H, Hao W, Wang T and Tang W 2017 *ACS Appl. Mater. Interfaces* **9** 12592
[62] Matsumoto T, Hatano T, Urata T, Iida K, Takenaka K and Ikuta H 2017 *Phys. Rev.* B **96** 205153
[63] Takenaka K, Shibayama T, Ozawa A, Hamada T, Nakamura T, Kodama K and Kinoshita T 2011 *J. Appl. Phys.* **110** 023909
[64] Lin J C, Wang B S, Lin S, Tong P, Lu W J, Zhang L, Zhao B C, Song W H and Sun Y P 2012 *J. Appl. Phys.* **111** 113914
[65] Lin J *et al* 2015 *Sci. Rep.* **5** 7933
[66] Hua L, Wang L and Chen L F 2010 *J. Phys. Condens. Matter* **22** 206003
[67] Qu B Y, He H Y and Pan B C 2011 *AIP Adv.* **1** 042125
[68] Antonov V N and Bekenov L V 2014 *Low Temp. Phys.* **40** 641

[69] Benmalem Y, Abbad A, Benstaali W, Bentounes H A, Seddik T and Lantri T 2018 *J. Comput. Electron.* **17** 881
[70] Khalladi R, Labrim H, Idrissi S, Mtougui S, Housni I E, Ziti S, Mekkaoui N E and Bahmad L 2019 *Solid State Commun.* **290** 42
[71] Garcia J, Bartolome J and Gonzalez D 1983 *J. Chem. Thermodyn.* **15** 1041
[72] Hamada T and Takenaka K 2012 *J. Appl. Phys.* **111** 07A904
[73] Sun Y, Wang C, Huang Q, Guo Y, Chu L, Arai M and Yamaura K 2012 *Inorg. Chem.* **51** 7232
[74] Sun Y S, Guo Y F, Wang X X, Tsujimoto Y, Matsushita Y, Shi Y G, Wang C, Belik A A and Yamaura K 2012 *Appl. Phys. Lett.* **100** 161907
[75] Zhao H, Sun Y, Shi Z, Zhang Q, Yao Y, Yu R and Wang C 2013 *J. Mater. Res.* **28** 23
[76] Qu B Y and Pan B C 2010 *J. Appl. Phys.* **108** 113920
[77] Masrour R, Jabar A, Bahmad L, Hlil E K, Hamedoun M, Benyoussef A, Hourmatallah A, Benzakour N, Rezzouk A and Bouslykhane K 2021 *J. Clust. Sci.* **32** 163
[78] Nabarro R, Rojo J A, Garcia J, Gonzalez D and Bartolome J 1986 *J. Magn. Magn. Mater.* **59** 221
[79] Song B, Jian J, Bao H, Lei M, Wang G, Xu Y and Chen X 2008 *Appl. Phys. Lett.* **92** 192511
[80] Feng W J, Li D, Deng Y F, Zhang Q, Zhang H H and Zhang Z D 2010 *J. Mater. Sci.* **45** 2770
[81] Matsunami D, Fujita A, Takenaka K and Kano M 2015 *Nat. Mater.* **14** 73
[82] Zhang X *et al* 2018 *J. Mater. Chem.* C **6** 13336
[83] Yuan Y, Liu Y, Xu C, Kang J, Wang W, Wang Q, Song B, Zhou S and Wang X 2020 *J. Appl. Phys.* **127** 073903
[84] Yoon I T, Kang T W and Kim D J 2006 *Mater. Sci. Eng.* B **134** 49
[85] Tashiro H, Suzuki R, Miyawaki T, Ueda K and Asano H 2013 *J. Korean Phys. Soc.* **63** 299
[86] Sakakibara H, Ando H, Kuroki Y, Kawai S, Ueda K and Asano H 2015 *J. Appl. Phys.* **117** 17D725
[87] Ishino S, So J, Goto H, Hajiri T and Asano H 2018 *AIP Adv.* **8** 056312
[88] Hajiri T, ishino S, Matsuura K and Asano H 2019 *Appl. Phys. Lett.* **115** 052403
[89] Hajiri T, Matsuura K, Sonoda K, Tanaka E, Ueda K and Asano H 2021 *Phys. Rev. Appl.* **16** 024003
[90] Nan T *et al* 2020 *Nat. Commun.* **11** 4671
[91] Miao M S, Herwadkar A and Lambrecht W R L 2005 *Phys. Rev.* B **72** 033204
[92] Lukashev P and Sabirianov R F 2010 *Phys. Rev.* B **82** 094417
[93] Lukashev P, Belashchenko K D and Sabirianov R F 2011 *Phys. Rev.* B **84** 133420
[94] Azouaoui A, Benzakour N, Hourmatallah A and Bouslykhane K 2021 *Philos. Mag.* **101** 1587
[95] Samathrakis I and Zhang H 2020 *Phys. Rev.* B **101** 214423
[96] Madar R, Gilles L, Rouault A and Bouchaud J 1967 *Comp. Rend. Acad. Sci.* **264** 308
[97] Xu T, Ji G P, Cao Z X and Ji A L 2018 *J. Phys. D: Appl. Phys.* **51** 055303
[98] Tan S, Gao C, Yuan H, Wu J, Wang C, Cao R, Sun Y and Solid J 2021 *State Chem.* **302** 122389
[99] Fruchart D 1976 *Thesis* University of Grenoble, Paris, No. 11922, at Centre de Documentation du CNRS

[100] Deng S, Sun Y, Yan J, Shi Z, Shi K, Wang L, Hu P, Malik M I and Wang C 2015 *Solid State Commun.* **222** 37
[101] Lu N P, Xu T, Cao Z X and Ji A L 2016 *J. Phys. D: Appl. Phys.* **49** 045308
[102] Takenaka K, Ozawa A, Shibayama T, Kaneko N, Oe T and Urano C 2011 *Appl. Phys. Lett.* **98** 022103
[103] Hadano M, Ozawa A, Takenaka K, Kaneko N, Oe T and Urano C 2012 *J. Appl. Phys.* **111** 07E120
[104] Lukose C C, Zoppi G and Birkett M 2022 *J. Mater. Sci. Technol.* **99** 138
[105] Sun Y, Guo Y, Li J, Wang C, Wang X, Sathish C I and Yamaura K 2014 *Sci. Adv. Mater* **6** 1
[106] Gomonaj E V and L'vov V A 1992 *Phase Transit.* **40** 235
[107] Mekata M 1962 *J. Phys. Soc. Jpn.* **17** 796
[108] Barberon M 1973 *Thesis* University of Paris-Orsay, Paris, No. 702, at Centre de Documentation du CNRS
[109] Sun Y S, Guo Y F, Wang X X, Yi W, Li J J, Zhang S B, Sathish C I, Belik A A and Yamaura K 2012 *J. Phys. Conf. Ser.* **400** 032094
[110] Yasuda T, Komori T, Mitarai H and Suemasu T 2022 *J. Cryst. Growth* **582** 126525
[111] Yasuda T, Komori T, Mitarai H, Honda S, Ghosh S, Vila L, Attané J P, Amemiya K and Suemasu T 2022 *J. Phys. D: Appl. Phys.* **55** 115003
[112] You Y, Bai H, Chen X, Zhou Y, Zhou X, Pan F and Song C 2020 *Appl. Phys. Lett.* **117** 222404
[113] You Y *et al* 2021 *Nat. Commun.* **12** 6524
[114] Fruchart D, Bertaut E F, Senateur J P and Fruchart R 1977 *J. Phys. Lett.* **38** 21
[115] Yasuda T, Komori T, Hirose T, Horiuchi T, Toko K and Suemasu T 2023 *AIP Adv.* **13** 015119
[116] Fruchart D, Bertaut E F, Madar R and Fruchart R 1971 *J. Phys.* **32** 876
[117] Takenaka K, Shibayama T, Kasugai D and Shimizu T 2012 *Jpn. J. Appl. Phys.* **51** 043001
[118] Kren E, Zsoldos E, Barberon M and Fruchart R 1971 *Solid State Commun.* **9** 27
[119] Chou R, Sun Y, Lu H and Lu G H 2018 *Int. J. Mod. Phys.* **32** 1850314
[120] Li D, Wang G, Li H, Wu S and Li S 2020 *Mater. Res. Bull.* **122** 110646
[121] Horiuchi T, Komori T, Yasuda T, Hirose T, Toko K, Amemiya K and Suemasu T 2023 *AIP Adv.* **13** 025107
[122] Li H *et al* 2018 *J. Appl. Phys.* **124** 093903
[123] Hou D, Su G, Tian Y, Jin X, Yang S A and Niu Q 2015 *Phys. Rev. Lett.* **114** 217203
[124] Meng M, Wu S X, Zhou W Q, Ren L Z, Wang Y J, Wang G L and Li S W 2015 *J. Appl. Phys.* **118** 053911
[125] Meng M, Wu S X, Zhou W Q and Li S W 2016 *Appl. Phys. Lett.* **109** 082405
[126] Evtushinsky D V, Kordyuk A A, Zabolotnyy V B, Inosov D S, Büchner B, Berger H, Patthey L, Follath R and Borisenko S V 2008 *Phys. Rev. Lett.* **100** 236402

Chapter 9

Recent progress

This chapter will discuss recent advances that have not been covered in previous chapters but which will broaden the horizon for spintronics application of Mn_4N. Recent results on ultrathin film growth, experiments on the Dzyaloshinskii Moriya interaction (DMI), SOT switching, anomalous Nernst effect, and modulation of the magnetic properties by light element doping are provided.

9.1 Ultrathin film growth

Current-driven magnetic DW wall motion is a promising technique with potential applications in nonvolatile devices. Mn_4N thin films meet the requirements for efficient DW motion, such as PMA, small magnetization, and thermal stability. Experiments have shown ultrafast DW wall motion driven solely by STT at room temperature (as described in chapters 4 and 7), making Mn_4N a strong candidate material for such applications. To further enhance the efficiency of current-driven DW motion in Mn_4N-based materials, exploring the use of SOT is necessary. Although SOT has been found to be more efficient than STT for driving DW motion [1, 2], it has not been achieved in Mn_4N. Achieving SOT-driven DW motion without an external magnetic field requires the creation of spin accumulation through the fabrication of multilayer structures and the formation of chiral Néel DWs in the magnetic layer [3, 4]. Stabilizing chiral Néel DWs commonly involves leveraging the DMI at interfaces [5, 6]. It is important to note that DMI occurs in systems with broken spatial inversion symmetry, and therefore requires a sufficiently thin magnetic layer. Therefore, as a first step towards achieving SOT-driven DW motion in Mn_4N, it is crucial to achieve ultrathin film growth and evaluate their properties. Mn_4N thin films have been successfully grown epitaxially on various substrates [7–9]. In particular, the perovskite-type STO(001) substrate has shown promise for synthesizing high-quality Mn_4N [7, 8] thanks to its similar crystal structure and a lattice constant matched to Mn_4N. However, there have been no reports on the formation of Mn_4N films thinner than 10 nm on STO substrates.

doi:10.1088/978-0-7503-5477-6ch9

An example of the formation of ultrathin (2–6 nm) Mn_4N films is introduced here. STO(001) substrates were immersed in an organic solvent and subjected to ultrasonic cleaning. Chemical etching using buffered HF (HF: 5 wt%, NH_4F: 36 wt%) was performed to expose a flat Ti-terminated surface [10]. After rinsing with deionized water and drying with N_2, the substrate was immediately transferred into the growth chamber. The background pressure in the chamber was approximately 1.0×10^{-8} Torr. The Mn_4N film was prepared using the plasma-assisted MBE method, where Mn and N atoms were simultaneously supplied. The nitrogen flow rate was set at 0.9 sccm, the input RF power supply was 132 W, and the pressure was maintained at 3.60–3.65×10^{-5} Torr during the growth. The thickness of the Mn_4N film was controlled by the deposition time. In situ RHHED patterns of the Mn_4N film were obtained, and then a SiO_2 film of approximately 3 nm was deposited at room temperature using RF sputtering to prevent oxidation of grown films. The film thickness of the sample was evaluated using XRR with pure Cu-Kα_1 radiation. The crystalline quality was evaluated through RHEED and x-ray diffraction (XRD) profiles.

Figure 9.1 shows the $\theta - 2\theta$ XRD patterns with RHEED images taken along the STO[100] azimuth for approximately 6 nm thick and 2 nm thick grown films. It was found that Mn_4N epitaxial films with RHEED streaks were formed on STO(001) in both cases; however, the magnetic properties, such as squareness ratio and the coercivity in the AHE curves, are quite sensitive to the growth conditions. It is also important to note that the growth conditions for ultrathin films are not the same as those for 10–20 nm thick Mn_4N epitaxial films. We need to take relatively 'N-rich' conditions. Surprisingly, the sign of the anomalous Hall resistivity of such ultrathin films of Mn_4N reverses with film thickness at room temperature [11]. XMCD measurements suggest that the magnetic structure of Mn_4N with a thickness of around 4 nm is different from that of conventional ferrimagnetic Mn_4N films [11]. The formation of ultrathin Mn_4N films is of great importance when considering the use of SOT and interfacial DMI in those films.

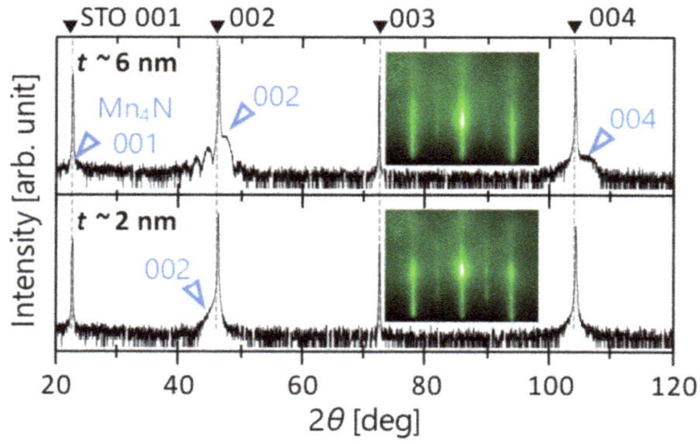

Figure 9.1. $\theta - 2\theta$ XRD profiles with RHEED images for 6 nm and 2 nm thick Mn_4N epitaxial films with RHEED patterns observed along the STO[100] azimuth. Open triangles and dashed lines indicate the peaks attributed to Mn_4N and STO(001) substrate, respectively.

9.2 Magnetic skyrmions and Dzyaloshinskii Moriya interaction

9.2.1 Observation of magnetic skyrmions

Magnetic skyrmions are topological spin textures, as shown in figure 9.2, and can be stabilized through DMI [12, 13] or stray field [14]. DMI-skyrmions occur from broken symmetry in bulk materials with a non-centrosymmetric crystal structure or in thin films because of the interfacial asymmetry. Néel skyrmions have been reported in various materials such as Pt/CoFe/MgO, Ir/Fe/Co/Pt, Pt/CoGd/Ta, Pt/CoGd/PtW, and Pt/Co/Ru [15–19], while Bloch skyrmions arises in bulk B20 crystals at low temperatures [20, 21]. For high-density data storage, small (10 nm in size) and stable skyrmions at room temperature are inevitable. The large saturation magnetization leads to the large stray field, and thereby large skyrmions (>50 nm) at room temperature [15–17]. In contrast, around 10 nm-sized skyrmions have been reported in amorphous ferrimagnets such as GdCo [18]. Thanks to thermal stability, small M_S, and PMA, Mn_4N is considered to be a potential material for skyrmion applications.

Observations of skyrmions in Mn_4N films in contact with heavy metals are reported in Refs. [22, 23]. The layered structure is MgO/Mn_4N(15 nm)/$Pt_{1-x}Cu_x$(30 nm)/Pt(3 nm) prepared by RF sputtering, wherein the top Pt layer is the capping layer to prevent oxidation. Figure 9.3 shows MFM images of skyrmions taken at room temperature in Mn_4N films with various capping layers. At $x = 0$, the average skyrmion size is 300 nm, while at $x = 1$, no evidence for any skyrmions is

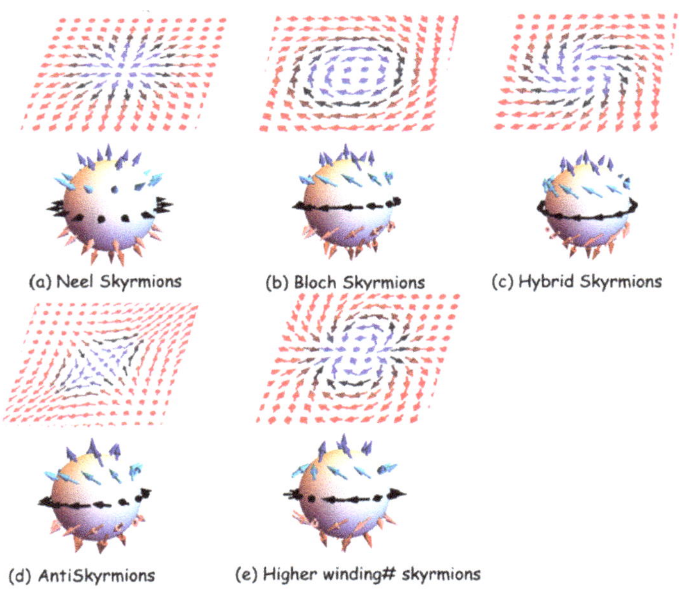

Figure 9.2. Three-dimensional spherical arrangement of spins that generate two-dimensional skyrmions through stereographic projection. The north pole becomes the skyrmion core while the south pole maps to the background spin texture. The figure shows skyrmions of various winding numbers with same polarization and various domain angles/helicities. Reproduced from [22]. CC BY 4.0.

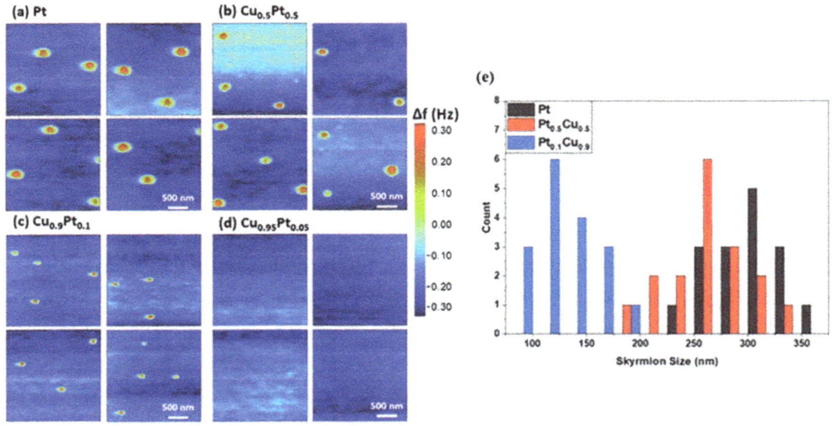

Figure 9.3. Room temperature MFM images of skyrmions in Mn_4N with various cap layers (a) Pt; (b) $Pt_{0.5}Cu_{0.5}$; (c) $Pt_{0.1}Cu_{0.9}$; and (d) $Pt_{0.05}Cu_{0.95}$. In each figure, four areas of the same sample have been shown. Blue colors correspond to regions of down magnetization, which are the ferrimagnetic backgrounds. Red colors correspond to up magnetization, which is the core of a skyrmion. (e) A histogram of distribution of skyrmion size in Pt (Black), $Pt_{0.5}Cu_{0.5}$ (Red), and $Pt_{0.1}Cu_{0.9}$ (Blue), these distributions correspond to all areas imaged, in addition to the four areas shown. Reprinted from [23], with the permission of AIP Publishing.

found, showing that Pt is inevitable for the formation of skyrmions. It is interesting to note that the skyrmion's size can be controlled by the Cu content x in $Pt_{1-x}Cu_x$(30 nm) layers, as shown in figure 9.3(e).

Figure 9.4 shows the simulated skyrmion DMI-K_u phase diagram of 15 nm thick Mn_4N films at room temperature [23]. It is noted that the dilution of capping layers using a lighter element like Cu is an effective means to stabilize smaller skyrmions by reducing the interfacial DMI (iDMI) strength.

9.2.2 DMI measurement

The averaged DMI was measured in Mn_4N films [24] using the method proposed in Reference [25]. Figure 9.5 shows a schematic of the measurement setup. Here, θ is the angle between the external magnetic field and the sample normal. The θ-dependent coercivity field H_c of the Mn_4N Hall cross-structure is measured. H_c is divided into the perpendicular component (H_z) and the in-plane component (H_x).

In the presence of DMI, there is a threshold point in the $(H_z)^{1/2}$ versus H_x curve. As shown in figures 9.6(a)–(d), $(H_z)^{1/2}$ begins to decrease with H_x at a certain value of H_x. This value is the effective magnetic field induced by DMI, and is denoted as H_{DMI}. The DMIs are calculated based on the following equation:

$$\mathrm{DMI} = \sqrt{\frac{A}{K_{\mathrm{eff}}}}\, M_S H_{\mathrm{DMI}}. \tag{9.1}$$

Here, A is the exchange stiffness, and K_{eff} is the effective PMA energy. The measured DMI for $MgO/Mn_4N/Pt$ was found to be 0.267 ± 0.065 mJ m^{-2}, and gradually decreased to 0.011 ± 0.01 mJ m^{-2} with increasing Cu concentration.

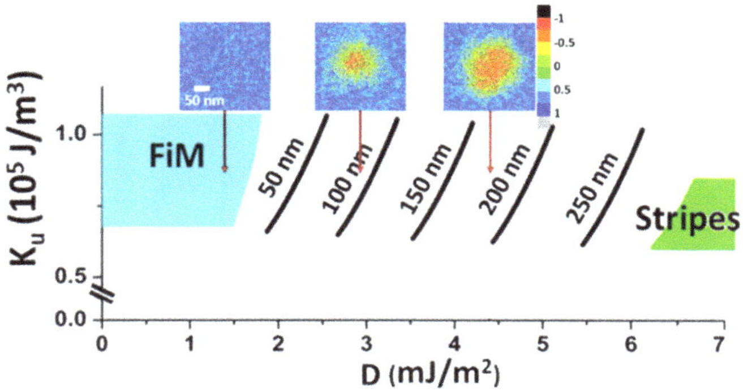

Figure 9.4. Simulated skyrmion DMI-K_u phase diagram of 15-nm Mn_4N at 300K. Dilution Pt cap layer by Cu reduces iDMI, which leads to smaller skyrmions. The figures inserted above show the mapping of out-of-plane magnetization from simulations. The red region corresponds to down magnetic moments, while the blue region corresponds to up magnetic moments. For an iDMI in the range from 3.5 to 4 mJ m^{-2}, a close to 200nm skyrmion is stabilized at 300K. If the iDMI is reduced to 3.0 mJ m^{-2}, then skyrmion size is reduced to 100nm. If the iDMI is further reduced to 1.5 mJ m^{-2}, then skyrmions are unstable at room temperature and collapse into ferrimagnetic state. Reprinted from [23], with the permission of AIP Publishing.

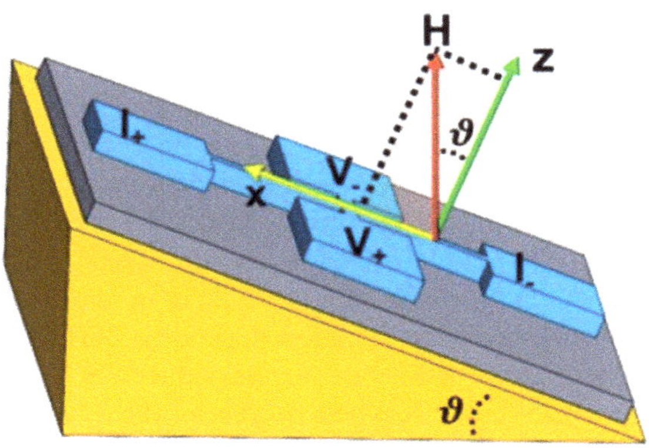

Figure 9.5. Schematic setup of the DMI measurement. The samples were put on a tilted holder while the external magnetic field is applied in the vertical direction. The out-of-plane component H_z and the in-plane component H_x of the coercivity field Hc were calculated with $H_z = H_c \cos(\theta)$ and $H_x = H_c \sin(\theta)$. Reproduced from [24]. CC BY 4.0.

Figure 9.6(e) demonstrates the control of the DMI through the $Pt_{1-x}Cu_x$ capping layers. The interfacial DMI constant, D_S, i.e., the averaged DMI multiplied by the magnetic layer thickness, is used to compare the DMI strength with other materials, as shown in figure 9.6(f). The obtained values are higher than Co-based materials [26]. First-principles calculation reveals that the DMI chemistry of $3d - 5d$ ultrathin films follows Hund's first rule and shows a similar trend to the magnetic moment of unsupported $3d$ monolayers or $3d - 5d$ interfaces [27]. For details, please see the review article about DMI [28].

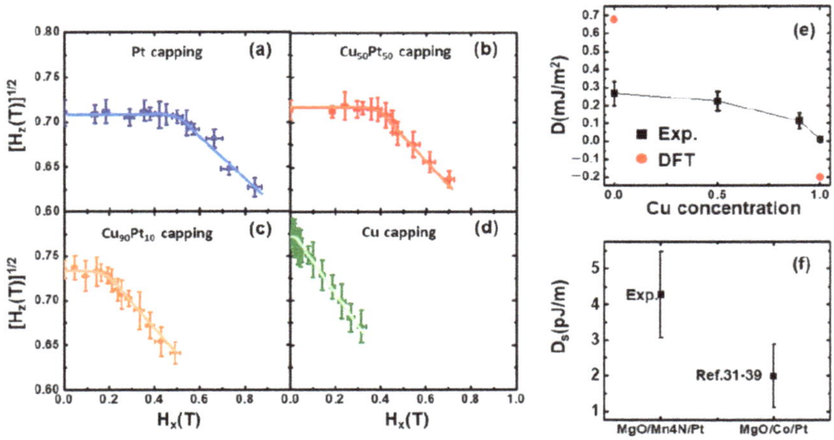

Figure 9.6. (a)–(d) $(H_z)^{1/2}$ versus H_x curves of $MgO/Mn_4N/Cu_xPt_{1-x}$ with different capping layers with $x = 0$, 0.5, 0.9, 1. The lines in the graphs are the fitted lines. The turning point H_x corresponds to H_{DMI} of each capping. (e) The measured DMI of $MgO/Mn_4N/Cu_xPt_{1-x}$ with different Cu content x in the capping layers. Red dots indicate the DMI from density functional theory calculations. (f) D_S comparison between $MgO/Mn_4N/Pt$ and MgO/Co/Pt. Reproduced from [24]. CC BY 4.0.

9.3 SOT switching

Current-induced SOI switching has been demonstrated to electrically switch the perpendicular magnetization vectors [29, 30], and thereby a substantially enhanced SOT efficiency has been reported [29, 31]. SOT-driven magnetization switching has been demonstrated in the Hall bar structure of Mn_4N(60 nm)/Pt bilayer structure, as shown in figure 9.7 [32]. The critical switching-current density (J_c) is $J_c = 6.17 \times 10^7$ A cm^{-2} in the presence of in-plane magnetic field in parallel with current flow $H_x = 300$ Oe. Figure 9.8(a) shows the thickness dependent SOT switching for Mn_4N/Pt (t nm/4 nm) bilayers, wherein a consistent current-induced SOT switching was clearly demonstrated. As the Mn_4N layer thickness decreases, the J_c was decreased, as shown in figure 9.8(b), implying the existence of a bulk-like SOT in the Mn_4N film.

This is an important demonstration which shows that rare-earth-free ferrimagnet Mn_4N may be used to develop spintronic devices as an alternative to widely used rare-earth transition metal ferrimagnets.

9.4 Anomalous Nernst effect

When a temperature difference is applied to both ends of a metal or semiconductor, a voltage is generated in a direction parallel to the heat flow due to carrier transfer caused by the heat flow. This phenomenon is called the Seebeck effect [33, 34], and has been studied for more than 50 years as an environmentally friendly thermoelectric (TE) power generation technology that directly converts thermal energy into electrical energy. The most commonly used TE materials are PbTe and bismuth chalcogenides such as Bi_2Te_3 [35, 36] because they show excellent dimensionless figure of merit values of approximately 1 at room temperature. The anomalous

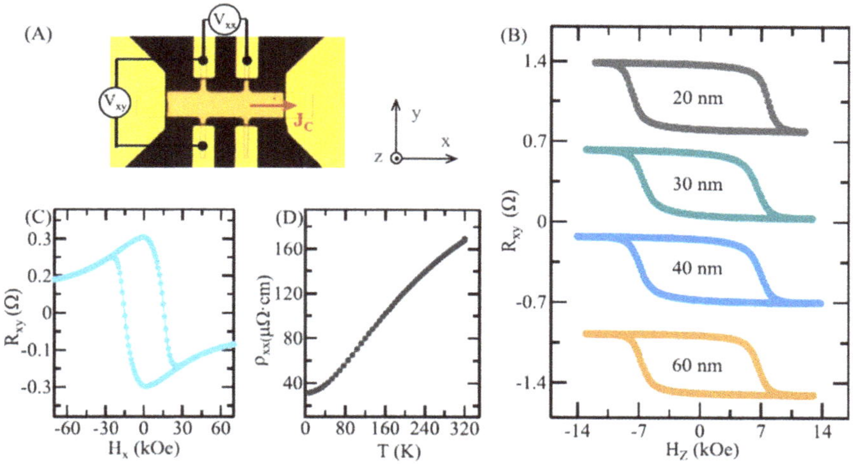

Figure 9.7. (A) The optical image of the Hall bar device, where the width of the Hall bar is 20 μm. (B) The AHE loops obtained in Mn_4N films with thickness ranging from 20 to 60 nm, confirming the existence of PMA in these films. (C) The corresponding PHE loop for Mn_4N/Pt (40 nm/4 nm), from which a H_k = 80 000 Oe is estimated. (D) The *R*–*T* curve for the Mn_4N (40 nm) film. Reproduced from [32]. CC BY 4.0.

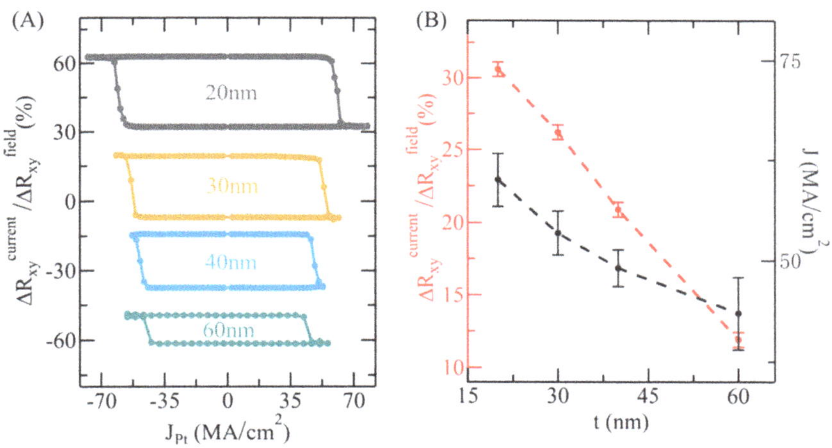

Figure 9.8. (A) The thickness dependent SOT-induced magnetization switching in the 20 μm-Hall bar device for Mn_4N/Pt (*t* nm/4 nm) bilayer with a fixed H_x = 300 Oe. The evolution of the critical switching-current density (J_C) and the switching ratio ($\Delta R_{xy}^{\text{current}}/\Delta R_{xy}^{\text{field}}$) as a function of the thickness (*t*) of Mn_4N is given in (B). Reproduced from [32]. CC BY 4.0.

Nernst effect (ANE) is a phenomenon in which a voltage is generated in the direction perpendicular to the magnetization of the material and the given heat flow when a thermal gradient is applied to a magnetic material. It is therefore an alternative technology to those based on the normal Seebeck effect. The transverse TE power based on the ANE has been reported on ferromagnetic Heuslter alloys [37–40], exhibiting a large value of around 6 $\mu V\ K^{-1}$. Materials possessing high K_u

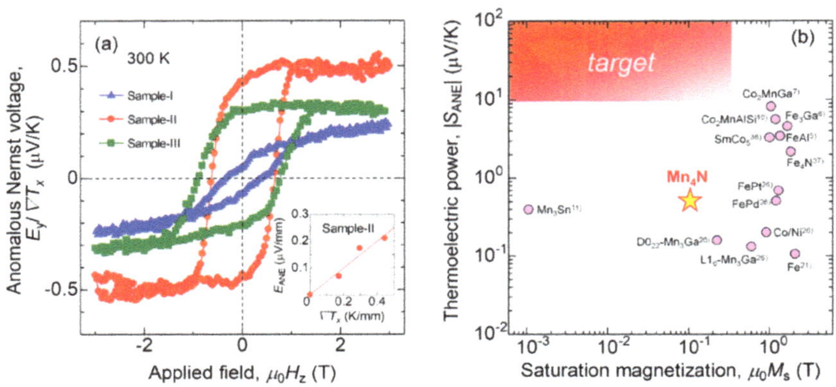

Figure 9.9. (a) Representative anomalous Nernst voltage (V_{ANE}) at 300 K as a function of the applied magnetic field, pointing in the z-direction (H_z). The inset represents saturated VANE as a function of temperature gradient. The solid line shows the least squares fitting with a linear function. The corresponding slope yields V_{ANE}. (b) Thermoelectric power (S_{ANE}) for various materials plotted against saturation magnetization ($\mu_0 M_S$). Red color represents the target values for heat flux sensors. [41] John Wiley & Sons. [original copyright notice].

are desirable to stabilize the spontaneous magnetization. Since Mn_4N shows PMA, a temperature gradient in the longitudinal direction of Mn_4N can be used to extract voltage in the width direction. Isogami *et al* demonstrated the ANE on sputter-deposited 20 nm thick Mn_4N films on MgO(001) substrate, as shown in figure 9.9 [41].

9.5 Modulation of the magnetic properties by light element doping

Reference [42] reviews the effect of doping light element (B, C, and N) into Mn_4N films, and demonstrates the possible modulation of its various properties. This is attributed to the fact that the material properties are governed by the transfer integrals, as shown in figure 9.10. For instance, the topological spin texture in lightly B-doped Mn_4N was found to become more stable at low temperatures from the extraction of the anomalous and topological Hall resistivity from the total transverse Hall resistivity [43].

9.6 Perspective

To develop spintronic applications of Mn_4N, it would be worthwhile to study whether the relatively high spin polarization of Mn_4N-based ferrimagnets can be exploited to obtain large tunnel magnetoresistance ratios. The room temperature double MCP found in $Mn_{4-x}Co_xN$ is also very attractive from an application point of view because the observed dual magnetic compensation mechanism may constitute a pathway for smaller spontaneous magnetization over a relatively wide range of Co concentrations. This is important because some magnetic quantities, such as the velocity of DW, are very sensitive to compositional changes around the MCP. Thus, slight differences in concentration can cause undesirable results. From a fundamental point of view, it is interesting to compare magnetic and spintronic

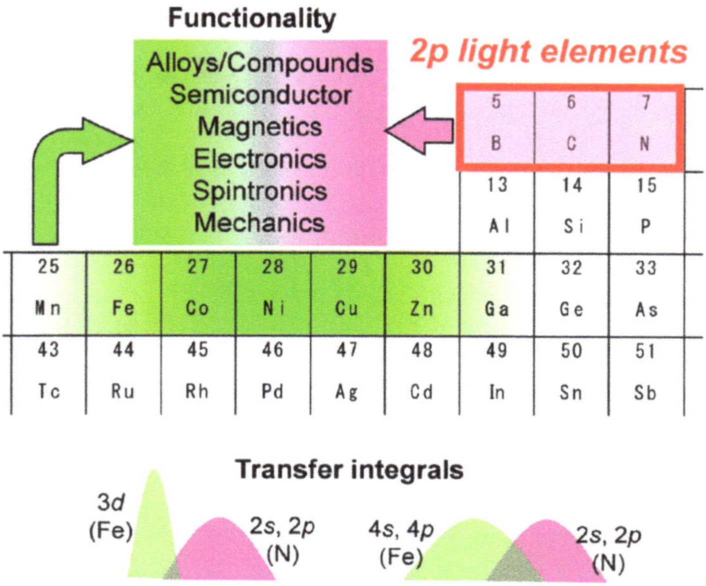

Figure 9.10. Overview of tailorable and multifunctional materials consisting of transition metals and 2*p* light elements. The material properties are governed by the transfer integrals as depicted by the lower schematics. Reproduced from [42]. CC BY 4.0.

properties such as SOT, STT, AHE, and polarization in the vicinity of two magnetic compensation concentrations. Although the DW mobility in Mn_4N is considerably higher than those achieved in SOT, the threshold current density for DW to start moving in Mn_4N is relatively high, probably due to the large damping constant α. From the line width of the microwave absorption spectrum and the oscillation frequency of the time-resolved MOKE signal, α was measured to be about 0.1 – 0.2 at room temperature [44]. This value is much larger than that reported for Fe_4N ($\alpha \sim$ 0.03) [45] and two orders of magnitude higher than that of the MnGa alloy [46]. Unfortunately, the reason for this high value of alpha is not made clear. Therefore, extensive research is needed to resolve this issue. In any case, the results presented in this book demonstrate the versatility of Mn_4N and Mn_4N-based mixed crystals and open up a rare-earth-free platform for ferrimagnetic spintronics.

Bibliography

[1] Chureemart J, Sampan-a-pai S, Boonchui S, Chantrell R W and Chureemart P 2021 *J. Magn. Magn. Mater.* **529** 167838

[2] Manchon A, Železný J, Miron I M, Jungwirth T, Sinova J, Thiaville A, Garello K and Gambardella P 2019 *Rev. Mod. Phys.* **91** 035004

[3] Martinez E, Emori S and Beach G S D 2013 *Appl. Phys. Lett.* **103** 072406

[4] Khvalkovskiy A V, Cros V, Apalkov D, Nikitin V, Krounbi M, Zvezdin K A, Anane A, Grollier J and Fert A 2013 *Phys. Rev.* B **87** 020402

[5] Franke K J A, Ophus C, Schmid A K and Marrows C H 2021 *Phys. Rev. Lett.* **127** 127203

[6] Conte R L *et al* 2017 *AIP Adv.* **7** 065317
[7] Gushi T *et al* 2018 *Jpn. J. Appl. Phys.* **57** 12
[8] Hirose T, Komori T, Gushi T, Anzai A, Toko K and Suemasu T 2020 *AIP Adv.* **10** 025117
[9] Zhang Z, Cho Y, Singhal J, Li X, Dang P, Lee H, Casamento J, Tang Y, Xing H G and Jena D 2020 *AIP Adv.* **10** 015238
[10] Kawasaki M, Maeda T, Tsuchiya R and Koinuma H 1993 *Science* **266** 1–3
[11] Yasuda T, Amemiya K and Suemasu T 2023 *Appl. Phys. Lett.* **123** 122404
[12] Dzyaloshinsky I 1958 *J. Phys. Chem. Solids* **4** 241–55
[13] Moriya T 1960 *Phys. Rev.* **120** 91–8
[14] Büttner F, Lemesh I and Beach G S D 2018 *Sci Rep.* **8** 4464
[15] Woo S, Litzius K, Kr€uger B, Im M-Y, Caretta L, Richter K, Mann M, Krone A, Reeve R M and Weigand M *et al* 2016 *Nat. Mater.* **15** 501
[16] Soumyanarayanan A *et al* 2017 *Nat. Mater.* **16** 898
[17] Raju M, Yagil A, Soumyanarayanan A, Tan A K C, Almoalem A, Ma F, Auslaender O M and Panagopoulos C 2019 *Nat. Commun.* **10** 696
[18] Caretta L *et al* 2018 *Nat. Nanotechnol.* **13** 1154
[19] Quessab Y, Xu J-W, Ma C T, Zhou W, Riley G A, Shaw J M, Nembach H T, Poon S J and Kent A D 2020 *Sci Rep.* **10** 7447
[20] Jena J, Göbel B, Ma T, Kumar V, Saha R, Mertig I, Felser C and Parkin S S P 2020 *Nat. Commun.* **11** 1115
[21] Zheng F *et al* 2018 *Nat. Nanotechnol.* **13** 451
[22] Vakili H *et al* 2021 *J. Appl. Phys.* **130** 070908
[23] Ma C T, Hartnett T Q, Zhou W, Balachandran P V and Poon S J 2021 *Appl. Phys. Lett.* **119** 192406
[24] Zhou W, Ma C T and Poon S J 2023 *Nanomaterials* **13** 1672
[25] Kim S, Jang P H, Kim D H, Ishibashi M, Taniguchi T, Moriyama T, Kim K J, Lee K J and Ono T 2017 *Phys. Rev.* B **95** 220402
[26] Kuepferling M, Casiraghi A, Soares G, Durin G, Garcia-Sanchez F, Chen L, Back C H, Marrow C H, Tacchi S and Carlotti G 2023 *Rev. Mod. Phys.* **95** 015003
[27] Belabbes A, Bihlmayer G, Bechstedt F, Blügel S and Manchon A 2016 *Phys. Rev. Lett.* **117** 247202
[28] Camley R E and Livesey K L 2023 *Surf. Sci. Rep.* **78** 100605
[29] Mishra R, Yu J, Qiu X, Motapothula M, Venkatesan T and Yang H 2017 *Phys. Rev. Lett.* **118** 167201
[30] Finley J and Liu L 2016 *Phys. Rev. Appl.* **6** 054001
[31] Liu L, Lee O J, Gudmundsen T J, Ralph D C and Buhrman R A 2012 *Phys. Rev. Lett.* **109** 096602
[32] Bai H, Xu T, Dong Y, Zhou H A and Jiang W 2022 *Adv. Electron. Mater* **8** 2100772
[33] Mahan G D and Sofo J O 1996 *Proc. Natl. Acad. Sci* **93** 7436
[34] Yang L, Chen Z G, Dargusch M S and Zou J 2018 *Adv. Energy Mater.* **8** 1701797
[35] Venkatasubramanian R, Siivola E, Colpitts T and O'Quinn B 2001 *Nature* **413** 597
[36] Vashaee D, Chen X, Liu J, Dresselhaus M S, Chen G and Ren Z 2008 *Science* **320** 634
[37] Sakai A *et al* 2018 *Nat. Phys.* **14** 1119
[38] Guin S N *et al* 2019 *NPG Asia Mater.* **11** 16
[39] Sumida K *et al* 2020 *Commun. Mater.* **1** 89
[40] Sakuraba Y, Hyodo K, Sakuma A and Mitani S 2020 *Phys. Rev.* B **101** 134407

[41] Isogami S, Mesuda K, Miura Y, Rajamanickam N and Sakuraba Y 2021 *Appl. Phys. Lett.* **118** 0924407
[42] Isogami S and Takahashi Y 2023 *Adv. Electron. Mater* **9** 2200515
[43] Isogami S, Ohtake M and Takahashi Y K 2022 *J. Appl. Phys.* **131** 073904
[44] Isogami S, Anzai A, Gushi T, Komori T, Takahashi Y K and Suemasu T 2019 *JSAP Spring Meeting 10a-M101-7 (Tokyo)*
[45] Isogami S, Tsunoda M, Oogane M, Sakuma A and Takahashi M 2014 *J. Magn. Soc. Jpn.* **38** 162
[46] Mizukami S, Wu F, Sakuma A, Walowski J, Watanabe D, Kubota T, Zhang X, Naganuma H, Oogane M and Ando Y 2011 *Phys. Rev. Lett.* **106** 117201

www.ingramcontent.com/pod-product-compliance
Ingram Content Group UK Ltd.
Pitfield, Milton Keynes, MK11 3LW, UK
UKHW051856100926
8556IPUK00005B/651